A GEOSCIENTIST'S GUIDE
TO PETROPHYSICS

IFP PUBLICATIONS

▶ **B. ZINSZNER**

▶ **F.M. PELLERIN**

A GEOSCIENTIST'S GUIDE TO PETROPHYSICS

Translated from the French by Trevor Jones (Lionbridge)

2007

t Editions TECHNIP 25 rue Ginoux, 75737 PARIS Cedex 15, FRANCE

FROM THE SAME PUBLISHER

- Basin Analysis and Modeling of the Burial,
  Thermal and Maturation Histories in Sedimentary Basins
  > M. MAKHOUS, Y. GALUSKIN

- Sedimentary Geology
  Sedimentary basins, depositional environments,
  petroleum formation
  > B. BIJU-DUVAL

- Geomechanics in Reservoir Simulation
  > P. LONGUEMARE

- Oil and Gas Exploration and Production
  Reserves, costs, contracts
  > CENTRE OF ECONOMICS AND ADMINISTRATION (IFP-SCHOOL)

- Integrated Reservoir Studies
  > L. COSENTINO

- Geophysics for Sedimentary Basins
  > G. HENRY

- Basics of Reservoir Engineering
  > R. COSSÉ

- Well Seismic Surveying
  > J.L. MARI, F. COPPENS

- Geophysics of Reservoir and Civil Engineering
  > J.L. MARI, G. ARENS, D. CHAPELLIER, P. GAUDIANI

Printed in France

ISBN 978-2-7108-0899-2

Acknowledgements

We are highly indebted to many of our colleagues for their contributions to this book.

Firstly, a word of thanks to the reviewers of the French manuscript who contributed directly, bringing numerous improvements to the text: F. Fournier, T. Le Meaux, B. Mouroux, M. Zamora and particularly J.P. Herbin for his long and careful contribution. Special thanks to B. Layan who played a significant role in this improvement, several figures being directly inspired from his lectures at the IFP School.

We are also grateful for the indirect but invaluable contributions made over the years during the numerous discussions with our colleagues at Boussens, Pau or Rueil. Our special thanks to: M.T. Bieber, R. Botton-Dumay, D. Bosie-Codreanu, P. Bousquié, D. Broseta, T. Cadoret, O. Coussy, J. Delclaud, P. Egermann, F. Ferfera, H. Greder, J. Guélard, C. Jacquin, D. Jeannette, Paul Johnson, B. Levallois, J.M. Lombard, N. Lucet, D. Marion, L. Nicoletis, C. Pabian-Goyeneche, D. Pavone, M. Poulet, M. Robin, E. Rosenberg, P. Trémolières.

At IFP, thanks to P. Boisserpe, B. Colletta and L. Montadert for their precious support right from the start of our project. Our thanks also go to D. Allinquant, M. Darthenay, P. Le Foll, M. Masson and J. Pirot who helped us, each in his own field of skill.

Lastly, BZ would like to give special thanks to T. Bourbié for the experience gained from him in writing book, which proved highly encouraging, and to P. Rasolofosaon for so many years of friendly cooperation in Rock Physics.

And FMP wishes to extend his warm and friendly thanks to B. Layan and G. Hamon, for the extremely fruitful exchanges over the many years shared at the ELF Scientific and Technical Centre.

Foreword

As its name implies, petrophysics is the study and measurement of the physical properties of rocks. As soon as you start to take an interest in this field, you are struck by the duality between the "Physics" aspects and the "Geology" aspects. In most publications, the "Physics" aspect prevails, in other words the quantitative description of the laws governing the phenomena affecting the rocks subject to various stresses: hydraulic, mechanical, electrical, etc. The rock itself is generally no more than a "black box" whose microscopic structure is never described, or at best very briefly, as a model sometimes far remote from reality. This is because the evident rich diversity of the natural porous media scares the physicist, as expressed so well by G. Matheron [in French, 1967] *"...ces millions de grains et la variété inépuisable de leurs formes et de leurs dimensions...."* ("...these millions of grains and the inexhaustible variety in their shapes and dimensions...".

Yet, how can we understand the anomalies sometimes observed in fluid flow or capillary equilibria within a rock without a vital piece of information, the description of a feature in the geometry of the pore space? But above all, if we are to scale up highly isolated petrophysical observations to an entire oil reservoir or an aquifer, it is essential to implement the powerful extrapolation tool of "geological interpretation". This is clearly based on a good understanding of the relations between the petrophysical parameters studied and the petrological characteristics of the rock considered. *This "Geological" approach of Petrophysics is at the hub of our project.*

Firstly, however, we must define our perception of the field of petrophysics and clarify the terminology, since two virtually identical terms, "Rock Physics" and "Petrophysics" coexist.

The term "Rock Physics" was popularised by Amos Nur, who developed a famous "Rock Physics" laboratory at Stanford University in the 70's. The field assigned to Rock Physics is basically both quite simple to define... and extremely vast: study the physical properties of geological materials, focusing in particular on the properties implemented in the various applications: electrical, hydraulic, nuclear, mechanical (static and dynamic) properties. The term "Petrophysics", which is older (G. Archie?), initially referred to the study of "reservoir" properties in an exclusively petroleum environment. At the present time, both terms coexist and are used in slightly different ways (although the distinction remains vague): "Rock Physics" is used mainly in university environments, often concerning mechanical or magnetic properties; "Petrophysics" is preferred by the oil exploration-production community, with a strong emphasis on hydraulics. A quick search for these two key words on an Internet search engine gives approximately the same number of hits for "Petrophysics" and "Rock Physics".

A relatively recent terminological evolution is noticeable, resulting in the inclusion of "Log Analysis" in "Petrophysics". In 2000, the journal "The Log Analyst" published by the Society of Petrophysicists and Well Log Analysts (SPWLA), was renamed "Petrophysics".

In this book we will only use the term "Petrophysics" and consider it in its usual common meaning, which is slightly restrictive: study of the physical properties of rocks focusing on the storage and flow of the fluids contained within them. We will investigate more specifically the properties related to the greater or lesser presence of a porous phase inside the rocks. This porosity will play a central role in our descriptions. The *raison d'être* of Petrophysics, as we will describe it, lies mainly in its relation to petroleum, hydrogeological and civil engineering applications.

The book is divided into two sections of unequal size:

The first section (by far the largest in terms of the number of pages) describes the **various petrophysical properties of rocks**. *Each property is defined, limiting the mathematical formulation to the strict minimum but emphasising, using very simple models, the geometrical (and therefore petrological) parameters governing this property.* The description of the measurement methods is restricted to an overview of the principles required for good communication between the geoscientist and the laboratory petrophysicist. For each property, we detail one or two aspects of the relations between petrophysics and geology which we feel are of special interest (e.g. the porosity/permeability relations in carbonate rocks for single-phase permeability, or irregular water tables and stratigraphic traps for capillary equilibria).

The various properties are classified into three subsets according to their main use in the study of oil reservoirs or aquifers (this organisation also allows us to consider the cases of perfect wettability and intermediate wettability separately, making it easier to expose capillary phenomena):

a) Calculation of fluid volumes (accumulations); **Static properties:** *Porosity* and *Capillary Pressure* in case of perfect wettability.

b) Fluid recuperation and modelling; Dynamic properties: *Intrinsic permeability, Wettability, Relative Permeability* and End Points.

c) Log and geophysical analysis; *Electrical properties*: Formation factor and saturation exponent, *Acoustic properties:* Elastic wave velocity, *Nuclear Magnetic Resonance* and its petrophysical applications. This third subset acts as a link with log analysis, a technique which is increasingly considered as being part of petrophysics. We provide a description of the most general principles without discussing log analysis as such.

The second section, which concentrates on methodological problems, provides a few notions which are at the root of the problems on the understanding and applicability of petrophysics.

It concerns, above all, **the representativeness of the measurements and the size effects,** so important for extrapolation of results and characterisation of reservoirs.

In this context we describe a few general points on the *effect of stresses and temperature* on the petrophysical characteristics. These effects are described separately for each property, however, in the first section. We therefore decided, perhaps rather over-simplifying matters in the process, to consider compressibility as "the effect of stresses" on the porosity.

The notions of *Representative Elementary Volume, Homogeneity, Anisotropy* provide a better understanding of the problems of *up-scaling*, for which a few examples are given (plug, core, log analysis, well test). After giving the most pragmatic definitions possible, we give a few examples concerning *permeability anisotropy* or the consequences of *saturation inhomogeneity* on the acoustic or electrical properties. The consequences of these size effects in reservoir geology are particularly spectacular in the contrast between matrix property (centimetric scale) and bulk properties (plurimetric scale) as observed in the fractured reservoirs, which will only be discussed briefly.

We then mention the extrapolation of petrophysical properties and characterisation of reservoirs using the *Rock Typing methods* which still remain to be defined, a clear consensus not having yet been reached.

Lastly, we provide a description of several **Porous Network investigation methods**. We describe in short chapters the methods used to observe these networks at various scales: *Thin sections, Pore Casts, Visualization of capillary properties, X-ray tomography*. Each method offers an opportunity to give a few examples of porous geometries. A brief overview of *X-ray diffraction* technique is proposed.

Concerning the **Bibliography**, we decided not to follow a certain inflationist trend. We restricted the list to the English texts explicitly referred to in the document. We made occasional exceptions to this language rule for some "reference" documents or for French documents from which we have extracted data.

We wanted to pay special attention to the **Index**, mainly since we know from experience just how important this is for readers. But also since a detailed index simplifies the plan. Rather than trying to produce a perfect logical order to introduce various notions, which rapidly turns out to be a very difficult task, we decided instead to insert numerous cross-references in the text, allowing readers to obtain all the details they need. The numerous porosity terms are often confusing, we try to clarify this in a "*porosity terms glossary-index*".

Nomenclature

Symbols

The nomenclature below does not include the multiple constants used in the text. These are generally represented by the Characters A, B, C... a, b, c... etc.

B	equivalent ion conductance in clay (W&S model)
B_0, B_1	magnetic fields
C, D	distance, dimension
C_P	pore compressibility, (C_{PP} C_{pc})
C_b	bulk rock compressibility (C_{bp} C_{bc})
C_w, C_o	water, oil compressibility
C_s	spreading coefficient
C_o	brine saturated rock conductivity
C_t	partially brine saturated rock conductivity
C_w	brine conductivity
C_{cw}	apparent conductivity of "clay bounded" water (S m^{-1})
cal	(subscript) calculated
d	diameter (e.g. grain)
d	d-spacing (X-ray diffraction)
e	thickness (e.g. crack)
f	frequency
f_L, f_{LA}, f_{LB}	Larmor frequency (NMR)
E	void ratio
F	formation factor
F^*	formation factor, W&S model
F_a	apparent formation factor
G	pore geometrical factor (Thomeer's hyperbola)
G_x	gradient (magnetic field)
G_0	geometrical factor
G	hydraulic gradient
g	(subscript) gas

H	Hirschwald coefficient
h	height (fluid column, etc.)
h_t	height above free water level
h_{td}	"displacement height", corresponding to the base of the water table
h	relative moisture
Hg	(subscript) mercury
Iw, Io	Amott's method (wettability measurement) index
I	radiation intensity
I	resistivity index
i	associated with w (subscript): "initial"
J	Leverett's transform (or function)
K	bulk modulus
K_{dry}, K_{sat}, K_{grain}, K_{fl}	bulk moduli used in Biot-Gassmann formula
K	intrinsic permeability
k	hydraulic conductivity
$KW_{(Sw)}$	water effective permeability at S_w saturation state
$KO_{(Sw)}$	oil effective permeability at S_w saturation state
$KRW_{(Sw)}$	water relative permeability at S_w saturation state
$KRO_{(Sw)}$	oil relative permeability at S_w saturation state
$KRG_{(Sw)}$	gas relative permeability at S_w saturation state
l	for liquid in e.g. Kl
l	length (plug, sample, etc.)
lab	(subscript) laboratory conditions
Ln	napierian logarithm
log	decimal logarithm
meas	(subscript) measured
M	elastic modulus
M	water molar mass
nw	(subscript) related to non-wetting fluid e.g. P_{nw}
m	(subscript) mean, average
m	cementation factor (formation factor formula)
*m*1	Humble Co. formula exponent
n	saturation exponent (Archie's formula)
o	(subscript) oil or hydrocarbon
p	compressional (P) wave symbol
pF	moisture potential
P	pressure (P_w, P_o, P_g in water, oil, gas)
P_{nw}, P_w	pressure in the non-wetting, wetting fluid respectively

P_d	displacement pressure
Pc	capillary pressure, by definition $= P_{nw} - P_w$
p_S	dry sample weight
p_{Sat}	fluid saturated sample weight
Q	volume flow rate
Q	quality factor
Q_v	cation exchange capacity per volume unit
R, R_c, R_{apc}	radius (capillary, apparent, capillary rise)
R_G	molar gas constant
R	not subscript: Resistivity; R_w brine, R_o brine saturated rock, R_t undetermined saturation state
r	associated with o or g (subscript): residual
res	(subscript), reservoir conditions
S	saturation (S_w, S_o, S_g, S_{nw}, S_w,) (water, oil, gas, nonwetting fluid, wetting fluid)
S_{or}, S_{gr}, S_{gi}, S_{li}	saturation (residual oil, residual gas, initial gas, initial liquid)
S_{wi}	initial or irreducible water saturation
Sw_t	saturation related to total porosity
Sw_e	"effective" saturation (related to effective porosity)
S_{Hg}	mercury saturation
S_{cal}	computed water saturation, as opposed to S_{meas} measured water saturation
S_{wR}	normalised (or reduced) water saturation
S	surface
ST	storage coefficient or storage capacity
s	specific surface
s	shear (S) wave symbol
T_1	relaxation time (longitudinal or spin laticce) (NMR)
T_2	relaxation time (tranverse or spin-spin) (NMR)
T	transition zone
T	tortuosity
T_K	absolute temperature
t_{go}; t_{gw}; t_{ow}	surface tension gas/oil, gas/water, oil/water
U	filtration velocity, specific discharge, or Darcy flux
u_a	air pressure
u_w	water pressure
V	velocity
Vp	P-wave velocity
Vs	S-wave velocity

V_T	total volume (sample)
V_S	solid volume
V_V	void volume
V_a, V_b, V_c, V_d	fluid volumes in Amott's method
V_w	water volume
$\mathcal{V}$	volume water content (related to solid volume)
V_q	coefficient related to the diffuse layer thickness (bound water) $(meq^{-1} cm^3)$
V_{cl}, V_{sh}	"clay mineral", shale, volume content
$v(i)$	i element volume content
w	(subscript) related to wetting fluid e.g. P_x
w	(subscript) water, brine
W	mass (sometimes weight)
$w(i)$	i element weight content
W_{USBM}	USBM wettability index
W_{IA}	Amott wettability index
z	depth
Z	impedance
Δ	indicates a difference e.g. ΔP
α	Clavier's model coefficient
β	Biot's coefficient
β_F	Forcheimer's coefficient
ε	strain
γ	acceleration (used also in γ-ray and in Formation Factor discussion)
γ_p	gravity acceleration
Λ	wavelength
λ	Lamé's parameter
ϕ	porosity
ϕ_t	total porosity
ϕ_e	effective porosity
ϕ_{co}	"conductive" porosity, used in resistivity discussion
Φ	total soil water potential
μ	shear modulus
η	viscosity
ρ	specific gravity
ρ_e	electron density
ρ_o	oil specific gravity

ρ_w	water specific gravity
ρ_{ma}	matrix specific gravity
$\rho_{M(i)}$	mineral i specific gravity
$\rho_{F(i)}$	specific gravity of the fluid i saturating the porous medium
ρ_S	"dry" specific gravity (solid phase)
ρ_A	porous medium ("apparent") specific gravity
σ	stress, σ_{diff}, σ_{eff}, σ_{conf} differential, effective, confining
τ	shear stress (used also for tortuosity)
θ	angle (wettability, X-ray diffraction)
θ_W	soil moisture (weigth water content)
θ_V	volume water content
ψ_z, ψ_m, ψ_o	potential (gravity, moisture or matrix, osmotic)
ω	angular velocity

Acronyms

AM	Available moisture
AS	Available Storage or Available Reserve
CEC	Cation Exchange Capacity
DT	Drilled Thickness
FWL	Free Water Level
GWC	Gas Water Contact
MSL	Mean sea Level
NEOM	Not Extractible Organic Material
NEOC	Not Extractible Organic Carbon
OWC	Oil Water Contact
REV	Representative Elementary Volume
SS	Sub Sea
TST	True Stratigraphic Thickness
TVD	True Vertical Depth
TVT	True Vertical Thickness
USBM	US Bureau of Mines
WSC	Water Storage Capacity
W&S	Waxman and Smits model
WWJ	Wright, Wooddy & Johnson

Unit conversion factors

Some conversion factors are rounded (r); in that case the resulting error is lower than 10^{-4}. [After M. Dubesset, in French, 2000].

From	To	Multiplication factor		
acre	m^2	4.047	E + 03	(r)
acre-foot	m^3	1.2335	E + 03	(r)
atmosphere (standard)	Pa	1.0132	E + 05	(r)
bar	Pa	1	E + 05	
barrel (US, Petrol)	m^3	0.159		(r)
barrel per day	m^3/s	1.84	E − 06	(r)
centimeter of water (4°C)	Pa	9.8064	E + 01	(r)
centimeter of mercury (0°C)	Pa	1.3332	E + 03	(r)
centipoise	Pa. s	1	E − 03	
centistokes	m^2/s	1	E − 06	
compressibility, usual, $(10^6 PSI)^{-1}$	GPa^{-1}	1.45	E − 01	(r)
cubic foot	m^3	2.8317	E − 02	(r)
cubic inch	cm^3	16.387		(r)
darcy	m^2	0.987	E − 12	(r)
darcy, for 1Pa. s fluid	cm/s (*NOT the same physical prop.*)	1	E − 03	
dyne par cm	N/m	1	E − 03	
foot	m	0.3048		
foot of water (conventional)	Pa	2.9891	E + 03	(r)
foot per second (ft/s)	m/s	0.3048		
gallon (imperial)	dm^3	4.5461		(r)
gallon (US)	dm^3	3.7854		(r)
inch	mm	25.4		
inch of mercury (0°C)	Pa	3.3846	E + 03	(r)
Inch of water (4°C)	Pa	2.4908	E + 02	(r)
millimeter of mercury (standard)	Pa	1.33323	E + 02	(r)
millimeter of water (standard)	Pa	9.8066		(r)
ounce (avdp)	g	28.3495		(r)
ounce (troy)	g	31.1035		(r)
(psi) pound force per square inch	Pa	6.895	E + 03	(r)

Contents

PART 1

PETROPHYSICAL PROPERTIES AND RELATIONS WITH PETROLOGY

PART 2

SCALE CHANGES AND CHARACTERISATION OF POROUS MEDIA: METHODS AND TECHNIQUES

PART 1

PETROPHYSICAL PROPERTIES AND RELATIONS WITH PETROLOGY

CHAPTER 1-1

Calculation of Fluid Volumes *In Situ* (Accumulations): Static Properties

This first chapter describes the properties used to calculate fluid volumes *in situ*: oil and gas for the oil industry, water for hydrogeology. This first step, known as "calculation of accumulations" leads to the calculation of reserves (fluid quantities recoverable under given financial conditions). This first phase involves the "static properties": **Porosity** and **Capillarity**. Since we decided to study the effects of temperature and pressure variations separately for each property, **Compressibility** is discussed under porosity, although traditionally it is not considered as being one of the "static properties".

1-1.1 POROSITY, MINERALOGY OF THE SOLID PHASE AND COMPRESSIBILITY

1-1.1.1 Porosity: definitions (connected, occluded, effective porosity, etc.)

By definition, porosity, a petrophysical quantity, is the ratio of void volume (V_V) to total volume (V_T) of the body considered. In everyday language, the term "porosity" is often used as a simplification of "porous space". In this case, we far exceed the notion of petrophysical value to include a morphological notion whose description makes the definition of porosity considerably more complicated. In this chapter, we will restrict ourselves to the study of the "petrophysical property" porosity. Numerous qualifiers are used to characterise the notion of porosity (in this respect, note that the language confusion between the terms "porosity" and "porous space" too often leads to many inaccuracies). Some qualifiers are specific to a particular technique and the same qualifier used in different fields sometimes refers to quite different notions (e.g. "secondary porosity" in petrology and "secondary porosity" in acoustics). We can therefore appreciate the degree of confusion possible. We will describe these various porosity qualifiers in the broad sense in the corresponding chapters (e.g. trapped porosity under the heading of capillary pressure) and attempt to summarise all these terms in a Glossary-Index.

Still in the field of "petrophysical" porosity, we must first clarify a few elementary notions:

A) Consolidated and unconsolidated materials, rocks and soils

The definition of porosity implies that the total volume and the void volume are precisely defined. This point is much less trivial than it might appear at first sight: in geological materials, the voids are generally saturated (either totally or partially) with water. In some cases (poorly consolidated materials, clays), water plays a major role in the definition of the material: it cannot be extracted without disturbing the structure. In this case, the notion of supposedly invariant total volume and void volume loses its precision.

Consequently, in order to define porosity and describe the measurement methods, we must make a distinction between:

- materials whose mechanical structure is determined by the continuity of the solid phase insensitive to the presence of water. These are consolidated rocks containing little or no clay;
- materials whose mechanical structure depends both on the structure of the solid phase and the presence of water. Clays are an example of this situation and we will discuss the definition of their porosity separately (§ 1-1.1.3, p. 12). This group includes a large number of non-consolidated rocks, but not all. The same applies for soils, which must often be classified in this group even though the definitions do not quite agree. Soil mechanics or pedology techniques are used to define and measure the porosity of these materials.

B) Connected and occluded porosity

In the case of consolidated rocks with little clay, for which the notion of porosity in terms of V_V/V_T is simple, there may be voids not accessible from outside the sample. For example, "fluid inclusions" in crystals correspond to **Occluded Porosity** inside the solid phase forming the rock. **Connected Porosity**, connected to the exterior by passages, even very small, corresponds to the complementary part. Occluded porosity, which does not affect the rock's storage or transport properties, is largely ignored by conventional petrophysics. It may have a significant impact on some physical properties, however, such as the solid phase density, a point which will be discussed later (§ 1-1.1.5, p. 34).

We will mention some points which we feel to be important:

- the notion of occluded porosity must be considered in the absolute: any pore connected to the exterior is part of the connected porosity, even if the passage is so small that it creates special phenomena. This is the case with pumice stone for example, whose porosity is largely connected but which is trapped during imbibition (see § 1-1.2.2D, p. 59), i.e. retaining the residual air when it is immersed in water;
- occluded porosity is rare in porous sedimentary rocks, generally less than a fraction of $1/10^3$. It can usually be ignored when studying oil reservoirs or aquifers. There are some exceptions to this general rule. A few examples of porous rocks in which the occluded porosity is important are given in § 1-1.1.5Bc, p. 34. We must also mention the case of poorly porous rocks, whose cement consists of crystals with well-marked faces (e.g. quartzite). Occasionally, some intercrystalline residual pores are only connected by very thin voids between the faces of the crystals (intercrystalline contacts).

Some of these joints may close up due to the effect of the confining pressure. In this case, we may therefore observe significant occluded porosity, in variable proportions depending on the confining pressure;
- for artificial materials it is more difficult to generalise. Porous bodies manufactured by sintering (glass, alumina, steel, etc.) sometimes have a large fraction of occluded porosity (up to 9% of the total volume in sintered nickel, for example).

C) Effective porosity

Although "effective porosity" (ϕ_{eff} or PHIE) is a fluctuating notion (whose details are unclear), it cannot be overlooked due to its central role in log analysis. One practice derived from the first electric log interpretation methods is to isolate in the total porosity a fraction known as the "effective porosity" (the confusion is even worse since this notion is presented as a synonym of the reservoir engineer's "effective porosity": proportion of the porosity that contributes to the storage and flow of fluids of interest).

Log analysis methods used to measure porosity are based on rock resistivity (resistivity tools), hydrogen atom content (H_2O) (hydrogen index-neutron tool) or the overall density (Density tool). In these three cases, some geological materials (mostly clays) have singular responses caused by the complex nature of the mineral-water relations (developed in § 1-1.1.3). If the basic interpretation methods were applied blindly, clay formations would sometimes exhibit high "porosity", since they contain a large amount of water.

In practice therefore, the implicit assumption defining the difference between total porosity "true", measured in laboratory) and the so-called "effective" porosity is the proportion of the porosity attributed to the clay phase.

This use is a major source of ambiguity in reservoir modelling since, in this case, if the data have not been carefully homogenised, there is a risk of mixing:
- data derived from well log analysis, possibly expressed in reference to this so-called "effective" porosity;
- data derived from laboratory experiments which always correspond to the total porosity, provided that the experimental conditions are good.

The same problem may arise in hydrogeology where the effective porosity defines the proportion of the porosity containing the volume of water available to flow during aquifer exploitation. We even speak of "usable porosity" or "kinematic porosity".

The calculations concerning effective porosity will be discussed again in § 1-1.1.5.B, p. 32.

1-1.1.2 Porosity measurement principles

A) Consolidated rocks

a) General principles

Porosity measurement consists in measuring the void volume and the total volume. Since the solid phase volume (V_S) is given by $V_S = V_T - V_V$, we simply measure two of these three parameters to calculate the porosity. Numerous methods have been devised to measure these various volumes (and the field is still open for more!). Amyx *et al.* [1960] and Monicard [1980] gives a pragmatic description of these methods. Table 1-1.1 summarises the principles of the most traditional methods.

Table 1-1.1   Summary of the main porosity measurement methods

Volume Measured	Reference in the text	Measurement Method
Total volume V_T	(1)	Buoyancy in mercury (mercury, a non-wetting liquid, does not penetrate without pressure in the common porous media).
	(2)	Volume occupied by the mercury between the sample and a chamber of known volume.
	(3)	Volume occupied by a powder of known density between the sample and a chamber of known volume.
	(4)	Direct measurement of the sample dimensions.
Void volume V_V	(5)	Compressibility of a perfect gas in a device treating only the pores as void volume (Hassler) or whose "dead volumes" are known.
	(6)	Intake of wetting fluid by total saturation under vacuum
	(7)	Mercury saturation under very high pressure (porosimetry test) and after in-depth degassing of the sample.
Solid volume V_S	(8)	Compressibility of a perfect gas in a chamber of known dimension.
	(9)	Buoyancy in a wetting fluid.
	(10)	Measurement of the solid density, after fine grinding of the porous material.
	(11)	Calculation of the solid density from quantitative analysis of the component minerals; or known density (monomineral rock).

Useful porosity calculation formulae:

We provide below some useful porosity calculation formulae connecting the various volumes (total, solid, pore): V_T, V_S, V_V; the masses of the dry porous body and the body saturated with fluid: p_S; p_{Sat} and the densities of the solid, the saturating fluid and the porous medium itself ("apparent" density): ρ_S, ρ_F, ρ_A.

$$\phi = V_V/V_T = 1 - V_S/V_T = V_V/(V_V + V_S)$$
$$V_T = V_V/\phi = V_S/(1 - \phi)$$
$$V_S = V_T(1 - \phi) = V_V ((1 - \phi)/\phi)$$
$$V_V = \phi V_T = V_S (\phi/(1 - \phi))$$
$$\rho_S = p_S/V_S = p_S/(V_T(1 - \phi)) = (\phi/(1 - \phi))p_S/V_V$$
$$\rho_A = \phi \rho_F + (1 - \phi) \rho_S$$
$$\phi = (\rho_S - \rho_A)/(\rho_S - \rho_F)$$

b) Practical remarks

Washing the sample

The measurements are taken on "clean" sample, i.e. after removal of the liquid hydrocarbons and the brine. There are numerous washing procedures, the main experimental concern being related to the problem of wettability state (§1-2.2, p. 175). The principle consists in sweeping the sample with mixtures of liquids dissolving the salts (e.g. methanol) and the hydrocarbons (e.g. chloroform, toluene, etc.). In practice, two main methods are used:

- To wash a large number of small samples quickly (routine core analysis), an apparatus with a still and a siphon is used to alternate the sample drying and immersion periods and purify the washing liquid by distillation. Obviously, eutectic mixtures must be used. The most well-known apparatus is the "Soxhlet extractor" (Fig. 1-1.1).
- For more thorough washing of larger samples, the cleaning mixture is forced through the sample held in a "Hassler" type apparatus (see Fig. 1-2.4a, p. 130).

If carried out according to professional standards, this cleaning should not extract the so-called non-extractable organic matter (e.g. kerogen and coal). Paragraph 1-1.1.5, p. 36, gives a detailed description of the consequences of this situation or that of incomplete washing (e.g. presence of salt in the sample). Note also that some solvents (e.g. alcohols) may react strongly with clay materials.

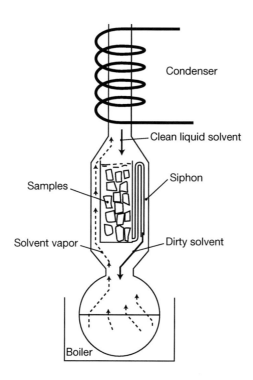

Figure 1-1.1 Schematic diagram of the Soxhlet extractor

Drying the sample

One of the most difficult practical problems to solve is that of the drying conditions. All the water must be extracted without any risk of damaging the clay phase (see § 1-1.1.3D, p. 16). Drying is generally carried out in a ventilated oven at a temperature of between 60°C and 150°C, depending on the laboratory. It is best to work between 85°C and 105°C and double this normal drying by drying at 150/160°C on a scrap plug in order to control the effect on a possible clay phase.

Drying operations under controlled humidity are not, in our opinion, recommended since by trying to "preserve" the clay phase and only accessing the so-called effective porosity (§ 1-1.1.1C, p. 5), a certain amount of liquid water remains in the porous medium, biasing the subsequent porosity-permeability measurements as well as various laboratory procedures: saturations, capillary pressure by mercury injection, etc. However, the – fortunately exceptional – presence of minerals dehydrating at low temperature (e.g. gypsum) poses a problem which can only be solved on a case by case basis.

Other techniques less damaging than drying are available for fragile, and especially clay, materials. The damage caused by drying, to a first approximation, is due to capillary tensions developed by the liquid-gas menisci. The method therefore consists in avoiding crossing this liquid/gas branch of the three-phase diagram (Fig. 1-1.2):

- Either the triple point is bypassed: this is the lyophilisation method. Two steps are required: 1) very fast freezing of the water by quenching effect so that crystallisation of the ice does not disturb the structure of the material; 2) sublimation of the sub-amorphous ice created in this way.

    This technique can only be applied to very small samples (a few mm3), otherwise the quenching effect is not guaranteed. It is highly suited to preparations for observations

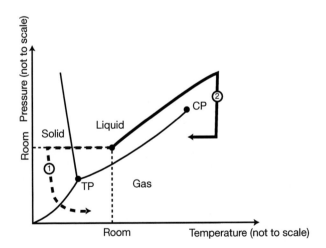

Figure 1-1.2 Diagram of dehydration techniques bypassing the triple point (TP)
(lyophilisation, dotted line ①) or the critical point (CP) (solid line ②)

CP water 22.4 MPa, 374°C
CP CO_2 7.4 MPa, 31°C

Table 1-1.2   Summary of the two special drying methods

	Advantages	Disadvantages	Use
Lyophilisation	Speed	Very small samples (less than a few mm³)	– Mainly SEM observation – Mercury injection
Bypassing the Critical Point	Samples of normal size	Relatively slow: Miscible exchanges and intermediate solvents	All petrophysical measurements

by scanning electron microscope (some SEMs have a preparation prechamber in fact), or for measurements such as capillary pressures by mercury injection.
–   Or the critical point is bypassed: the liquid/gas fluid remains single-phase throughout the process and the destructive capillary pressures are unable to develop. To avoid the need for the high-pressure high-temperature apparatus which would be required to bypass the critical point of water (P = 22.4 MPa; T = 374°C), however, a series of miscible exchanges is carried out, using intermediate solvents of decreasing polarity in order to replace the water in the porous medium by liquid CO_2. The critical point of CO_2 is then bypassed under pressure and temperature conditions quite accessible to conventional laboratory techniques (P = 7.4 MPa; T = 31°C).

Measurement of total volume

This operation is less trivial than it would appear. Direct measurement of the sample dimensions (method 4) is only accurate for perfectly cylindrical samples. Vernier calliper devices interfaced with a computer are available on the market, which provide a fast way of calculating the mean of numerous measurements.

The most accurate measurement is that of buoyancy in mercury (method 1). Mercury, a non-wetting liquid, does not penetrate without pressure in the common porous media and, due to its high density, the measurement is extremely accurate (Fig. 1-1.3a). The presence of

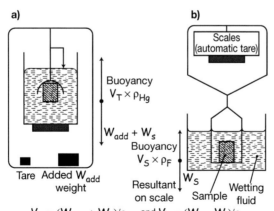

$$V_T = (W_{add} + W_s)/\rho_{Hg} \text{ and } V_s = (W_s - W_i)/\rho_F$$
where W_s and W_i represent the weights of the dry and immersed sample, respectively

Figure 1-1.3   Schematic diagram of the buoyancy measurement;
a) in mercury; b) in the fluid saturating the porous medium

vugs in contact with the sample surface may seriously disturb the measurement. This method, which involves the use of mercury "in the open air", is gradually being phased out due to increasingly stringent safety rules.

The differential measurement technique (method 2) can also be used to determine the volume of mercury required to fill a calibrated chamber containing the sample. A graduated pump fills the chamber up to a precise mark. To avoid the use of mercury, it can be replaced by powders of fixed, known apparent density: this powder is used to fill the voids left in the calibrated chamber by the sample (method 3). The volume of powder used is determined by weighing. Although practical, this method is far less accurate than that based on the use of mercury.

Measuring the void volumes

To measure the void volume, the most common method in routine use is probably that based on the ideal gas law (Mariotte-Boyle law) (method 5). By measuring the pressure difference after injecting a known volume of gas into an isothermal system, the total volume accessible to the gas can be calculated. The experimental techniques consist in either directly plotting the graph of volume against the pressure of gas (helium, since its behaviour is closest to that of an ideal gas) contained in a sample holder allowing as void only the pores and a minimum dead volume (Hassler core holder, see Fig. 1-2.4a, p. 130), or measuring the pressure drop occurring when a reservoir of known pressure and volume is opened to the holder containing the sample. Helium compressibility measurements may generate experimental errors (extremely sensitive pressure sensors) difficult to check *a posteriori*. They require strict quality control.

In another family of methods, the sample is saturated with a quantity of liquid whose volume (or mass and density) will be measured. The simplest method is to carry out total saturation under vacuum with a wetting fluid (method 6). The difference between the dry mass and the saturated mass gives the volume of saturating fluid directly. One might expect that wiping off the saturated plug could lead to measurement uncertainty. Experience has shown, however, that if the operator is careful the uncertainty is very low. The disadvantage with this reliable method, which requires no expensive equipment, is that it involves saturation by a wetting fluid, with possible consequences on the clay fractions.

The sample can also be totally saturated with mercury, under very high pressure (method 7). This (destructive) method is in fact only used as a by-product of porosimetry measurement (§1-1.2.4D, p. 79). Obviously, the methods proposed above can only be used to measure the volume of connected voids (connected porosity).

Measuring the solid volume

The method which is probably the most widely used for routine measurements (method 8) is, like method 5, based on the compressibility of ideal gases (helium). In this case, the sample is placed in a chamber of precisely known volume and "compressibility". Using the same principles as in (5), this time the total volume of the voids remaining in the chamber in the presence of the sample is calculated, then the solid volume is obtained by subtraction.

One reliable method, which requires no expensive equipment, involves measuring the buoyancy on the fully saturated sample (method 9). It is the complement of method (6). The buoyancy is measured directly by the weight loss of the saturated sample when immersed in the saturation liquid (Fig. 1-1.3b).

In the solid volume measurement methods described above, the occluded porosity is included with the solid. This is not the case with method (10) where the sample is ground finely enough (down to the order of the micrometer) to destroy every porous structure then the density of the powder measured. To our knowledge, this (totally destructive) method is rarely used. It may be worthwhile if abnormally high occluded porosity is suspected. The density of the solid (method 11) could also be calculated from a mineralogical analysis and the volume deduced directly using the dry weight. This is the corollary of the methods detailed in § 1-1.1.5, p. 26. In practice, this method is probably never used. For a precisely identified monomineral rock (quartzitic sandstone, limestone), this method may be applied to "correct" experimental errors.

Abstract

In current practice, the methods most often used by the Service Companies for routine studies involving thousands of measurements are those based on helium compression (5 and 8). They are ideal for automation and the gas permeability can be measured at the same time. The accuracy of the pressure sensors must be checked very carefully.

In research laboratories, when fewer measurements are taken, the saturation methods (6 and 9) give reliable results and do not require expensive equipment. Some laboratories of operating companies have automated these techniques, achieving productivity similar to that obtained with the previous methods (5 and 8). If saturation is a problem, the mercury methods (1 or 2 and 8) are used. When the use of mercury is not desirable, the powder method (3) is generally preferred to the notoriously inaccurate direct measurement (4).

B) Unconsolidated rocks and soils

The parameters V_S, V_V and V_T of consolidated materials are generally independent of the saturation state. For porosity measurements, these samples can therefore be dried or saturated without too much disturbance (taking care, nevertheless, to protect the clay phase) and implement the methods described above.

This is not the case for unconsolidated materials and soils in the meaning of surface formations or pedology.

In the oil industry, considerable care is required in order to obtain a representative sample of unconsolidated rock (see also § 2-1.3D, p. 298) both during the coring phase (double-envelope and low invasion core barrel) and during the sampling phase in laboratory (specially designed cutting ring). From coring to measurement, the sample is never removed from its sleeve (see diagram of Hassler cells Fig. 1-2.4, p. 130). The sleeve must therefore be able to withstand attack by hydrocarbons and solvents. The washing and drying operations are exclusively carried out by miscible exchanges at controlled temperature. The pore volume variations are acquired by balances on the incoming and outgoing volumes, the total volume variations are estimated using strain gauges. Obviously, these measurements have only become truly operational, leaving the domain of research laboratories, with the emergence of electronic sensors and automatic acquisition programs. Lastly, the current trend is to systematically acquire these data directly on the sample subject to a stress regime similar to that found *in situ*.

In the fields of pedology and especially geotechnics, the notion of void ratio $E = V_V/V_S$ is generally used rather than that of porosity $\phi = V_V/(V_V + V_S)$. This notion offers the advantage of having a parameter with constant denominator. In soil mechanics, in contrast with petroleum petrophysics, a rough estimation of the absolute porous volume is often sufficient to try to accurately evaluate the variations of this volume.

In the hydrogeology field, the notion of porosity is virtually never used under ordinary circumstances. Instead, the notion of "storage coefficient" (or "storage capacity") is used. The physical meaning of the term depends on whether the water table is free (in this case equivalent to effective porosity) or captive (in this case the parameter depends on the compressibility).

1-1.1.3 Clay porosity

The extremely vast subject of clay environments has been discussed in numerous publications. We will restrict ourselves to pragmatic considerations, mentioning only a few special aspects which may lead to misunderstandings or experimental protocol errors.

We decided not to deal with the specificity of clay environments in a single paragraph, but instead as each of the various petrophysical parameters is discussed.

A) Clay minerals

a) Originality of clay minerals

To a first approximation, clay minerals are characterised both by tabular or fibrous shapes, which develop large specific areas, and by a crystalline structure involving for some of them a lack of positive electrical charges, inducing the presence of compensating cations. These cations are mobile and undergo exchanges with those in the brine saturating the porous space. The clay phases therefore display a specific electrical conductivity, discussed in § 1-3.1.3., p. 213. Although they do not play a role in the crystalline formula, these cations intercalate between the crystallite sheets of some clay species such that these interlayer spaces may vary: this feature is used to define smectites.

b) Main families of clay minerals

Details of the mineralogical classification of clays will not be discussed here. To avoid certain interpretation errors, however, we must remember that the simplified classification into three broad families (kaolinites, illites and smectites), and sometimes four if we include the chlorites, is only a practical shortcut concerning the minerals encountered most frequently. Strictly speaking, clay minerals and phyllosilicates in the broad sense, consist of an extraordinarily wide range of compositions and structures. They include:

1. The kaolinite-serpentine group, characterised by electrically uncharged sheets. The basal distance (i.e. between two points in crystallographically equivalent positions) is about

7 Å when they contain no interlayer water and about 10 Å in the presence of interlayer water;

2. The group of uncharged talc-pyrophyllite minerals;
3. Micas and illites, of basal distance 10 Å;
4. Chlorites, of basal distance 14 Å;
5. Smectites and vermiculites, characterised by basal distances which vary depending on the content of the interlayer (relative humidity, solvation liquid, nature of the interlayer cation);
6. The group of fibrous sepiolite-palygorskite (or attapulgite) silicates.

We can see from this list that one of the first classification criteria is the value of the interlayer charge, a relevant feature for the applications since the log analyst often derives useful information on porosity and saturation from the electrical properties.

Lastly, these clay species may hybridise to generate the so-called interstratified clays, where the interlayer structure may either be regular (e.g. corrensite: regular chlorite/smectite interstratified clay) or irregular (e.g. the illites/smectites, and the family formerly known as hydromicas: interstratified mica/vermiculite or mica/smectite).

We will make a few practical remarks:

– In operational practice, a distinction is generally made between the clays and the micas. Implicitly, this separation is based on a distinction of particle size and therefore properties. Micas are reputed to behave as "ordinary" minerals whereas clays are considered as very fine media, reactive from the physico-chemical and mechanical points of view. We must bear in mind, however, that this is only a general and empirical rule. It does not define by principle the clay medium, and forgetting it could lead to errors.
– The confusion between the mineralogical definitions of clays (those to which we refer here) and the granulometric definitions is sometimes dangerous. The minerals have a continuous size distribution, for example:
  • There is no clear boundary between a mica and an illite;
  • Some kaolinites (or dickites, from the same family) are very large (see for example the photograph of Figure 1-1.16, p. 32), much larger than the 2 microns generally considered as the conventional limit of granulometric clays;
  • Chlorites also exist in an even wider range of sizes, which often depends on the conditions of their genesis.
  • In contrast, smectites are virtually always small.
– Similarly, there is no continuity break between the porous space of the clay phase and the intergranular porous space.

B) Distribution of clays in rocks

a) Organisation of clay minerals in rocks

The organisation of the clay materials and their porous space must be seen as a nesting of structures at decreasing scales. It is important to remember this, to avoid rashly extrapolating a characteristic or measurement acquired at one scale to another. The worst example, although one which is sometimes given, is to compare the macroscopic swelling properties of clay soils (geotechnics, pedology, etc.) with the intimate properties of a class of minerals (the smectites) whose interlayer swelling during a well-defined experimental procedure is used precisely to characterise this mineral family with respect to the others.

Since the aim is to understand and establish relations with the physical and mechanical properties, we must consider that it is the textural and structural (micro-fabric) aspect which controls these properties, the intimate mineralogical composition of the clay only being of secondary importance. The genetic aspect (i.e. study of the formation of these clay fractions during the history of the rock) is only important insofar as it determines these textural/ structural aspects and allows them to be distributed in a geological model of the reservoir. This allows us to hierarchise the data acquisition requirements which, in such a complex field, might appear to be unlimited.

At macroscopic level, log interpretation has naturally led to a classification into three broad classes (schematised on Figure 1-1.4) where the organisation of the clay phase and the differentiated petrophysical responses are implicitly equated. Although this classification has stood the test of time and must be retained, we need nevertheless to be aware of its limitations and approximations.

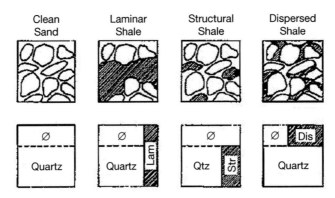

Figure 1-1.4 Forms of shale sorted by manner of distribution in the rock formation and their volumic representations (Schlumberger Co. document [1989])

Two inaccuracies must be clarified:

– There is still a confusion between the "claystones" in the lithological sense, whose mineral clay content may be no more than 50% to 60%, and the mineralogical clays in the strict sense of the term. In the remainder of this document, we will reserve the term clay to the latter and the term shale to the former.
Consequently, "laminar clays" are always shales, "dispersed clays" are generally mineralogical clays and "structural clays" can be either one or the other.

– "Laminated clays" and "structural clays" are generally considered to be inherited whilst "dispersed clays" would be authigenic. In practice, we end up with the previous distinction: detrital "allogenic clays" are interspersed in shales and cohabit with a very fine quartzo-feldspathic detrital phase, whilst "authigenic clays" consist of clay assemblages, sometimes associated in shallow reservoirs with amorphous, more or less hydrated, phases (silica, alumina, iron).

b) The two types of texture: "mud supported" and "grain supported" and the consequences on porosity

The distinction which induces fundamental behavioural differences separates the mud supported and grain supported textures.

- **In mud supported textures**, the grains are dispersed in a clay or silt-clay matrix. In this case, the hydromechanical properties are "soil" type with, in particular, the existence of shrinkage phenomena during the first drying phase (or draining phase in the sense of Capillary Pressures) (Phase A of Figure 1-1.5). Loss of the wetting fluid occurs at the expense of a reduction in Total Volume. The porous volume drops accordingly and the material remains saturated up to a limit known as the "shrinkage limit". We will return to this mechanism in the chapter dedicated to capillary pressures. In practice, the claystones and some of the log analysts' "structural" shales belong to this category. At the depths investigated by oil drilling, the stresses are such that these materials are well beyond this "shrinkage limit".
- **In grain supported textures**, the grains are interconnected and the structure is rigid. We then speak of argillaceous sandstone. Some very simple examples of sphere packings are given in the next paragraph (§1-1.1.4)

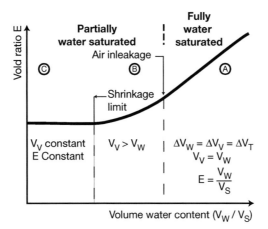

Figure 1-1.5 Shrinkage curve diagram (correlative evolution of water content and void ratio). We can identify three phases as the water content decreases

Ⓐ the reduction of total volume is equivalent to the loss of water. We speak of plastic or liquid "soil" type behaviour.

Ⓑ from the "air inleakage point", simultaneous desaturation of the sample and reduction in total volume.

Ⓒ from the "shrinkage limit", the total volume of the sample is constant. The water loss corresponds entirely to desaturation. We speak of rigid type behaviour (e.g. consolidated rock).

C) Geological classifications for the description of clays

Mixing genetic considerations (geological history) with descriptive considerations often leads to later confusion and on principle must be avoided. A practical exception can be made, however, to distinguish for classification purposes the allogenic clay phases (deposited at the same time as the associated detrital phase) and the authigenic clay phases (crystallised during diagenesis in the intergranular space).

- The facies of allogenic clay minerals are synthesised according to Willson and Pittmann [1977] on Figure 1-1.6 which is frequently used.
- The "authigenic" clays can be described by referring to one of the two classifications of Neasham [1977] or Willson and Pittmann [1977] summarised on Figure 1-1.7. These descriptions are complementary. The terms *"grain coating"* and *"pore lining"* are synonyms.

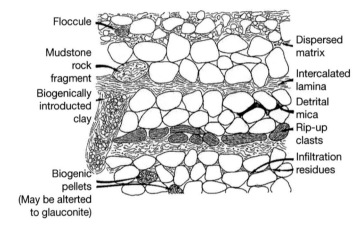

Figure 1-1.6 Mode of occurrence of allogenic clay in sandstones, according to Willson and Pittmann [1977]

D) Influence of drying techniques on modifications of the clay phase in rocks

We have seen that the clay phase collapses in on itself during ordinary drying (see shrinkage limit).

- In shales ("mud-supported" texture), all the material shrinks. The sample cracks (see aggregate structure) and its porosity decreases.
- In reservoir rocks with low clay content (e.g. "grain-supported" texture), the clay phase shrinks on the detrital skeleton. The total porosity remains constant but we observe a redistribution between the microporosity (which decreases) in favour of the intergranular macroporosity which increases, like the effective permeability.

In a first analysis, this damage is due to the considerable capillary forces exerted on the clay phase. This is therefore related to the presence of the two-phase gas-liquid system in the porous medium. To dehydrate the material without disturbing its texture, the two techniques used to bypass the gas-liquid branch were described in § 1-1.1.2.A.

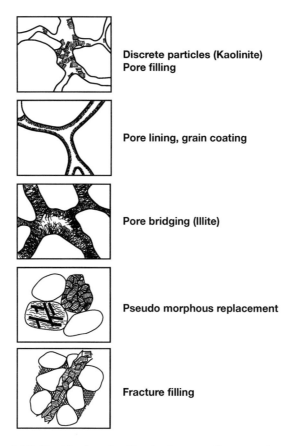

Discrete particles (Kaolinite)
Pore filling

Pore lining, grain coating

Pore bridging (Illite)

Pseudo morphous replacement

Fracture filling

Figure 1-1.7 Classification of authigenic clays according to Neasham [1977]
and Willson & Pittmann (1977)

The result on materials may be quite spectacular, for a fragile clay phase: authigenic phase (smectite, "fibrous illite" arranged in pore-lining (or grain coating). It is less clear when the pore-lining phase is more rigid (chlorite cover). Although drying precautions are essential with muds or poorly consolidated shales and surface formations (soils in the geotechnical and/or pedological meanings), in contrast, the effect is imperceptible in shales which have already been subject to high consolidation pressures, or in some special cases of allogenic grain-coating.

The effects of drying are illustrated on Figure 1-1.8 which concerns an argillaceous sandstone (tertiary continental delta; France, drilling). The clay phase (Fig. 1-1.8a) is inherited from the deposit. This is a special case where the grain-coating clay phase is allogenic and not authigenic. This is clear, especially since the crystallites are arranged parallel, not perpendicular, to the surface of the detrital grains. This structure has been acquired by the exundations of the sediment before burial.

a) Scanning Backscattered Electrons Microscopy
Polished section

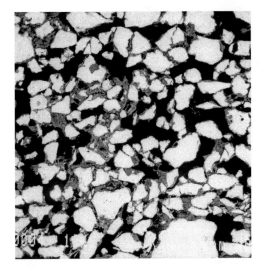

White: Quartz
Grey: Clay
Black: Porous medium
(Sample conventionally dried)

b) Porosimetric Spectra

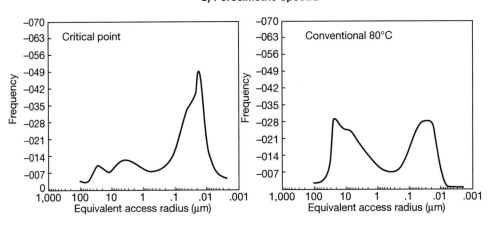

Drying method	Macroporosity >0.4 µm	Microporosity <0.4 µm	Gas Permeability (mD)
Critical point	0.084	0.185	75
Conventional 80°C	0.145	0.124	115

Figure 1-1.8 Example of the effect of drying techniques
on modifications of the clay phase of an argillaceous sandstone

After its deposition, over the course of geological time, the sediment has been desaturated and the clay phase has "frozen" the shape of the capillary menisci.

a) Microphotograph of the sandstone studied; SEM, backscattered electrons.

b) Porosimetry spectra (see § 1-1.2.4.D, p. 79) of two samples subjected to different drying methods.

The effect of the drying method can be clearly seen on the porosimetry curves (Fig. 1-1.8 b) where, with traditional drying, we observe an increase in the macroporosity due to the collapse of the clay phase onto the detrital quartz skeleton.

E) Summary on clay porosity: various "water types" and concordance with log analysis terminology

Summing up, on Figure1-1.9, the previous definitions are mapped against those of the terminology generally used in petroleum log analysis, comparing them with the fields investigated by the various laboratory measurement techniques. On Figure 1-1.10, the laboratory measurement results are compared against those obtained using log analysis.

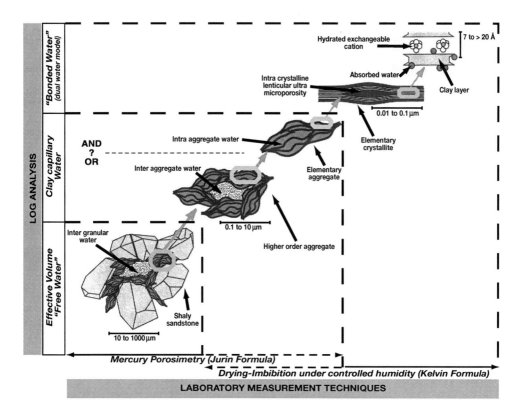

Figure 1-1.9 Various "water types" found in shaly rocks, comparison between petrophysics laboratory and log interpretation terminologies. [according to B. Layan, private correspondence]

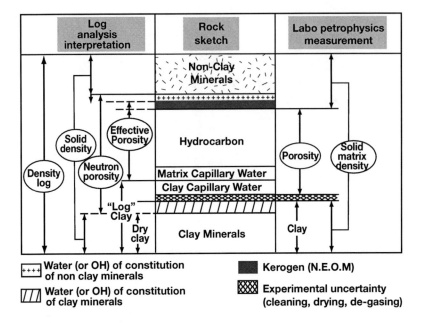

Figure 1-1.10 Comparative summary of the results of laboratory measurements and log analysis, concerning the porosity of clays

(As a complement to this diagram, note that advanced log analyses can be used to express the porosity in terms of total porosity) [according to B. Layan, private correspondence].

1-1.1.4 Order of magnitude of porosity in geomaterials

A) Theoretical data on intergranular spaces

The porous space which is simplest to model, while remaining close to the true situation of geological materials, is the intergranular space existing between spheres of the same diameter. The characteristics of this space depend on the type of packing chosen for the spheres. Extensive studies have been conducted on this subject, an overview can be obtained in Cargill [1984]. Some classical results are listed in Table 1-1.3 [Bourbié *et al.* 1987].

It is extremely important to note that regular packing can only exist in nature on volumes of a few grains. Concerning porosity therefore, there is little point in considering them and the extreme porosity values of 0.26 and 0.48 are both quite unrealistic as regards "true" packings of spherical grains of the same diameter.

In contrast, Dense Random Packing of Hard Spheres is closer to reality. It consists of random packing corresponding to the result of light settling (grains not deformed at the points of contact). They are studied on computer-generated models and soon turn out to be highly complex. We will take the average porosity value of 0.36, which corresponds to the maximum porosity that can be observed in the very well sorted sandstones.

Table 1-1.3   Characteristics of some sphere packings

Packing type		Solid phase structure	Pore structure [3]	Porosity (%)	Number of contact points per sphere	Void type	Radius of maximum inscribable sphere [1]	Radius of maximum sphere passing through narrowest pore channels	Fraction of porosity contained in the maximum inscribable sphere (%)
Regular packings	Simple cubic	Cubic void		47.6	6		0.732	0.414 curvilinear-square pore access	43
	Simple hexagonal	Simple rhombus void		39.6	8	2 trigonal	0.528	0.414 and 0.155 curvilinear-triangular pore access	45
	Compact hexagonal or tetrahedral	Tetrahedral void	Separated view	25.9	12	2 tetrahedral + 1 octahedral	0.225 0.414	0.155	27
Dense random packing of hard spheres [2]				about 36	around 9 on average	at least five main types Bernal's canonical holes	most frequent radius 0.29		

The unit of length is the radius of the sphere.
(1) From Guillot (1982).   (2) From Cargill (1984).   (3) From Graton and Fraser 1935.

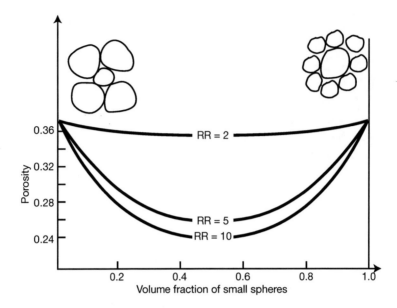

Figure 1-1.11 Porosity of a packing of spheres of different radii (two values in ratio RR) as a function of the value of this ratio and the volume fraction of smaller spheres. According to Guyon *et al.* [1987]

The packings of isogranular spheres only correspond to the relatively rare case of very well sorted sandstones. If we consider packings of spheres of different diameters, porosity is much more variable. On the still quite simple example of Figure 1-1.11, we can observe the porosity variation of an aggregate containing two types of sphere. The porosity value is always less than the maximum of 36% mentioned above, the minimum corresponding to the intermediate volume fractions.

B) Sedimentary rocks

In sedimentary rocks, the porosity varies considerably under the effect of compaction during burial. Two main phenomena are responsible for the reduced porosity: rearrangement of grains under the effect of stresses and cementation, in which the pressure-solution mechanism (increased solubility of minerals under the effect of stress) may play a major role, especially in limestone rocks.

a) Clastic formations

In clastic formations, the porosity near the surface corresponds to that of sands. Sand porosity varies around 0.35, as we have seen above for the well sorted sands. It can be significantly increased by the presence of irregularly shaped grains (shell fragments, micas) or decreased by granulometric dispersion.

During diagenesis (modification of the rock structure under the effect of pressure or temperature during burial), the porosity reduction will vary to such an extent, depending on the geological formation and the sedimentary basin, that it would be pointless giving any trend whatsoever. We will just say, simplifying matters to the extreme, that sandstones of porosity greater than 0.25 and 0.10 are rare below depths of 2 500 m and 5 000 m respectively. At these extreme depths, sandstones generally only retain porosity values of a few percent, but this porosity is sometimes sufficient to generate large reservoirs of gas at high pressure, known as tight sands.

b) Carbonates

The change in porosity of carbonate rocks during burial is even more complex than with sandstones, due to the high chemical sensitivity of calcite and dolomite. At shallow depth, we find limestones which are almost compact (porosity less than 0.05), as is often the case in lacustrine series, or extremely porous (over 0.60) for shell and coral limestones. Attempting to plot a curve to illustrate the trend between depth and porosity would be futile, even for a given basin or zone. The porosities of carbonates are statistically even lower than the orders of magnitude indicated for the sandstones in the previous paragraph. There are, however, a few lucky exceptions, such as some large carbonated reservoirs in the Middle East whose porosity exceeds 0.25 at a depth of 2 500 m.

We must nevertheless point out one exception to the general rule of "chaotic" variation of limestone porosities: the chalks, which are extremely thick in some areas (Paris Basin, North Sea). Their porosities are in the range of 0.35 to 0.45. We may sometimes observe a regular reduction in porosity with burial, however, corresponding to a compaction phenomenon comparable to that in the clay series.

c) Compaction laws for shale-sand formations

The shale-sand series represent almost 90% of the sedimentary terrains. Oil drillings therefore cross them at more or less the same proportion and numerous log analysis records are available which provide a fairly accurate picture of the change in porosity (and in the related change in compressional wave velocity, see § 1-3.2.6C, p. 243) against the depth of burial. These porosity vs. burial (z) relations are known as compaction laws or profiles. They are generally exponential type ($\phi = \phi_0 e^{cz}$, where z is the depth, c is a constant and ϕ_0 is the initial porosity of the surface sediment). Figure 1-1.12 gives two highly diagrammatic examples of extreme profiles, corresponding to the Oklahoma primary series (highly compacted) and to the Gulf of Mexico tertiary series (poorly compacted [according to Magara, 1978]. Note that the porosity values are given in total porosity, not in effective porosity. These compaction profiles serve as data for mathematical modelling of the evolution of sedimentary basins. They are also used in the detection of "under-compacted areas" exhibiting abnormally high porosity (and correlatively abnormally high pore pressure).

C) Porosity of crystalline and volcanic rocks

Although petroleum geologists are unfamiliar with the porosity of crystalline and volcanic rocks, this subject is nevertheless very important. We must make a distinction between unweathered crystalline rocks (granite, gneiss, etc.), leached crystalline rocks and volcanic rocks (lava).

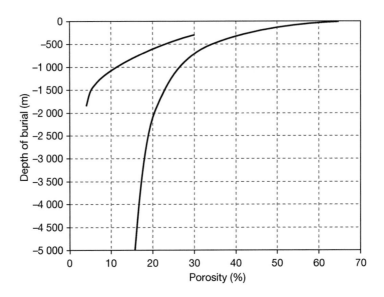

Figure 1-1.12 Compaction profile of shale-sand series according to Magara [1978]

The top curve corresponds to the Oklahoma primary series (highly compacted).
The bottom curve corresponds to the Gulf of Mexico tertiary series (poorly compacted).

Crack porosity may play a very important role in crystalline rocks, although its fractionary value never exceeds 10^{-2}. This special type of porosity is discussed in § 2-1.1.2, p. 267.

a) Unweathered crystalline rocks

Unweathered crystalline rocks exhibit very low porosity ($\phi < 0.01$ in granites). It consists mainly of occluded porosity (fluid inclusion in crystals) and, to a lesser extent, of microcrack porosity. As regards the "matrix" properties (centimetric or metric scale, see § 2-1.3.2, p. 311), unweathered crystalline rocks can therefore be considered as compact for everything concerning storage or circulation of fluid at human time scale. The situation may be quite different if we consider the bulk properties, including possible networks of open fractures.

b) Leached or hydrothermalised crystalline rocks

Crystalline rocks near the surface may be subject to percolation of chemically aggressive meteoric water (e.g. in tropical climate) and undergo very deep weathering of some minerals (e.g. feldspar). There is little change to the external appearance of the rock, but this leaching could create up to 0.2 porosity. This phenomenon may affect the rock for depths of several hundred metres. On the surface, these areas may form large aquifers (Figure 1-1.13, weathered diorite, north eastern USA). Phenomena corresponding to an upflow of geochemically active fluids (hydrothermalism) may produce the same results. When, over

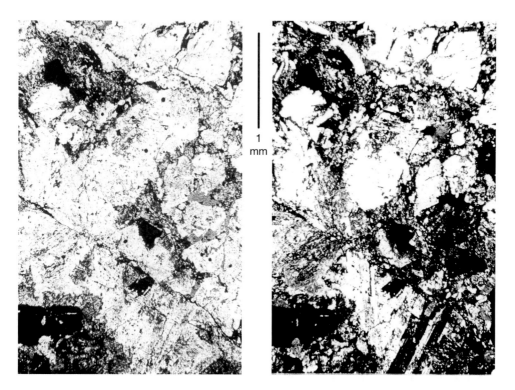

1
mm

Figure 1-1.13 Photograph of a thin section of weathered crystalline rock (diorite?)
"natural" light on the left and polarised light on the right. Blue resin is injected to highlight the porosity

the course of geological history, these leached or hydrothermalised rocks are buried under sedimentary covers, they may become oil reservoirs (e.g. Sumatra, Venezuela, Vietnam).

c) Volcanic rocks

There are numerous types of volcanic rocks and this diversity is reflected in their porosity.

Consolidated volcanic rocks (lavas) are rarely compact (e.g. some basalts or andesites) but are generally porous (in particular, vug porosity due to surface magma degassing (Fig. 1-1.14). The most well known example is pumice stone, whose porosity can reach very high values ($\phi > 0.80$).

The porosity of unconsolidated, or poorly-consolidated, volcanic rocks (ashes, volcano-sedimentary) is also often very large. The study of porosity in volcanic rocks has not received the strong financial backing which, for more than half a century, has focused on the study of porosity in oil-bearing sedimentary rocks. The petrophysical study of these rocks is in its infancy and should prove to be highly interesting.

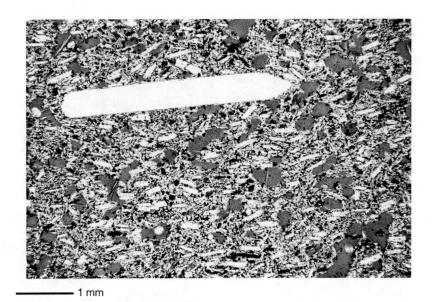

—————— 1 mm

Figure 1-1.14 Photograph of a thin section of porous lavas,
Volvic andesite ($\phi = 0.23$, Kmax = 60 mD, Kmin = 15 mD)

1-1.1.5 Solid phase density and mineralogy

When measuring the porosity it is always possible, with no need for additional experimental work, to measure or calculate the solid phase density (derived from the sample mass and solid volume). It is strongly recommended during the porosity-permeability measurement programme (routine core analysis, § 2-1.1.2, p. 269) to simultaneously request a density calculation.

Unlike porosity, permeability or saturation, density is not a variable directly required for reservoir evaluation. This parameter, however, at the boundary between petrophysics and petrography, is essential data throughout the reservoir description process.

We will develop here the advantage of implementing a combined "porosity/mineralogy/density" process. We will mention its application to calibration of the quantitative (petrophysical) log analysis often expressed in "effective" porosity.

A) Advantage of the mineralogical description for petrophysical evaluation of materials

In addition to the purely geological objectives, mineralogy processing presents the advantage of allowing:

- cross-checking of petrophysical data and geological data (verification/adjustment of solid densities);
- cross-checking of geological data;
- the preparation of "mineralogical poles" of the multi-mineral quantitative interpretation of well logs.

The first of these checks is still necessary, the second is recommended, and the preparation of "mineralogical poles" is essential if multi-mineral quantitative interpretation of well logs is planned. In these processes, the "true" mineralogical assays resulting from analytical techniques (X-ray diffraction cf. § 2-2.4, p. 352; elementary chemical analyses, etc.) must be used, not the "mineralogies" resulting from visual interpretation of thin sections.

a) Cross-check of petrophysical data and geological data

In the general case where the mineralogical and petrophysical data are acquired independently, sometimes using techniques whose procedure is difficult to control, this first step is unavoidable and mutually validates mineral data and physical data. We summarise below the main types of solid phases encountered and the data on their densities. These densities are summarised in Table 1-1.4

Table 1-1.4   Density of the main components in rocks

Mineral or Material	Density Standard value kg/m^3	Realistic extreme value kg/m^3
Quartz	2 650	
Potassium feldspar	2 560	
Albite	2 610	
Plagioclase		2 760 (Anorthite)
Calcite	2 710	
Dolomite	2 870	2 820
Siderite	3 960	
Anhydrite	2 960	2 920
Barite	4 500	
Halite	2 160	
Pyrite	5 010	4 800
Muscovite	2 790	
Kaolinite	2 600	
Illite	2 780	
Smectite	2 630	
Non-extractable organic matter	1 100	1 200
Amorphous silica	2 160	

Minerals

During the process proposed, the mineral densities can be varied in a "reasonable" proportion, by following several physically and geologically realistic paths. Some minerals must be considered *a priori* to have constant density, e.g. quartz, the potassium feldspars, calcite (to a lesser extent), pyrite and other sulphides. The plagioclase is typically denoted by its most common and lightest sodium-rich end member (albite $2\,610$ kg/m^3) but its density can increase up to $2\,760$ kg/m^3 depending on the exact composition of the plagioclase. The density of dolomite can drop down to $2\,820$ kg/m^3 due to the common presence of fluid inclusions and associated organic matter. The same applies for anhydrite and the evaporite minerals.

Amorphous phases

The "amorphs" are impossible to determine using X-ray diffraction techniques in the strict sense. Two types of low-density amorphous phases are commonly encountered:

- Non-extractable organic matter: kerogen, lignite and coals, etc. (NEOM). This organic matter – excluding hydrocarbons – is always present. The inexpensive measurement of "non-extractable organic carbon" (NEOC), should be carried out systematically. It is present in average quantities of 0.1% to 0.5% by weight in sedimentary rocks. Even when the content is low, it has a noticeable impact on the density calculation. If no special precautions are taken, this organic matter will be included in the hydrocarbons by conventional quantitative log analysis. The log analyst must be notified if it is relatively abundant. Figures 1-1.15 *a* and *b* show a quantitative aspect of this effect. Figure 1-1.15a shows the impact of the non-extractable organic matter on the "dry solid" density ρ_S, corresponding to the solid fraction of the material (sometimes called ρ_{ma}, for "matrix", but this term may be confusing). ρ_S is the result of the petrophysical measurement. Figure 1-1.A15b shows the error made when estimating the porosity by log analysis if this NEOM is not taken into account. The raw log analysis, in fact, considers the NEOM as a hydrocarbon and therefore includes its volume in the effective porosity calculation.
- Amorphous silica, quite difficult to detect, is relatively rare. It may be suspected when a tridymite/cristobalite assemblage is detected. In this case, the log analyst must be informed since this silica is light and hydrated (hydrogen index). Failure to take this amorphous silica into account may induce a large overestimation of the initial water content of a reservoir. In contrast, the "false amorphs" detected during analysis of petrographic thin sections could be misleading. The maximum resolution of the optical microscope is deceiving in this context.

"Artefact" minerals

Present in small quantities, some minerals are artefacts:

- rock salt (halite) or other "soluble salts", resulting from the crystallisation of brine saturating the sample;
- the presence of smectic clay (bentonite) in permeable materials may indicate invasion. The same applies for the usual mineral loads of sludges: barite, calcite;
- some minerals produced by open air oxidation of the sulphides in the reservoir: gypsum, natrojarosite, etc. Ideally, they should be converted (mole for mole) into pyrite type sulphite.

Some quantitative aspects are discussed in § *Bd*

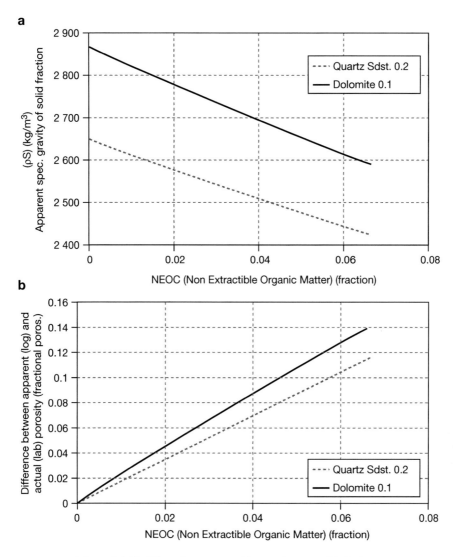

Figure 1-1.15 Effect of non-extractable organic matter (NEOM) on:
a) the solid phase density (including the NEOM);
b) the log analysis porosity calculated by incorrectly considering the NEOM as a hydrocarbon.
The values of the x-axes are expressed in content by weight of non-extractable organic matter (NEOM). The density of NEOM is 1 100 kg/m^3.

b) Cross-check of various geological data

This cross-check must be carried out in three stages for optimum control of the scale effect:

Mineralogy versus description/quantification on thin section:

Description of the thin section is used in microtextural (micro-fabric) analysis which, in terms of petrophysical interpretation, is not without interest (e.g. distinction between detrital quartz and quartz cement in sandstones, Dunham/Folk classification in carbonates, etc.). Mineral counting in the strict sense (or its visual estimation) allows coherence analysis, but cannot be considered, under any circumstances, as a quantitative mineral analysis.

The comparison will be made by plotting on a graph, for each mineral, the volume contents resulting from mineralogical analysis and analysis of thin sections after correcting the microporosity effect on the "thin section" mineral phases. An advanced analysis consists in also comparing the "thin section porosity" and the macroporosity derived from specific processing of the porosimetry curves by mercury injection (§ 1-1.2.4D, p. 79). Whilst it is impossible to obtain absolute equality, a blatant inconsistency could be the sign of:

- failure to respect the sampling procedures;
- presence of mineral phases too difficult to determine on thin section (type or content) under the observation conditions generally practised in petroleum sedimentology analysis;
- incorrect estimation of the microporosity distribution.

In oil reservoir levels, observation of thin sections is not always reliable. The hydrocarbons are frequently not (or poorly) washed off. The residual oil is often interpreted incorrectly.

Mineralogy *versus* plug description

The "sandstone, limestone, dolomite, etc." description must be coherent with the mineralogical analysis: The quantitative mineralogy is compared with the lithology (mineralogical crossed graphs, box-and-whisker plot, variance analysis, etc.). Software products are available to carry out more advanced analysis easily, creating mineralogical groups by statistical classification applied to mineralogical parameters and performing a mapping analysis with the lithofacies.

An inconsistency is symptomatic of a sampling problem.

Mineralogy *versus* elementary layer description (log 1/40 on core: interval data)

The method is the same as that above, but is not redundant. It is used to control the start of the "upscaling" logic: from the sample to the "unit" geological bed of thickness equal to the vertical resolution of conventional log analysis tools. An inconsistency may be due to:

- failure to respect the sampling procedures;
- heterogeneousness of the level (organised: e.g. laminated, or not: e.g. bioturbated, pyrite, anhydrite, siderite diagenetic nodules, etc.). This information will be critical when constructing the "petrophysical log" or the "core log" in general (§ 2-1.3.1F, p. 305).

c) Preparation of the "mineralogical end-members" for quantitative interpretation

General case

The log analyst often needs to reduce the number of mineralogical end-members used in the interpretation. Data analysis can be used as one of the grouping guides, by:

- building the correlation matrix to identify the minerals whose contents are correlated or anti-correlated;
- if necessary, executing this operation by depth intervals based on geological units;
- checking that these associations are physically/geologically realistic.

It is therefore recommended to associate – clay assemblage/pyrite/organic matter –. This set is an electrical conductor. The organic matter participates both in cation exchange and radioactivity. The extreme densities of sulphide and organic matter will tend to compensate for each other.

If zeolites (clinoptilolite, heulandite, analcime, etc.) are detected, they can also be associated with a clay phase (hydrogen index, cation exchange capacity, etc.).

Potassium feldspars offer the advantage of "consuming" some of the potassium radioactivity. When the contents of these feldspars exhibit good correlation with those of quartz, it may be worthwhile to suggest grouping. The same applies for plagioclases. Creation of this "clastic" end-member should be avoided, however, if log analysis is also used by the sediment-diagenetic analysis.

In contrast, it is best not to group the carbonates (calcite, dolomite and siderite).

Special case of the clay phase

Interpretation of the clay phase in thin section may lead to a great deal of confusion. For example, the geologist's "glauconite" is not the same as crystallographic glauconite (it often consists of an illite/ferriferous chlorite association, etc.). Consequently, a "clay-mineralogy" derived from thin section analysis should never be used quantitatively in log analysis.

The new techniques combining X-ray diffraction and X-ray fluorescence (XD-XF, § 2-2.4, p. 352) provide reliable quantification of the clay mineral phase. Generally however, they only investigate the fraction below 5 μm. "Large" clays such as some authigenic kaolinites (dickite – photo on Figure 1-1.16; see also SEM photo on Figure 1-1.46, p. 90) will be underestimated.

Whenever possible, both as regards data analysis and log analysis, it may be worthwhile separating two "clay" end-members: an "authigenic" type end-member (which cannot be extrapolated to inter-reservoirs and cap-rock) and an "allogenic" type end-member (which, for simplification purposes, can be considered as common to the reservoirs, the inter-reservoirs and the cap-rock). These end-members must be built on the basis of the clay-XD analysis but must be corroborated by thin section observation and especially by scanning electron microscopy (SEM).

After completing these operations, the dispersion of the "calculated solid density" ($\rho_s cal$) *versus* "measured solid density" ($\rho_s mes$) graph is a good indication of the practical uncertainty of this parameter after reconciliation.

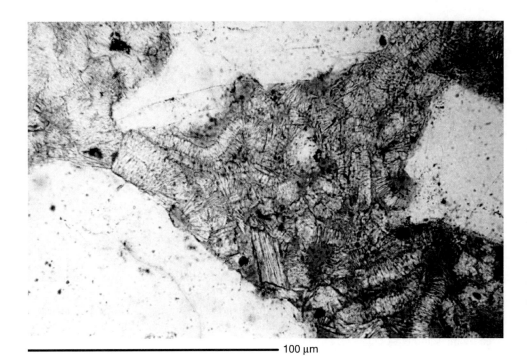

100 µm

Figure 1-1.16 Large crystals of authigenic kaolinite (dickite)

B) Example of process used to reconcile various data

In this section, we provide some examples of processes used to reconcile various "porosities", in particular "petrophysical porosity (ϕ)" (laboratory measurement) and "effective porosity (ϕ_{eff} or PHIE)".

a) Volume content – weight content reconciliation

Especially in quantitative log analysis, it is important to make a clear distinction between the two notions of volume content and weight content. The confusion between these two types of content is sometimes overlooked since the resulting differences on the calculation of total sample density are negligible provided that the densities of the minerals present are similar (general case). This can prove disastrous, however, if there are large differences between the densities of the mineral phases. Contents determined using analytical techniques on samples are generally expressed in weight contents, reduced to a dried weight according to an accurately defined procedure (example: 105°C in ventilated oven; sometimes 80°C or 150°C in ventilated oven, etc.).

In contrast, the mineral contents of quantitative log analyses are expressed in volume content, which is consistent with the expression of porosities and saturations, also given in terms of volume.

Where: $v_{(i)}$ i element volume content, of volume $V_{(i)}$, of weight content $w_{(i)}$ of weight $W_{(i)}$ and density $\rho_{M(i)}$; and W_s the dry weight of the sample of density ρ_s, hence:

$v_{(i)} = V_{(i)}/V_t$ and $w_{(i)} = W_{(i)}/W_s$, noting that $W_s = \rho_s V_s = \rho_s V_t(1 - \phi)$

$V_t = W_s/\rho_s(1-\phi)$ and that $V_{(i)} = W_{(i)}/\rho_{M(i)}$, we obtain the correction formula:

$$v_{(i)} = (w_{(i)}/\rho_{M(i)})\, \rho_s(1 - \phi)$$

where we can see that if $\rho_{M(i)}$ is close to ρ_s, the difference between the two contents is low.

b) Notion of porosity associated with minerals (clays)

The porosity associated with a mineral is directly related to *the log analysis notion, not the physical notion,* of effective porosity. In this paragraph, we will not try to define this porosity associated with a mineral (exclusively clay in fact). This point has been discussed earlier (§ 1-1.1.3). We will simply give a few examples of results obtained after taking into account a "porosity associated with a mineral".

This data can be expressed in three different ways:

- $\phi_{M(i)}$; the proportion of the porous volume internal to the phase (i) – as a fraction of the total sample volume –;
- or $\phi_{P(i)}$; the "internal" porosity of the phase (i) – as a fraction of the volume of this phase (i) –;
- or $\rho_{A\ (i)}$; the "apparent" density (i.e. including the internal porosity). This is known as the "wet density" of the phase (i).

As throughout this section, we assume that the porosity ϕ of the petrophysical file corresponds to the "actual" ("total") porosity. We therefore obtain the "effective" porosity expression (principle formula): $\phi_{eff} = \phi - \Sigma(\phi_{M(i)})$

We use the new parameters $v_{phm(i)}$ volume content of the mineral phase "i", *including the associated porosity* and $w_{phm(i)}$ weight content of the mineral phase "i" *including the fluid saturating the associated porosity,* fluid of density $\rho_{F(i)}$

Note that, for convenience, the sum of the clays is generally represented by a "mineral-equivalent". The total volume content of the clay minerals is written v_{cl} and the volume content of the clay phase together with the porous volume, to which it is associated by convention, is written v_{sh}

$$\Sigma(v_{(clay)}) = v_{cl} \text{ and } \Sigma(v_{phm(i)}) = v_{sh}$$

The "apparent" density of the clay fraction together with its associated (or "internal") saturated porosity is called the "wet" density ρ_A. Figure 1-1.17 represents a graph of this density against $\phi_{P(i)}$

During a quantitative interpretation, it is important to specify the types of content/porosity used. A quantitative example based on the case of a kaolinitic sandstone is given below.

Petrophysical data (lab)

quartz weight content, $w_{(quartz)} = 0.75$

"150°C dry" clay weight content (kaolinite), $w_{(clay)} = 0.25$

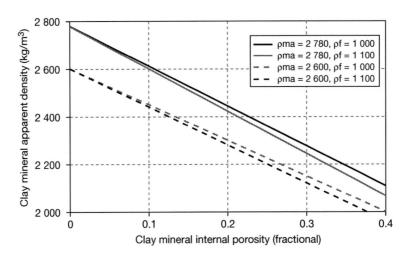

Figure 1-1.17 "Wet" density of two types of clay, against the internal porosity of the clay and the density of the saturating fluid (ρ_F)

e.g. Illite, $\rho_M = 2\ 780$ kg/m^3 and Kaolinite, $\rho_M = 2\ 600$ kg/m^3.

total porosity (petrophysical), $\phi = 0.20$

solid density (petrophysical) $\rho_S = 2\ 640$ kg/m^3

First log analysis

apparent density of the "wet" clay phase $\rho_A = 2\ 100$ kg/m^3

with, by assumption, a fluid density $\rho_F = 1\ 100$ kg/m^3

Using the formula given in the previous paragraph (b), we can calculate the dry clay volume content v_{cl} directly, $v_{cl} = v_{(arg)} = 0.203$ as well as the quartz volume content $v_{(quartz)} = 0.598$.

Considering clay as an "intrinsically" porous medium and using the formula of § 1-1.1.2A, p. 6, giving the porosity against various densities, we obtain $\phi_{P(clay)} = 0.333$. Using the formula in the same paragraph giving V_V as a function of ϕ and V_S, we calculate $\phi_{M(clay)} = 0.101$ since v_{cl} can be assimilated to a solid volume.

We have therefore recalculated the volume contents of quartz (0.598), clay together with its own internal porosity (0.203 + 0.101) and the "effective" porosity (0.099 = 0.2 − 0.101).

As expected, the sum of all these contents is equal to 1, apart from rounding errors.

c) Impact of actual occluded porosity

This is the exception that confirms the rule: Some of the porosity is actually occluded in the sense of the petrophysical measurement. In practice, and if the experimental conditions of the petrophysical measurements are correct, this case is limited to the intracrystalline porosity (fluid inclusions). Since this phase is not taken into account by the petrophysical measurement, the density of the crystalline species concerned turns out to be abnormally low.

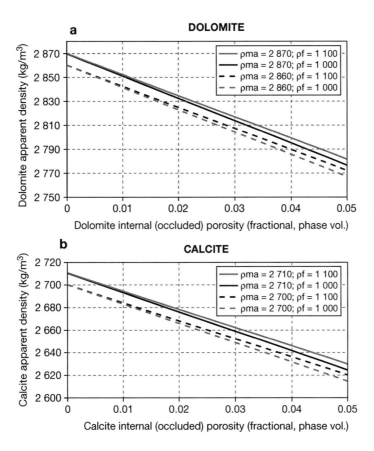

Figure 1-1.18 Effect of occluded porosity on the density of
a) Dolomite with fluid inclusions (two assumptions regarding ρ_M of the dolomite without occluded porosity and ρ_F).
b) Crinoidal limestone with syntaxic cementation (two assumptions regarding ρ_M of calcite and ρ_F).

In concrete terms, it should therefore be possible to "force" this density below the reference crystallographic minimum value. Strictly, the user must simultaneously enter a Hydrogen Index HI(i) consistent with the reduction in mineral density since some of the fluid is included in this mineral phase.

Figures 1-1.18a and b illustrate a few orders of magnitude. The dolomites are a relatively classical case (e.g. the Lacq gas reservoir in France or oil reservoirs from the Pinda in Angola where $\rho_{eff} \approx 2.83$ to 2.84).

An example of crinoidal limestone very similar to the case schematised on Figure 1-1.18b, is given on Figure 1-1.19. Obviously, the occluded porosity does not correspond to the microporosity visible on the pore cast (micritisation crown of the crinoid fragments, under the effect of bacterial attack) since this microporosity, being saturated with epoxy resin (§ 2-2.1, p. 325), is connected porosity. At other places however, this microporosity is occluded.

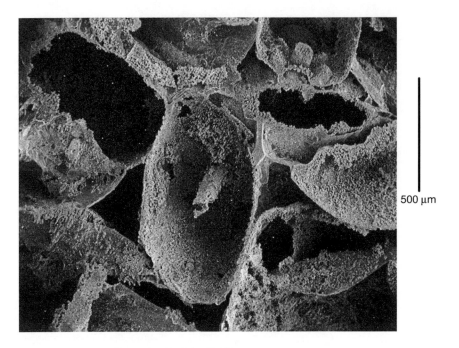

500 µm

Figure 1-1.19 Crinoidal limestone, epoxy pore cast

d) Impact of artefact minerals

Depending on the operating conditions of the mineralogical analysis and/or the core conservation conditions, the mineralogical assemblage may include minerals acquired after extracting the core: the artefact minerals (see § A):

It may be desirable to take these artefacts into account. There are two possible cases:

- the mineral concerned is added to the initial assemblage with no influence on the petrophysical porosity measured, since only present in trace quantities, in which case it is simply removed from the mineralogical analysis;
- the mineral concerned fills the porous medium: after calculating its volume content, the porosity must be corrected.

A quantitative example is illustrated on Figure 1-1.20 (a pure quartz sandstone containing crystallisation salt). We see that a content of 1.5% salt (halite) lowers the density of pure sandstone (quartz) from 2.65 to 2.64 g/cm^3. The apparent porosity would be measured at 0.185 for an actual porosity of 0.20.

These differences are not negligible since a good petrophysical laboratory easily achieves, in "routine analysis", precision of better than 5 kg/m^3 for density and 0.005 for porosity.

Another quantified example of these effects (artefacts and/or omission of minerals present in low quantities) is given in Table 1-1.5. We will start with a kaolinitic sandstone

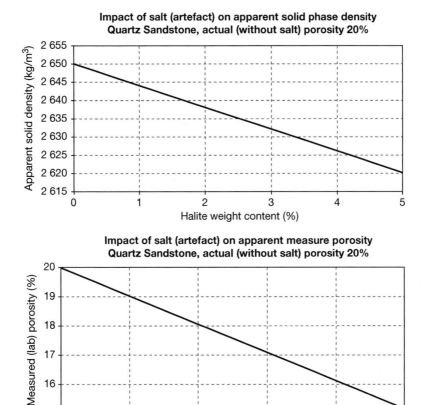

Figure 1-1.20 Impact of salt deposited during drying on the solid phase density (ρ_S) and the apparent porosity of a sandstone of actual (without salt) porosity of 0.20

containing pyrite and non-extractable organic matter (NEOM). The weight contents ($w_{(i)}$ "lab" data) are indicated in column 1, the volume contents $v_{(i)}$ (column 3) and the solid density (ρ_S) of this example are calculated.

Note that taking the assumption that only the major minerals (quartz, kaolinite and illite) would be taken into account, the apparent ρ_S calculated would be equal to the previous one (NEOM – Pyrite compensation) but since the NEOM (whose volume content exceeds 2%) is therefore considered as a hydrocarbon in the log analysis, there would be a significant overestimation of the porosity and saturation of this poorly permeable sandstone.

If the pyrite is detected and correctly evaluated but the NEOM is not measured (most likely situation in many laboratories) the apparent ρ_S is 2686 kg/m3. We may therefore observe the inconsistency with the "measured" value which is much lower, indicating the possible presence of organic matter.

Table 1-1.5   Example on calculating the impact of an "artefact" mineral (bentonite) on the porosity and apparent density of a sandstone

Material	1	2	3	4	5	6	7
	Initial data Sandstone 10% porosity			Bentonite artefact (Smectite) 150°C		Bentonite artefact (Smectite) 85°C	
	Weight contents $w_{(i)}$	Material density $\rho_{M(i)}$ kg/m^3	Volume contents $v_{(i)}$	Weight contents	Volume contents	Weight contents	Volume contents
Quartz	0.750	2 650	0.6745	0.711	0.6745	0.707	0.6745
Kaolinite	0.160	2 600	0.1467	0.152	0.1467	0.151	0.1467
Illite	0.050	2 780	0.0429	0.047	0.0429	0.047	0.0429
Pyrite	0.030	5 010	0.0143	0.0285	0.0143	0.0285	0.0143
NEOM	0.010	1 100	0.0217	0.0095	0.0217	0.0095	0.0217
Smectite (150°C)		*2 610*		*0.052*	*0.050*		
Smectite (85°C)		*2 310*				*0.057*	*0.0625*
Porosity	**10%**			**5%**		**3.75%**	
ρ_S **in kg/m^3**		**2 648**			**2 646**		**2 626**

The same table shows the results for the same material, where half of the porous medium is invaded by bentonite (smectite) from the drilling mud (columns 4 to 7). We can consider two different cases. Firstly, the sample has been dried at "high" temperature (150°C) and the smectite no longer contains any water inducing porosity associated with clay (§ 1-1.1.2, p. 12). The smectite density is then 2610 kg/m3, resulting in the values shown in columns 4 and 5. Secondly, if the rock has been dried at "low" temperature (80°C), we may consider that the porosity associated with bentonite is 20%, resulting in the values shown in columns 6 and 7.

e) Conclusion: Application of the approach to other parameters

The approach described is used to reconcile the various density and mineralogical content data, while checking parameters important for log analysis (NEOM, presence of hydrated amorphs). In addition however, this approach offers the advantage of being able to process other physico-chemical parameters at the same time, such as:

- Cation Exchange Capacity (CEC): This parameter is directly related to the presence of clays and is used in resistivity log analysis (Waxmann, Sushash methods, etc.). Generally, the total CEC is measured in the laboratory on a limited number of samples from a cored series. To extrapolate this result to the entire series, the calculation is based on this batch of samples presenting simultaneously the mineralogical content analysis, NEOM included (after check and validation). Multivariable statistical analysis methods can therefore be used to calculate the CECs of all the mineralogically identified samples.

– Specific Areas: in argillaceous-sandstone, some interpretations methods, or petrophysical considerations, use the notions of (external or total) Specific Areas. Calibration by mineralogical analysis can be carried out in the same way as that described above.

– The Hydrogen Index (HI) characterizing the hydrogen atom "content" of the material is extremely useful when interpreting the Neutron and RMN log analysis (§1-3.3.1, p. 248). A theoretical Hydrogen Index can be calculated by adding together the HIs brought by each mineral (mineralogical table), including the HI brought by the porosities associated with the mineral phases, and not forgetting the organic (NEOM) and possibly mineral (e.g. hydrated silica) amorphous phases. If a compositional analysis of the hydrocarbons is available (see PVT analyses), the theoretical HI of the oil can be calculated.

1-1.1.6 Effect of stresses on porosity, Compressibility

A) Effect of stresses on porosity

When a porous medium is subject to a stress variation, the geometry of the porous space varies due to two main phenomena:

– rearrangement of the solid elements leading to bulk deformation of the skeleton (example of the crushed sponge);
– variation in volume of the solid elements themselves under the effect of their own compressibility. This solid compressibility (corresponding to $1/K_{grain}$ of § 1-3.2, p. 222) is generally much lower than that related to the previous rearrangement, especially for rocks of average or high porosity.

The formalisation and quantification of these relative variations are not discussed in this book, since they pertain to the field of porous medium mechanics. Studies have been conducted by numerous authors (Terzaghi and Biot are the most well-known) on these subjects and we will refer the reader to the specialised literature on mechanics [Coussy, 2004]. We will simply give a few definitions which will be useful in the remainder of the discussion. The purpose of this chapter is to explain the impact of stresses on the porous medium, before the rupture mechanisms appear. For further information, readers can refer to books on Rock and Soil Mechanics and Civil Engineering Geology [e.g. Jaeger and Cook, 1969; Lambe and Whitman, 1979; Fjær *et al.*, 1992]. Tiab and Donaldson [1996] give detailed information on the petrophysical aspects of the effects of stress.

a) Confining, pore and differential pressure

A porous material may be subject to several types of stress (or pressure).

– The *confining stress* which is transmitted by the solid phase. In the materials of the Earth's crust, this stress is the result of the combined action of the overburden (lithostatic stress) and the tectonic stresses induced by the relative movement of the continental plates (e.g. compression). More generally, it is a true state of stress (tensorial representation).

– The *pore pressure*: the pressure of the fluids saturating the pore space. In sediments, the pore pressure value is generally equal to the value of the hydrostatic column, with sometimes major exceptions in the zones where hydraulic communications with the rest of the basin are reduced or non-existent. This is the "undrained" case of soil mechanics and the pore pressure can "absorb" a significant proportion of the tectonic and lithostatic stresses: they are over-pressure zones (thick shale formations, gas reservoirs, etc.).

– The *"capillary" pressure* (§ 1-1.2.1, p. 54) which can be transmitted to the solid phase. It becomes extremely important in soil mechanics, in water/air two-phase state, when the water saturations are low ("sand pile" effect). Soil physicists prefer to substitute the notion of potential, in order to generalise the logic. This notion is discussed in Chapter 1-1.2.3, p. 67.

We can see intuitively that the pore pressure (P_{pore}) opposes the confining stress (σ_{conf}) and that the parameter which is central when quantifying the mechanical state of the porous medium depends largely on the difference between these two types of stress. The differential stress (σ_{diff}) or differential pressure (P_{diff}) is defined as their difference: $\sigma_{diff} = \sigma_{conf} - P_{pore}$. This parameter was originally defined by Terzaghi [1943]. Another expression of the result of these two stresses is given in Biot's poromechanical theory [1941]: the effective stress: $\sigma_{eff} = \sigma_{conf} - \beta P_{pore}$ where $\beta = 1 - K_{dry}/K_{grain}$ (K_{dry} and K_{grain} are the bulk moduli of the rock saturated with infinitely compressible fluid and of the mineral forming the grains, cf. 1-3.2.5, p. 236). β is often known as Biot's coefficient, see below § B).

One point is often a source of confusion: in studies concerning the effect of stresses on the petrophysical properties, the expressions of "effective" stress used by Biot and Terzaghi are sometimes incorrectly compared.

Biot's approach, which considers a "solid + fluid" set, can be used for example to calculate the "distribution" between these two media of a stress variation induced by an external disturbance. Biot's approach is also used to evaluate the incompressibility variation of a porous system (solid + fluid) according to the variation of conditions at the limits (see § 1-3.2. Biot-Gassmann's equation).

In contrast Terzaghi's approach, through the notion of differential stress, can be used to characterise the "geometric" state of a porous medium, independently of the characteristics (modulus, pressure) of the saturating fluid. The "Terzaghi" parameter should be used when studying the effect of stresses on properties dependent on the geometry of the porous network.

Using a theoretical approach, one can demonstrate that for a homogeneous porous medium whose solid phase consists of a single isotropic element, the pore pressure variation, which affects the compressibility of the solid, induces a homothetic variation in the geometry of the porous space. The "geometric" structure itself only depends on the differential stress state.

b) *Overview on porosity evolution: Relaxation, Compaction, Pore collapse*

The broad trends in the evolution of porosity against differential stress are represented highly schematically on Figure 1-1.21.

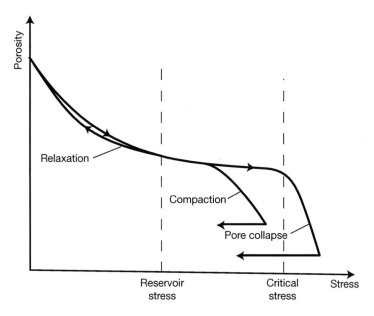

Figure 1-1.21 Schematic diagram of various trends of the evolution of porosity against differential stress

Relaxation

Note firstly that most of the samples used when studying porosity evolution are core samples which, when extracted, undergo sudden stress relaxation. This relaxation is sometimes associated with expansion of the gas bubbles present in more or less well-connected pores. The gas expansion, although initially absent *in situ* (dissolved gas), may have a dramatic effect when the core is brought up too quickly and if the medium is insufficiently permeable.

Relaxation affects the microstructure of the rock. This is clearly indicated by the change in sound wave velocities under the effect of pressure (Hertz coefficient, see § 1-3.2.4, p. 233). These studies tend to demonstrate that this damage is not totally reversible during the application of a differential stress. The increased porosity induced by relaxation often has little effect on the volume but, being concentrated on specific zones of the porous space (intercrystalline joints, microcracks), it may have a major impact on the petrophysical characteristics.

Characteristic cracks may sometimes appear in the core due to macroscopic effects of relaxation. "Discing" is the most spectacular example of this (Fig. 1-1.22). It is a frequent phenomenon in which the core is cut into saddle-shaped discs from one to several centimetres thick, roughly perpendicular to the core axis.

Figure 1-1.22 Example of core discing, observed in longitudinal cross-section.
This favourable example shows the morphological evolution
of the discontinuities near the edge of the core, therefore suggesting their "non-geological" origin

Compaction

Compaction is a very broad term, which in geology designates a reduction in rock porosity under the effect of burial.

In the much more restricted field of petrophysics, compaction corresponds to an irreversible reduction in porosity under the effect of reduced fluid pressure during production, inducing an increase in differential stress. One of the most famous examples is that of chalk from the Ekofisk oil field in the North Sea. Further to seabed subsidence of more than ten metres caused by a rapid drop in porosity, the production installations had to be built again. This phenomenon, frequently encountered in the case of highly porous limestones, is outside the traditional context of compressibility and must be considered separately.

Sandstone compaction will be briefly discussed below (§ Dc).

Pore collapse

Sometimes the differential stress is strong enough to break elements of the solid structure, resulting in "pore collapse". The limiting stress is known as the "critical stress". In consolidated rocks of average porosity (<0.2), the critical stress is often greater than 100 MPa. This value may be much lower in highly porous rocks (e.g. limestone). Obviously, if the petrophysical experiments are to remain significant, the differential stress must never reach this critical value. In limestones, pore collapse often occurs suddenly. In sandstones, collapse is sometimes more progressive and therefore more difficult to detect in the first phases.

c) Order of magnitude of the porosity variation under the effect of differential pressure

Figure 1-1.23 shows some examples of porosity variation under the effect of differential pressure. There is less data available in the public literature than might have been expected. The total porosity variation under the effect of a differential stress is in fact less important, for reservoir modelling applications, than the rate of change of this variation, i.e. the pore compressibility.

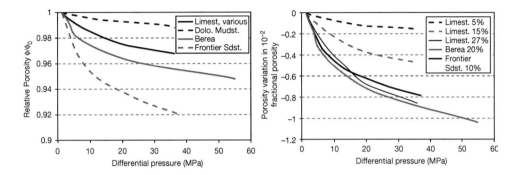

Figure 1-1.23 Example of porosity-differential pressure relation
according to data obtained by Harari *et al.* [1995] (and Jones and Owens [1980] for Frontier sandstone)

On the left, data expressed in relative porosity variations (ϕ/ϕ_0, ϕ_0 being the porosity without confinement). The "limestone" curve corresponds to numerous limestone samples of porosity between 0.05 and 0.27, with very similar relative porosity variation values.

On the right, the same data expressed as fractional porosity variations (unit 10^{-2} of fractional porosity, i.e. 1% porosity).

Estimating the compressibility values causes numerous problems which are summarised in the next paragraph. In practice, it is difficult to integrate the compressibility over a large differential stress interval. The values shown on Figure 1-1.23 must therefore considered as estimations (approximation probably on the high side). They can be summarised as follows: between zero and 30 MPa of differential pressure, the relative porosity variation is about 1% (10^{-3}) for slightly porous limestones, 3% (5.10^{-3}) for porous carbonates, 5% (10^{-2}) for porous sandstones and up to 10% (5.10^{-2}) for poorly consolidated sandstones or very porous carbonates (suspected of having a critical pressure less than 30 MPa). The previous values between parentheses correspond to an estimation of the fractional porosity variation.

Numerous measurements in low-porosity sandstones (about 10%) [Jones and Owens, 1980] indicate variations up to 5% (5.10^{-3}).

B) Various expressions of compressibility

We will focus on pore compressibility.

Compressibility is the variation in volume of a material divided by the stress variation. In view of the numerous definitions, however, it is more appropriate to speak of "compressibilities". For simplification purposes, we can identify the following compressibilities:

- rock mechanics compressibility (used in geomechanical modelling): this is the variation in total sample volume under the effect of a variation in pore or confining pressure;
- petrophysics compressibility (used in reservoir modelling): this is the variation in porous volume, mainly under the effect of variation in pore pressure due to the production of fluid;
- acoustic and log analysis compressibilities: this is the result of a calculation using the measurement of P and S wave velocities of ultrasonic frequency (sonic frequency for log analyses). In a perfectly elastic medium, they would be the same as the previous ones. The bulk moduli K_{dry} or K_{sat} (reciprocal of compressibility) discussed in Chapter 1-3.2, p. 237, correspond to this type of compressibility.

Depending on whether we are considering the variation in pore volume (V_P) or in total sample volume (V_B) under the effect of a variation in pore pressure (P_P) or in confining pressure (P_{conf}), we obtain the traditional expressions:

Compressibility concerning the total volume (rock mechanics):

$C_{BC} = (-1/V_B)(\partial V_B/\partial P_{conf})$ at constant P_P and $C_{BP} = (-1/V_B)(\partial V_B/\partial P_P)$ at constant P_{conf}

Compressibility concerning the porous volume (petrophysics):

$C_{PP} = (-1/V_P)(\partial V_P/\partial P_P)$ at constant P_{conf} (this is the compressibility measured in the oil laboratories) and $C_{PC} = (-1/V_P)(\partial V_P/\partial P_{conf})$ at constant P_P

Obviously, these compressibilities are directly inter-related for a given mechanical model. With a linear poroelastic medium, for example, the compressibilities so defined can be expressed as a function of the dry bulk modulus K_{dry} and of Biot's coefficient β [Boutéca, 1992] (used in petroacoustics in § 1-3.2):

$C_{BC} = 1/K_{dry}$ (by definition) and $C_{BP} = \beta/K_{dry}$

$C_{PC} = \beta/\phi K_{dry}$ and $C_{PP} = [\beta - \phi(1-\beta)]/\phi K_{dry}$

Biot's coefficient $\beta = 1 - K_{dry}/K_{grain}$ characterises the relative incompressibility of the porous space (K_{dry}) in the absence of any variation in pore pressure during the stress (for example, in the total absence of a saturating fluid) compared to K_{grain}, the bulk modulus of the skeleton-forming grain. If we consider that the solid is incompressible, $\beta = 1$ and we obtain the simple expressions:

$$C_{BC} = C_{BP} \text{ and } C_{PC} = C_{PP} = C_{BC}/\phi$$

In this configuration, we see that whether the pressure variation concerns the pore pressure or the confining pressure is unimportant. In this situation, the important parameter is the differential pressure $P_{diff} = P_{conf} - P_p$, as mentioned previously in § A.

C) Pore compressibility (effective rock compressibility)

We will restrict ourselves to the "pore compressibility" (C_P) called "effective rock compressibility" by Hall [1953] (more precisely C_{PP}) which relates to the modelling of flow in reservoirs (oil fields, aquifers). The definitions given above correspond implicitly to hydrostatic distribution of the confining stress (confining pressure). This is obviously a simplification since confinement generally corresponds to a stress state in the strict sense. It is important to note that the compressibility value measured in an experiment may vary significantly (up to a factor of 2) depending on the "stress path" used, i.e. how the stresses (horizontal, vertical) vary with respect to each other during the experiment.

a) Measurement of pore compressibility

In order to measure pore compressibility as accurately as possible in the laboratory, petrophysicists generally try to set up a configuration similar to the situation in a reservoir, where the pore pressure decreases under a constant vertical stress. The sample placed in a Hassler type apparatus (see Fig. 1-2.4) is brought, using closed loop control pumps, to conditions P_{conf} (or σ_{conf}) and P_{pore} corresponding to those of the reservoir to be investigated. The pore pressure is then released, maintaining the confining pressure at the fixed value. The important parameter measured is the fluid volume saturating the pore space produced for a given variation in pore pressure. Obviously, it is very important to correct the liquid compressibility as well as the compressibility specific to the experimental apparatus (cell, tube, etc.) so that only the actual variations in porous volume are taken into account. Equally, the temperature stability must be very carefully controlled. Note in this respect that the compressibilities considered here are "isothermal" compressibilities, i.e. measured so that the temperature remains constant during the pressure variation. The "dynamic" compressibilities, measured by wave propagation (§ 1-3.2.2, p. 229), are "isentropic" compressibilities, i.e. measured in the absence of any thermal energy exchange with the exterior, the wave passage time being too short for such an exchange. With compressible fluids, there may be a significant difference. For example, with hexane under ambient conditions, the values are respectively 2.4 and 1.9 GPa^{-1}. In rocks however, the differences are unimportant.

It is quite common to find compressibility measurements acquired by varying the total stress and by maintaining the interstitial pressure constant, the latter being close or equal to atmospheric pressure. This method, which is much simpler, produces more data at lower cost. Nevertheless, we have just seen that we must be very cautious when handling the values obtained from these experiments, above all by checking that the data have been suitably corrected. Similarly, and to simplify the experiment, service laboratories frequently use a completely (or virtually) isotropic stress regime. However, reservoirs are generally considered to exhibit zero lateral strain (oedometric type strain). In experiments with non-zero radial strain, correction is also required.

Although experiments of this type appear sufficient to estimate the porosity corrections required depending on the stresses ("laboratory conditions" versus "*in situ* conditions"), especially with consolidated rocks, their application in terms of compressibility must be used with caution.

An approximate value of the compressibility can also be obtained by measuring directly the variation in porosity of a sample subjected to a variation in stress state. With the apparatus used to take porosity measurement by helium compressibility, it is quite easy to apply stress to the sample (see 1-1.1.2, p. 10). Although the measurement accuracy is probably too low to obtain a true measurement in differential form (true compressibility), we can nevertheless estimate the value of the rock porosity parameter under confinement.

b) Remark on the evolution of occluded porosity under the effect of stress

All the previous considerations implicitly assume that the porosity remains connected in spite of the geometric variations of the porous space induced by the stress variations. In "ordinary" rocks with three-dimensional throats to the pores, it is highly likely that this condition is nearly always satisfied. In contrast, it is possible that in poorly porous rocks with abundant cement formed from crystals with clearly marked faces (e.g. quartzite), the intercrystalline joints can easily close up, thereby disconnecting some pores from the main network and generating a fraction of occluded porosity. Depending on the method used to measure the porosity variation during a stress variation, the compressibility may either be underestimated (measurement of the expelled saturating fluid) or highly overestimated (helium measurement under various stress conditions). In the first case, the fluid "trapped" in the occluded porosity will not be taken into account in the compressibility calculation whereas, in the second case, the occluded porosity fraction will be considered as having completely disappeared under the effect of the stress.

D) Factors affecting pore compressibility; its order of magnitude

Pore compressibility is an important parameter when modelling reservoirs devoid of free gas (e.g. to reconcile matter balance and pressure variation). We would therefore expect to find extensive public documentation on compressibility values. This does not seem to be the case. One possible explanation is that in the past the "compressibility" parameter was studied in less detail, considering its secondary importance in case of production where the pressure is maintained by water injection. Numerous studies are currently being conducted in laboratories to measure it and the scarcity of publications should not be considered as a lack of interest in this parameter.

a) Effect of differential pressure on compressibility

The compressibility of a sample varies with the differential pressure: under the "squeezing" effect induced by differential pressure, the compressibility generally drops. In some cases the opposite behaviour has been observed, generally in highly porous unconsolidated sediments, under low initial pressure.

The experimental relation C_p *vs.* P_{diff} seems fairly well represented by a power law with negative exponent (in the interval −0.6; −1). On Figure 1-1.24, we provide some experimental data from the literature. Note that for the data from Fatt [1958], the pressure value is supplied as net overburden pressure equal to $P_{conf} - 0.85\,P_{pore}$. This is probably an example of using (incorrectly, in our opinion) the "Biot" approach, but it does not have a significant impact on the results.

We learn from these experimental data, and remaining quite prudent in the generalisation, that:

– At low differential pressure (less than 10 MPa, for example), pore compressibility may vary considerably with pressure. For the very low pressures (<5 MPa) in fact, we may sometimes doubt whether the measurements are representative. These low differential pressures are not found in hydrocarbon reservoirs (except in the extreme cases of over-pressurised or very shallow reservoirs). This problem concerns the aquifers.

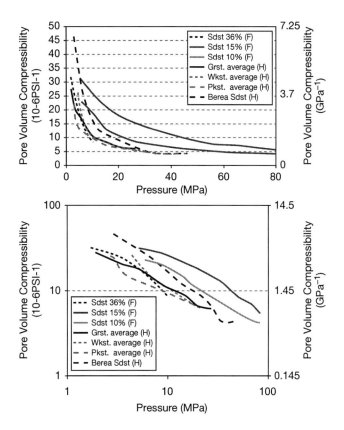

Figure 1-1.24 Example of pore compressibility (C_P)-differential pressure relation according to data from Fatt [1958] (**F.**) and Harari *et al.* [1995] (**H.**). Fatt's data are not specifically supplied as "differential pressure", but the final result does not appear to be substantially changed
Linear scale (top) and logarithmic scale (bottom).

– At the differential pressures found in oil reservoirs, the effect of differential pressure is relatively less important and experimental values can be compared even if acquired at different differential pressures, provided that they are high enough. By analogy with the seismic wave velocities, we could speak of "terminal" compressibility. We must not forget, however, the cases where compressibility increases with effective stress.

b) Effect of temperature on compressibility

Note that several of the authors mentioned in this chapter have studied the effect of temperature on compressibility. This effect is low enough to be neglected, obviously after correcting the direct effects of temperature on the fluids (fluid viscosity, compressibility) and, to a lesser extent, on the solid phase (mineral compressibility).

c) Effect of time on compressibility (loading rate)

This is an important effect which cannot be developed within the scope of this book. We will restrict ourselves to a brief summary. Remember that the loading rates (σ') have a major effect on the results of a compressibility experiment. Consequently, in addition to the corrections mentioned above and induced by the difference in "stress paths" between the laboratory and the reservoir, it is essential to estimate the loading rate corrections.

In practice, there are three orders of magnitude:

- laboratory experiments (σ'_{lab}), whose loading rate (order of magnitude 10^5 Pa/minute to 10^5 Pa/hour) is close to that of well tests;
- loading rates induced by reservoir exploitation ($\sigma'_{in\ situ}$), whose orders of magnitude are measured in 10^5 Pa/month or in 10^5 Pa/year;
- geological compactions, which are of the order of the Pa/year.

The resulting compaction decreases as the loading rate increases (Fig. 1-1.25).

De Waal and Smits [1988] propose a method to estimate these differences in the case of sandstones (and especially poorly consolidated or unconsolidated sands):

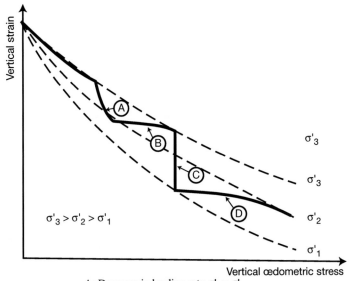

A: Decrease in loading rate $\sigma'_Z = \sigma'_2$
B: Increase in loading rate $\sigma'_Z = \sigma'_3$
C: Interruption of loading rate $\sigma'_Z = 0$
D: Re-start of loading rate $\sigma'_Z = \sigma'_2$

Figure 1-1.25 Effect of the loading rate (σ'_Z) on compressibility (expressed by the vertical strain Δh) in a schematic oedometric compaction experiment. According to de Waal and Smits [1988]

Considering the term compressibility in the broad sense, the relation between compressibility in the reservoir during production ($C_{m, \, in \, situ}$) and compressibility measured in the laboratory ($C_{m, \, lab}$) is of type:

$$C_{m, \, in \, situ} = A.C_{m, \, lab} \, (\sigma'_{lab}/\sigma'_{in \, situ})^{w}$$

where w = de Waal's material constant, between 0.007 and 0.03 which varies roughly the same as compressibility.

If we consider loading rate ratios of 10^4 or 10^5 (ratio of the hour to the year or the decade), application of this formula gives variations of 10 to 50% on the average to high compressibilities.

d) Porosity/Lithology – Compressibility relations; the paradox of Hall's plot

For the practical applications, it is important to understand the relations between lithology, porosity and compressibility under high differential pressure. Although this approach is used when studying all petrophysical properties, the situation is different in the case of pore compressibility: if we consider the usual relations, compressibility would be independent of lithology and the evolution as a function of porosity would be opposite to what one might expect. The origin of these relations can be found in a comparatively old publication [Hall, 1953] which became extremely popular due to its simplicity. These results are reproduced on Figure 1-1.26. Only the average curve given by Hall is represented, but we can see that the (relatively few) experimental points are grouped around this curve corresponding to a power law such as $C_{pp} = 1.7665\phi^{-0.4413}$, when the compressibilities are expressed in 10^{-6}PSI^{-1}. Some simulation programmes use an empirical formula giving similar results, in 10^{-6}PSI^{-1}:

for $\phi < 0.3$ $C_{pp} = 2.6 + 78 \, (0.3 - \phi)^{2.415}$ and for $\phi > 0.3$ $C_{pp} = 2.6$

The fact that compressibility decreases with increasing porosity represents a paradox. We must beware of intuition in this field. While it is true that increasing porosity leads to greater compressibility of the "skeleton", reducing the effect of compressibility on the solid itself, at the same time compressibility is given as a ratio of the porous volume, which becomes very low for low porosities. The only way to obtain a more reliable conclusion is therefore by calculation.

In paragraph B above, we indicated the relations between the various compressibilities (grain, skeleton) in the assumption of a poroelastic medium ($C_{pp} = [\beta - \phi(1 - \beta)]/\phi K_{dry}$). K_{grain} is quite accurately known (see Table 1-3.2, p. 240) and there are empirical relations between ϕ and K_{dry} based on the experience of sonic logs (§ 1-3.2.6E, p. 247). Although they consist of dynamic modules deduced from wave propagation (§ 1-3.2.2, p. 229), they can nevertheless be used in a first analysis. The result of the calculation (Fig. 1-1.26a) corroborates the intuitive impression: in a poroelastic assumption, pore compressibility decreases with porosity.

Other data from the literature do not help clarify this paradox. An exhaustive study conducted by Newman [1973] describing his own experimental results (over 200 samples) together with a compilation of results from other authors (Fig. 1-1.26b) led to the following conclusions:

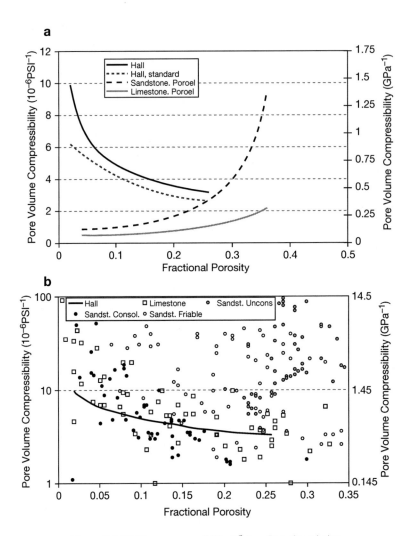

Figure 1-1.26 Pore compressibility (C_P) *vs.* Porosity relation
a) two expressions of the Hall plot [1953] compared with the results of a poroelastic calculation for sandstones and limestones using the sonic velocity *vs.* porosity "empirical relation" (§ 1-3.2.6.E, p. 247).
b) experimental results based on data from Newman [1973].

- the results of laboratory measurements are incredibly dispersed, the "brittle" sandstones or the unconsolidated sands extend over more than one order of magnitude and display no compressibility *vs.* porosity relation;
- the consolidated sandstones are also dispersed and their pore compressibilities extend over one order or magnitude. The results tend to confirm the inverse evolution between Porosity and Compressibility. Sandstones and limestones do not demonstrate different trends;

– the compressibility values of the Hall plot correspond to a "low trend" for the low porosity values. Generally, the compressibility values found in the literature are significantly higher than could be expected by the poroelastic calculation, using skeleton compressibilities deduced from the log analysis velocities.

Lastly, note that the orders of magnitude of oedometric compressibilities (Table 1-1.6) given by de Waal and Smits [1988] and based on a large number of samples, follow a trend opposite to that obtained by Hall, therefore corresponding quite closely to the poroelastic estimation.

Table 1-1.6 Order of magnitude of oedometric compressibilities (in the two usual units), according to de Waal and Smits [1988]

Type of sandstone	Unconsolidated	Brittle	Consolidated	Tight
Oedometric compressibility 10^{-6}PSI^{-1}	30-7	7-3	3-0.5	0.5-0.3
Oedometric compressibility GPa^{-1}	4-1	1-0.5	0.5-0.1	0.1-0.05

We have no convincing explanation for these apparent contradictions. We will simply give a few indications:

– the experimental uncertainty on the pore compressibility measurement increases as the porous volume (and therefore the porosity) decreases. In low porosity material, even greater experimental rigour is required;
– the measurements reported in the literature are generally taken on reservoir samples: reservoirs (i.e. permeable rocks) of low-porosity (<0.1) always correspond to special facies. Measurements taken on rock outcrops could possibly give different results;
– the reservoir rocks have all been cored. In poorly porous rocks, relaxation may induce microcracks likely to produce a significant increase in pore compressibility. Application in the laboratory, for a necessarily very short duration, of a differential pressure equal to that present *in situ*, may not be sufficient to close up all the microcracks, possibly resulting in atypical behaviour.

Conclusion on pore compressibility

When evaluating oil reservoirs [Bourdarot, 1998], the well test is essential. Its interpretation is based on the compressibility of a unit volume in the porous medium: C_t

where: $C_t = C_o S_o + C_w S_w + C_p$ and C_t "total" compressibility; C_o, C_w, C_p respectively oil, water, pore compressibility and S_o, S_w oil and water saturation. The orders of magnitude of the three compressibilities are very approximately equal [decade 3-30 $(10^6 \text{PSI})^{-1}$], which demonstrates that a good knowledge of pore compressibility is essential.

In hydrogeological technique, the usual parameters largely derived from pumping tests are firstly Transmissivity, which is the product of the hydraulic conductivity and the thickness of the aquifer (or of the level tested?) (§ 1-2.1.1, p. 123) and secondly the "Storage Coefficient" (or "storage capacity"). The Specific Storage Coefficient represents the ability of the aquifer to produce water under the effect of a reduction in hydraulic head. The Total Storage Coefficient is the product of the specific storage coefficient and the thickness of the aquifer (or of the level tested?). In practice, for a free aquifer it corresponds to the effective porosity. For a confined aquifer, the term Specific Storage Coefficient ($ST_{s)}$ is equivalent to the term total compressibility used by petroleum engineers.

ST_s = g(($1 - \phi$) C_{pv} + ϕ C_w) where C_{pv} is the "vertical compressibility of the porous medium".

It is therefore clear that compressibility is an important parameter. The most striking aspect concerning pore compressibility is the contrast between the apparent interest in this parameter for reservoir calculations, the abundance of theoretical developments concerning the mechanics of porous media (and therefore compressibility) and the uncertainty regarding the experimental results published.

To produce a good model of the reservoir, pore compressibility measurements should be taken whenever possible (core samples available), according to a precise protocol including:

  – Accurate estimation of the initial stress *in situ* (before exploitation).
  – Experiment under oedometric conditions on a sample, checking its verticality if the medium is anisotropic.
  – Control of the sample strain and integration of the loading rates.

If no cores are available, the only solution is to rely on the results obtained on reservoir rocks found to be similar to the case concerned after careful examination.

1-1.2 CAPILLARY PRESSURE IN CASE OF PERFECT WETTABILITY

Two or more immiscible fluids frequently cohabit in a porous medium. Water and gas in the surface areas of interest to the agronomist and the hydrogeologist, water, oil and gas in the oil reservoirs (and in polluted soils). A number of capillary phenomena, which have a major impact on the equilibria between these fluids, occur at the interfaces between the fluids and the porous solid. These capillary effects are the key to explaining numerous types of petrophysical behaviour.

A knowledge of wettability, i.e. of the stronger affinity of one of the fluids for the solid, is essential to understand the capillary phenomena. Wettability may give rise to a great deal of confusion (see § 1-2.2, p. 167) if we move away from the case of perfect wettability, where one of the fluids has a very strong affinity for the solid. *For a better understanding, we decided to deal with the cases of perfect and intermediate wettability separately.* Capillarity in case of perfect wettability, which is relatively simple and unambiguous, will allow us to explain numerous phenomena. And a thorough grasp of this first concept will considerably simplify the description of capillarity in case of intermediate wettability (§ 1-2.2).

1-1.2.1 Definition of perfect wettability and capillary pressure

We will discuss the two-phase case which corresponds to the presence, inside the porous network considered, of two immiscible fluids.

A) Wettability

Wettability characterises the relative affinity of one of the fluids for the solid. This parameter is traditionally quantified by the value of the contact angle θ between the fluid interface and the solid surface (Fig. 1-1.27a). Note that this notion only really applies to smooth surfaces. We can easily imagine the problems which will arise when attempting to define the contact angle for the extremely rough surfaces found in natural porous media (see Fig. 1-2.33, p. 177).

If one of the fluids exhibits a very strong affinity for the solid (wetting fluid), the angle θ will tend to zero and the wetting fluid will cover the entire solid surface. In the presence of gas, most liquids are wetting. The most well-known exception is mercury which forms small globules on a surface. Mercury is the classic example of a non-wetting liquid (Fig. 1-1.27b).

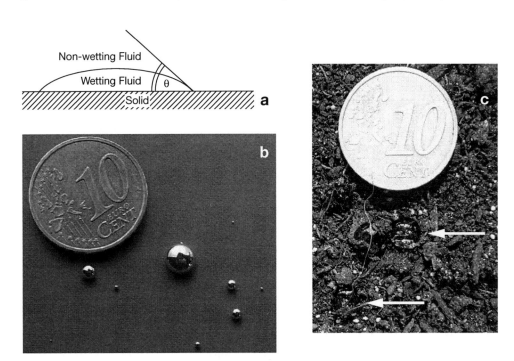

Figure 1-1.27 Wettability
a) Definition of the contact angle θ between the fluid interface and the solid surface.
b) Classic example of non-wetting liquid in the case Liquid/Air/Solid: mercury globules on glass.
c) Example of temporarily non-wetting water in the Water/Air/Dry Soil system: water "globules" on a very dry soil (compare with the mercury "globules").

For the water/air pair, although water is generally the wetting fluid, interesting exceptions may be observed. If the porous medium is contaminated or very dry, water may be non-wetting, at least temporarily. This phenomenon is easily observed by spraying water onto very dry ground: the water runs and is not absorbed (Fig. 1-1.27c).

We must emphasise the fact that wettability must be defined in the context of the assembly including the fluids and the solid surface. A fluid will only be wetting or non-wetting with respect to the other components of the system. In addition, it is the state of the solid surface and not the solid itself which plays the central role. Deposition of molecules of a foreign body, even in very small quantities, may have a considerable effect on wettability. This explains why we always refer to "clean" solids. In contrast, deposition of bodies contaminating the surface of a porous medium often creates wettability anomalies. This situation is systematically encountered in natural porous media where the mineral surfaces have evolved according to the conditions of deposition (sedimentary rocks) and/or of diagenesis (history of interstitial fluids).

B) Capillary pressure

As an illustration, we could say that the fluid/fluid interface behaves as a solid membrane, capable of opposing a certain force (the surface tension t_s,) to an increase of the area of this surface. This surface tension induces a pressure difference between the two fluids. The pressure in the non-wetting fluid is greater than that in the wetting fluid. This difference called the Capillary Pressure (Pc) is a function of the average curvature of the interface.

The capillary pressure is governed by the *Laplace* or *Plateau equation* (the latter name being used in particular by Anglo-Saxon authors):

$$Pc = t_s(1/R_1 + -1/R_2)$$

R_1 and R_2 are the main radii of curvature of the interface. The + sign corresponds to the general case where the centres of curvature O1 and O2 are located on the same side of the interface. The – sign corresponds to the case where the centres of curvature are located on both sides, as is the case with saddle-shaped interfaces (Fig. 1-1.28a). In practice, we generally use the term mean curvature $1/R_m$ such that $2/R_m = 1/R_1 + -1/R_2$.

The dimension of surface tension t_s is a force per unit length, t_s is therefore expressed in N/m (Table 1-1.7). In practice, the milliNewton per metre (mN/m) is often used, which corresponds to the dyne/cm, still quite frequently used.

The most common illustration of capillarity is the wetting fluid rising up a capillary tube (this is the best configuration to illustrate quantitatively the affinity of the wetting fluid for the solid).

Note that the classical expression (sometimes called Jurin's formula) relates the capillary rise height h to the radius R_c of the capillary tube by expressing the equilibrium of gravitational forces and capillary forces:

$$h = (1/\Delta\rho\gamma) (2t_s \cos \theta/R_c)$$

where t_s and θ are the capillary constants, $\Delta\rho$ the density difference between the two fluids and γ the gravitational acceleration.

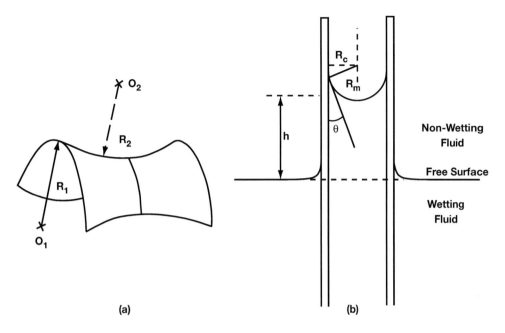

(a)   (b)

Figure 1-1.28

a) Example of saddle-shaped interface.
b) Diagram of the capillary tube.

Table 1-1.7 Order of magnitude of the surface tension for some typical fluid pairs, at ambient temperature

Fluids	Pure water/Air	Water/n-Heptane or n-Octane	Water/Gas	Water/Oil	Mercury/ vapour
t_s mN/m	70	50	50	20 to 50	470

By combining the hydrostatic equation and the Laplace equation, we see that the capillary pressure, by definition equal to the pressure difference on each side of the interface (with P_{nw} and P_w the pressures in the non-wetting fluid and wetting fluid, respectively) is such that:

$$Pc = P_{nw} - P_w = \Delta\rho\gamma h = 2t_s/R_m$$

The interface formed inside a cylindrical capillary tube can be compared to a spherical cap of radius equal to $R_c/\cos\theta$. The average curvature is therefore equal to this value and we can see that Jurin's law is the expression of a particular case of the Laplace-Plateau equation.

1-1.2.2 Capillary pressure curve

A) Definition

The previous formulae show the proportionality between capillary pressure and the average interface curvature or, restricting ourselves to the simple case of the capillary tube, to the reciprocal of the radius of this capillary. To progressively invade the porous space, i.e. to penetrate in the smaller and smaller anfractuosities, the non-wetting fluid must be under greater and greater pressure (capillary pressure). We can see that there is a relation between a given capillary pressure and the fraction of the porous space filled by the non-wetting fluid (non-wetting fluid saturation, S_{nw}).

For a given porous medium therefore, the capillary pressure curve will be the function relating the capillary pressure to the corresponding wetting fluid saturation (obviously equal to the complement to 1 of the non-wetting saturation). Since the wetting fluid considered is frequently water, the conventional form of the capillary pressure curve is: $Pc = f(S_w)$ (see Figure 1-1.29 and after).

This curve represents an extremely practical way of predicting the behaviour of the porous medium in two-phase saturation. It provides a means of quantifying certain geometric characteristics of the porous space, due to the proportionality between capillary pressure and average curvature. This property is put to advantage in the case of mercury porosimetry, for example (§ 1-1.2.4D, p. 79).

B) Effect of hysteresis, drainage, imbibition

It is important to note that non-wetting fluid saturation depends not only on the capillary pressure but also on the history of the experiment which led up to the pressure considered (hysteresis). We will return to this point from time to time. There are two broad types of fluid movement:

- if the non-wetting fluid expels the wetting fluid, further to an increase in capillary pressure, we speak of drainage;
- otherwise, if the non wetting fluid leaves the porous space further to a decrease in capillary pressure, we speak of imbibition.

This terminology is only clear in case of perfect wettability. It is still used in case of intermediate wettability, but in this case "drainage" replaces the expression "water content decrease" ($S_w \searrow$) and "imbibition" corresponds to a "water content increase" ($S_w \nearrow$).

We must therefore describe the capillary pressure curves during drainage and imbibition separately.

C) Capillary pressure curve during first drainage

a) Access (or displacement) pressure

A capillary pressure curve during drainage (Fig. 1-1.29, 1-1.31) starts with zero capillary pressure and at $S_w = 1$ (considering a case of perfect water wettability and using the

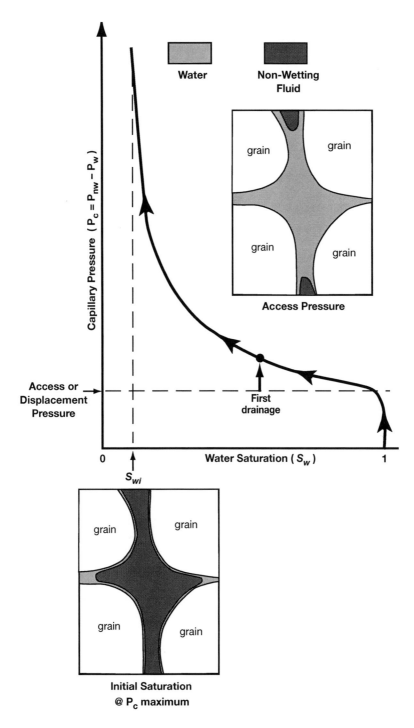

Figure 1-1.29 Capillary pressure curve during first drainage

fractional notation for the saturations). When the pressure increases, the water saturation remains equal to 1 until the capillary pressure is high enough to allow the non-wetting fluid (oil or gas in our example) to cross the pore throats of larger diameter: this is the "access pressure" or "displacement pressure"

b) "Transition zone"

As the capillary pressure continues to increase, the non-wetting fluid invades pores with smaller and smaller throats: the water saturation progressively decreases. In this case, we observe a relative "proportionality" between capillary pressure variation and water saturation variation. This section of the drainage curve is sometimes qualified as "transition zone" (the term originates from the distribution of fluids in oil reservoirs, § 1-1.2.6, p. 104). The non-wetting fluid occupies the pores whose capillary throats (the term equivalent capillary radius of the throat is used, applying Jurin's law bluntly) are greater than those corresponding to the capillary pressure considered. We will return on numerous occasions to one simple but fundamental point, namely that the parameter governing the drainage is the absolute value of the equivalent capillary radius of the throat, which may bear no relation to the "radius" of the pores as it appears on a thin section, for example. Various methods are available to visualize the relative locations of the wetting and non-wetting phases at various drainage states (§ 2-2.2, p. 336) and we can see, for instance, the number of large pores (vugs) which can only be accessed via throats of very small equivalent radii. One quite common example of this is clearly shown on Fig. 2-2.11, p. 338, where the macropores (> 100 μm) can only be accessed via throats of equivalent capillary radii of less than 1 μm.

c) Initial saturation at maximum capillary pressure and irreducible saturation (Swi)

If the water saturation is to continue decreasing as the capillary pressure increases, the water must be able to escape. We must therefore introduce the notion of relative permeability, a point which will be discussed in detail below (relative permeabilities, § 1-2.3, p. 179). While remaining on a highly qualitative level, we can say that from a certain (low) water saturation point, the water loses its mobility inside the porous network (decreasing towards "hydraulic" discontinuity of the water phase) and that, consequently, only a slight variation in saturation will be observed despite a large increase in capillary pressure. In the past, this observation led to the definition of a minimum (wetting fluid, water) saturation called the *Irreducible saturation* S_{wi}. The parameter so defined would be an intrinsic characteristic of the porous medium. When trying to produce an accurate definition, this notion of irreducible saturation turns out to be highly imprecise, to such a point that there may be some doubts as to its very existence. We will return to this problem later (§ 1-2.3.3, p. 186). At this stage, we will simply rely on a pragmatic definition of irreducible saturation: it would be the wetting fluid saturation value beyond which an increase in capillary pressure no longer induces any variation in saturation observable at "laboratory" time scale. At the scale of an oil reservoir, it would be the saturation such that its variation according to the height above the "water level" is less than the resolution of the deferred logging tools and their interpretation.

In this region of high capillary pressure the wetting fluid, whose saturation is often low to very low (0.05 - 0.15) in reservoir rocks, is located in the least accessible areas of the porous network and, in particular, in the microporosity (carbonate) or the shaly layers and the weathered feldspars (sandstones). Sometimes however, "macropores" are connected to the exterior by throats which are so small that they retain the wetting fluid, even at very high capillary pressures. Carbonates, for example, may contain vugs produced by selective dissolution followed by partial recementation or intergranular pores "forgotten" during cementation. The latter case occurs quite frequently in sparite grainstone. In highly cemented sandstones, similar phenomena of "forgotten" macropores may be observed. It is highly likely that this phenomenon is amplified by closure of intercrystalline films under the effect of high differential pressures during burial (§ 2-1.1, p. 267). We must not forget the possible existence, at very high capillary pressure, of water pockets of non-negligible size, especially during analysis of resistivity measurements (§ 1-3.1, p. 199).

The notion of irreducible saturation is too imprecise and may lead to significant errors. It will not be used in this book. We will only consider the notion of initial saturation at maximum capillary pressure which is clearly defined for any experiment and which, in the reservoirs, is directly related to the structural position (height above the free water level, see § 1-1.2.6, p. 104).

D) Capillary pressure curve during imbibition

a) Spontaneous imbibition

After a drainage phase, if the capillary pressure is released, the wetting fluid may once again penetrate inside the porous space. This is known as spontaneous imbibition. No external energy input is required, the fluid movement is maintained by some of the mechanical energy stored during the drainage phase. The fundamental point is the system hysteresis: at a given capillary pressure, the water saturation is lower (and generally much lower) than it was during the drainage phase (Fig. 1-1.30, 1-1.31). When the capillary pressure has dropped to zero, spontaneous imbibition stops since we are in a situation of perfect water wettability. In intermediate wettability regime the situation would be quite different. We will return to this experiment at the start of § 1-2.2.

b) Residual saturation, "trapped porosity"

When spontaneous imbibition stops, at zero capillary pressure, a variable amount of non-wetting fluid is trapped by the capillary forces in the porous space. This fraction of trapped fluid is known as *Residual non-wetting fluid saturation* (S_{or} or S_{gr}, oil or gas residual saturation). This terminology mainly applies to petroleum petrophysics. In fields dealing with water/air regimes (pedology, civil engineering), we sometimes speak quite simply of "trapped porosity". When studying porosity (§ 1-1.1.1), we mentioned the relative danger of these porosity qualifiers.

This fraction is of large practical importance since the capillary or moderate hydraulic mechanisms (flows) cannot release the trapped fluid. In oil reservoirs, for example, the oil trapped during waterflooding must undergo special physical or chemical treatments, implemented during tertiary oil recovery operations [Latil *et al.*, 1980], to be unblocked. We will also see how this phenomenon helps explain the water-air equilibria in partially saturated surface rocks.

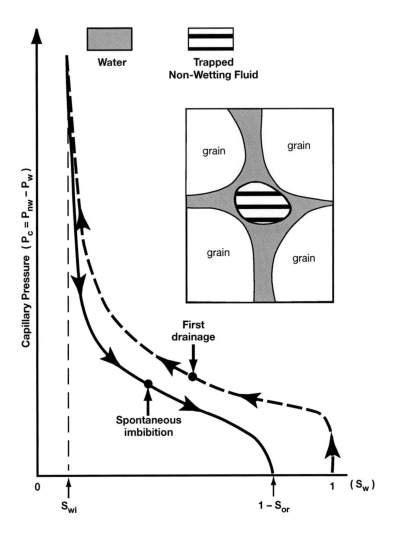

Figure 1-1.30 Capillary pressure curve during imbibition

The location of fluids during imbibition phase is far less intuitive than during the drainage phase. We do not have the simple capillary tube model or the criterion of absolute value of the equivalent throat radius. The visualization methods (§ 2-2.2.2, p. 336) provide a fairly accurate idea of the location of this "trapped" fluid: The residual non-wetting fluid consists of "globules" or "ganglions" occupying the central areas of the pores.

The characteristic parameter is no longer the absolute value of the throat radius, but the relative value of the ratio of the "maximum pore radius" to the pore-throat radius. As soon as this ratio increases, the non-wetting fluid is trapped. Since the absolute value of these radii is relatively unimportant, we may observe quite similar images at scales from about a hundred microns up to several millimetres. This character clearly appears on the preparations used to

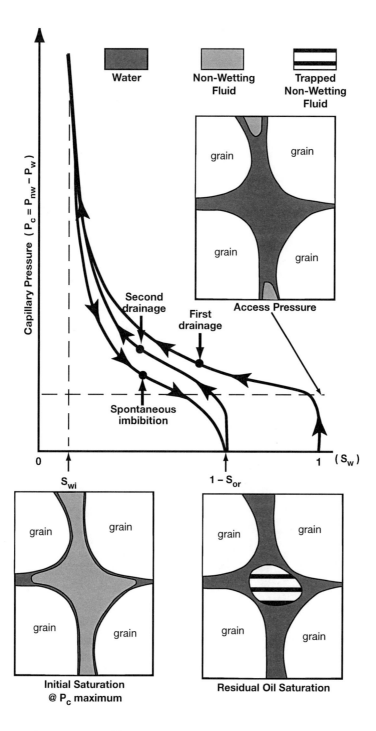

Figure 1-1.31 Capillary pressure curve and schematic location of fluids

visualize the fluids (Fig. 1-1.32, Fig. 2-2.13, p. 340), as soon as the maximum pore radius decreases, the trapping tendency disappears.

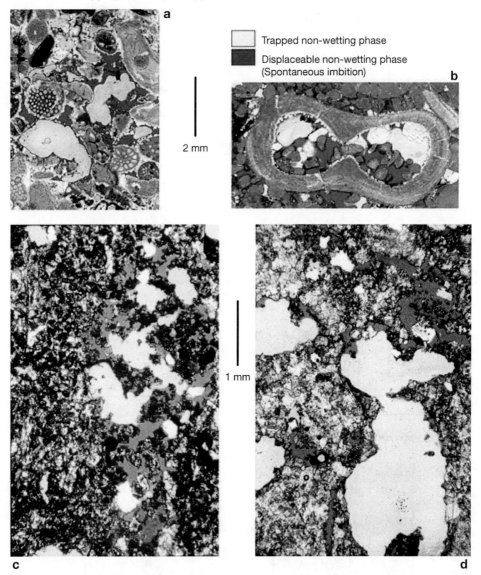

Figure 1-1.32 Example showing the location of the non-wetting fraction trapped
by imbibition in natural porous media

Photograph of thin sections prepared according to the resin spontaneous imbibition method (see § 2-2.2.2). The red resin corresponds to the fraction of non-wetting fluid displaceable by spontaneous imbibition. The yellow resin corresponds to the fraction trapped during imbibition.

a) Bioclastic grainstone; b) large bioclast in an oolitic grainstone; c) and d) vuggy dolomite. Notice the role of residual crystallisation (c) in limiting the capillary trapping.

The photographs on Figure 2-2.13, correspond to the same visualization method.

c) Attempt to explain the trapping phenomenon

We will now try to provide a qualitative explanation of the mechanism trapping a fraction of the non-wetting phase during spontaneous imbibition. (Note that there is no consensus of opinion on this explanation.)

We will consider a fraction of the porous space consisting of a pore (Fig. 1-1.33, on the left of the diagram) accessible by a pore throat of average radius of curvature R_{PT}. This pore throat will therefore be unable to accommodate a wetting fluid/non-wetting fluid interface of average radius of curvature larger than R_{PT} and we may therefore consider that, if the average radius of curvature increases up to R_{PT} under the effect of a reduction in capillary pressure, then the interface will break.

• State 1 on the diagram corresponds to a situation of high capillary pressure after drainage. The average radius of curvature is small, the interface follows quite closely the roughness of the pore wall and the wetting fluid saturation (S_w) is low.

• State 2 corresponds to the start of imbibition (reduction of Pc and correlative increase of S_w). The non-wetting phase is assumed to escape out of the right hand side of the diagram. Since the average radius R_2 corresponding to this phase is much smaller than R_{PT}, the phenomenon can continue normally. Once the wetting fluid saturation has increased

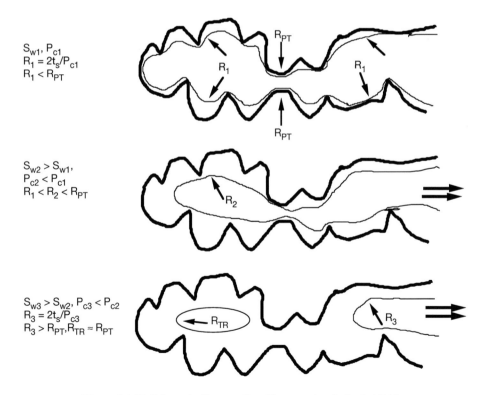

S_{w1}, P_{c1}
$R_1 = 2t_s/P_{c1}$
$R_1 < R_{PT}$

$S_{w2} > S_{w1},$
$P_{c2} < P_{c1}$
$R_1 < R_2 < R_{PT}$

$S_{w3} > S_{w2}, P_{c3} < P_{c2}$
$R_3 = 2t_s/P_{c3}$
$R_3 > R_{PT}, R_{TR} \approx R_{PT}$

Figure 1-1.33 Schematic diagram of capillary trapping during imbibition

sufficiently (and the corresponding capillary pressure has decreased sufficiently) for the average radius to reach the value of the pore throat radius, the interface breaks and we obtain state 3.

• State 3 is characterised by the presence of an isolated ganglion of non-wetting fluid in the pore on the left hand side. The average curvature of this ganglion is close to that of the pore throat. As imbibition continues, the pressure in the ganglion will remain constant but will never be high enough to pass through the pore throat. The ganglion is permanently trapped.

This explanation is rather brief (it does not, for example, take account of any dynamic effects which could "help" the ganglion to pass through the pore throat). However, it justifies the fact that the geometric parameter governing the trapping is the ratio of the pore radius to the pore throat radius: if the pore radius is much larger than the pore throat radius, the ganglion will still occupy a large volume when the non-wetting phase continuity breaks, independently of any notion of absolute value of these radii.

E) Capillary pressure curve during second drainage

After spontaneous imbibition, the capillary pressure can be increased again to start another drainage experiment. We have seen that at the end of spontaneous imbibition, the sample contained a large proportion of non-wetting fluid (S_{nwr}). The capillary pressure curve of this second drainage experiment is therefore different from the first drainage.

Hysteresis is observed once again, since the capillary pressure required to obtain the same non-wetting fluid saturation is larger during second drainage than during spontaneous imbibition (Fig. 1-1.34). It is "true" hysteresis since if the cycle of drainages/imbibitions was to be continued, the same spontaneous imbibition/second drainage curves would be described. This hysteresis corresponds to the energy difference required between "entry" and "departure" of the non-wetting fluid. It is sometimes called "drag hysteresis".

We should not really speak of hysteresis between first drainage and spontaneous imbibition, since the difference between the curves largely corresponds to the irreversible trapping of some of the non-wetting fluid. We sometimes use the term "trap hysteresis" to characterise the difference between first and second drainage, although it is not true hysteresis.

1-1.2.3 Capillary phenomena in soils

The description of capillary pressure curves in Chapter 1-1.2 implicitly concerns the aquifer or petroleum reservoir rocks, but this property can be generalised to all porous media. The purpose of this paragraph is to demonstrate the similarities, not always obvious at first sight, between the approaches adopted in petroleum and soil science techniques.

The soils, in the pedological sense and the geological sense, are worthy of special attention since capillary phenomena in soils are extremely important for all aspects of hydrology, agronomy and civil engineering. These disciplines have their own methods and terminology and readers interested either one can refer to the corresponding literature [e.g. Brady and Weil, 1999; Maidment, 1993].

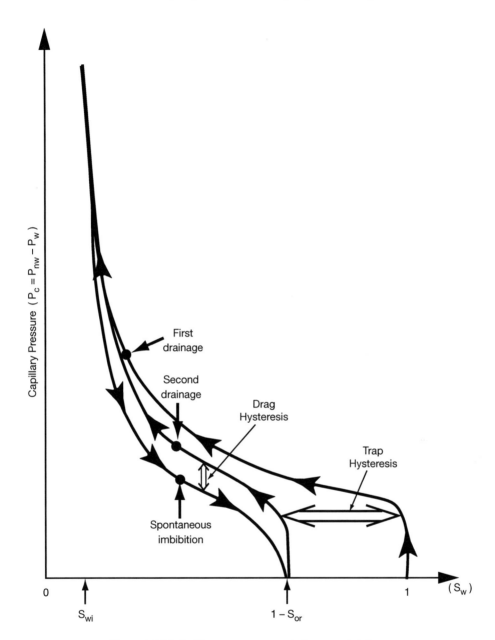

Figure 1-1.34 Capillary pressure curve during second drainage

A) Soil specificity

a) Specificity concerning the physical processes

Regarding the physical process, soil specificity is fundamental since the total volume of a soil sample may vary depending on the water content. This is not the case for reservoir rocks, which contain water and/or hydrocarbons and whose total volume (and *a fortiori* the volume of the solid phase) is considered as invariant: any loss of liquid in terms of volume leads to desaturation, and vice versa.

In a soil, loss of water first results in a reduction in total volume, the saturation index remaining equal to100%. Then, beyond a water content limit known as the "shrinkage limit", the soil behaves as a rigid rock and loss of water leads to desaturation of the porous space (see Fig. 1-1.5, p. 15).

In practice therefore, the water volume is either compared with the solid volume or to the dry mass in order to obtain an invariant. This leads to an abundance of definitions, sometimes a source of confusion, which will be summarised below, using the volume and mass parameters described in § 1-1.1.2 (p. 5): V_V, V_S, V_T, V_W respectively the void, solid and total volumes ($V_T = V_V + V_S$) and the water volume contained in the porous medium; W_S the dry mass and W_W the water mass:

- Porosity: $\phi = V_V/V_T$ The letter "n" is used in soil physics and soil mechanics (and sometimes "f" in hydrogeology, instead of "ϕ").
- Void ratio: $E = V_V/V_S$ where $E = \phi/(1 - \phi)$, often written "e".
- Saturation index $S_W = V_W/V_V$ ("saturation" in petroleum terminology). In soil physics and soil mechanics, S_R is often used instead of S_W.
- Soil moisture: $\theta_W = W_W/W_S$. The term "water content" is frequently employed instead of "soil moisture" and written "w" or "ω"
- Volume water content: $\theta_V = W_W/V_T$ This term is often confused with the expression V_W/V_T since these notions are used in soil mechanics and in soil physics where the assumption is made that $\rho_w = 1000$ kg/m^3 (freshwater).

b) Specificity concerning the methodological approach

There are a certain number of specificities in the study of soil capillarity which correspond to methodological differences:

- In petroleum petrophysics, due to the specialisation of the technical teams, petrophysicists mostly disregard fluid movement related to capillary forces. These studies belong to the field of reservoir modelling. Consequently, it is the capillary pressure curve as such which is studied. In soil sciences however, this specialisation is less obvious and the movement of fluids (water in fact) is an integral part of the studies on capillarity. This leads to the definition of potentials (moisture, gravity, etc.). Since our objective is primarily to highlight the similarities between petroleum methods and soil sciences, independently of terminology differences, we will first describe the equivalent of the capillary pressure curve. We will then briefly discuss the movements of water in soil.
- One of the main applications of soil science is agronomy. It is not surprising that plants (especially their roots) play a major role, extending as far as the definition of

capillary properties. The maximum capillary pressure that can be "overcome" by the roots to extract water from the soil is one of the major parameters of the capillary pressure curve (wilting point).

B) Curve of capillary pressure in soils

a) Matrix potential curve

Figure 1-1.35, found in various forms in hydrogeology or pedology books [e.g. de Marsily, 1986] is the equivalent of the capillary pressure curve described in the previous paragraphs. The terminology differences must nevertheless be taken into account.

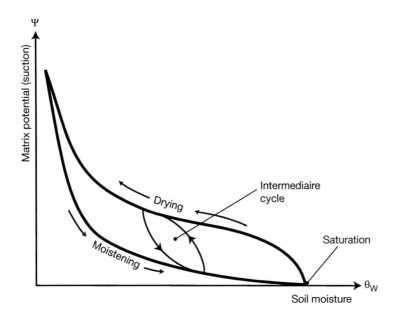

Figure 1-1.35 Curve of suction matrix potential against water content
(equivalent, in soil science, of the capillary pressure curve)

The y-axis does not strictly correspond to a pressure (Pc) but to a moisture potential (ψ_m), (also called "suction matrix potential", "matrix potential" or "capillary potential". It corresponds to the energy required to "extract" a unit mass of water fixed by the capillary forces. Note that:

- this physical quantity is expressed in practice as equivalent water height. We therefore obtain the same hydrostatic formula Pc = $\Delta\rho\gamma h$ as that used in § 1-1.2.6 (p. 102) to explain the saturation profiles in hydrocarbon reservoirs;
- in soil science applications, the air is virtually always at atmospheric pressure, which is taken as zero for pressure difference measurements. Since the capillary pressure is equal to the pressure difference between the non-wetting fluid (air, reference pressure

zero) and the wetting fluid -water), the pressures in water are negative. This explains why we speak of "suction";
- For historical reasons (in reference to the former CGS system of units), these heights are converted into centimetres. Since the range encountered extends over several orders of magnitude, the heights are expressed in logarithmic form. The moisture potential "pF" is therefore equal to the decimal logarithm of the capillary suction expressed in centimetres of water, taking into account the sign inversion inherent to the use of the term "suction".

Other sources of terminology confusion are given below: In the field of petroleum, and in the case of water-wet media, we have defined the two hysteresis branches:
- *drainage*: branch describing the decrease in water saturation;
- *imbibition*: branch describing the increase in water saturation;
  In soil science, the term "drainage" is reserved for the water evacuation process during consolidation (water-saturated soils).
- the branch describing the decrease in water saturation is often called *drying*;
- the branch describing the increase in water saturation is called *humidification, moistening or rewetting*.

b) Order of magnitude of moisture potentials

We have defined the potential "pF" as the decimal logarithm of the capillary suction expressed in centimetres of water. The orders of magnitude observed in soils are listed in Table 1-1.8. Note that the range of values is much broader than that generally considered for reservoir rocks.

c) Specific retention, Wilting point

By convention, pedologists identify two singular points on the pF curve:
- A point corresponding to pF = 2 (100 cm of water) corresponding to the "specific retention" of the soil expressed in volume content, equivalent to the hydrogeological notion of field capacity (weight content): if the pF is less, the soil is assumed to dry spontaneously (i.e. allow the water to flow by gravity).
- A wilting point corresponding to the energy beyond which plants cannot extract water from the soil, and set by convention at pF = 4.2 (16 000 cm of water, i.e. 1.57 MPa).

d) Soil water reserve

The difference between the volume water content at pF = 2 ("specific retention" or "field capacity") and the volume water content at pF = 4.2 (wilting point) defines the soil "Available Storage" (**AS**) (or available reserve). The "Field capacity" is also called the "Water Storage Capacity" (**WSC**). We sometimes even speak of an "Available Moisture" (**AM**) limited by the water content to pF = 3. Some authors define it even more empirically: 1/3 or 2/3 of the **AS**, or 1/3 of the "Water Storage Capacity" (**WSC**).

These characteristics are schematised on Figure 1-1.36. We observe a certain similarity with the distinctions made by petroleum petrophysicists (Fig. 1-1.10, p. 20). The most

Table 1-1.8   Order of magnitude of moisture potentials "pF" and pore throat equivalent radii
(data from Baize [2004, in French] and Banton and Bangoy [1999, in French]

Suction (Equivalent negative pressure)		Hydraulic head		Pore throat equivalent radius	Remarks
kPa $(10^{-2}$ Bar)	atm	Water column (cm)	pF		
0.1	≈ 0.001	1	0	1.5 mm	Pedologist's macroporosity
1	≈ 0.01	10.2	1.1	0.15 mm	
10	≈ 0.1	102	2	0.015 mm	Specific retention
98	0.967	1 000	3	1.5 µm	Limit of classical tensiometers *in situ*
1 000	≈ 10	10 200	4.0	0.15 µm	*In situ* measurements: – limit of new-generation tensiometers – start of the field of psychrometric measurements
1 569	15.48	16 000	4.2	0.094 µm	Wilting point
7 000	69	71 400	4.85	0.021 µm	Limit of psychrometric measurements *in situ* (95% humidity)
9 804	100	100 000	5	0.015 µm	Air-dried soil (relative moisture 92%)
98 039	1 000	1 000 000	6	0.0015 µm	Air-dried soil (relative moisture 48%)

interesting similarity is that between the notion of "wilting point" used in soil physics and that of "irreducible water saturation" used in petroleum petrophysics. This similarity will also apply below, when we examine polyphase flows.

In these notions of soil water reserves, we find ambiguities similar to those observed in the definition of petroleum reserves. Initially, these concepts were determined to quantify the water available in the soil for vegetation. The capacity of plants to extract water obviously varies from one species to another. In addition, the limit defining the "field capacity" is purely empirical. Determination of "AS" or "AM" based on a range of pF values is therefore conventional and corresponds to a usual value, obtained by observation and experiment and mainly involving crops.

We must also mention two other sources of ambiguity:

– Soil physicists working in the field of pedology generally express volumes in water height by analogy with pluviometries and in order to make comparisons. The difficulty arises not so much in the conversion of volume into height per unit area, but in the determination of the depth of soil concerned. Strictly speaking, it can only be the

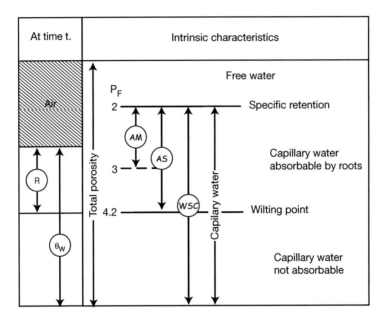

Figure 1-1.36 Occupation of soil porosity by water. Diagrammatic representation of various types of reserve

depth reached by the roots of the plants (cereals, trees, etc.) concerned and not the total thickness of the pedological soil or the geological surface formation considered.
– This leads to the other conceptual difficulty: a difference must be made between the potential total volume of a pedological horizon and the water actually available at a given time. This explains why modern agronomist make a clear distinction between the notion of **storage** (AS) and the notion of **reserve** (R).

It is interesting to point out the analogy with a notion also sometimes unclear in oil exploration, but which the petroleum engineers have had to strictly define by rigorously identifying the notion of hydrocarbon accumulation and the notion of reserves.

C) Total soil water potential and water movement in soils

The study of water movements in soils is clearly fundamental for soil science specialists and hydrogeologists. However, petrographers studying diagenesis (cementation, compaction) of sediments in the vadose zone, i.e. the zone very close to the surface in two-phase water/air saturation, also require a knowledge of these movements. Vadose diagenesis is of paramount importance in the study of numerous carbonate rocks, for example.

The generalised notion of total soil water potential is used to describe these movements. It is a physical quantity expressing the potential energy of water per unit mass, volume or weight of soil considered. Its dimensions vary depending on the definition used.

We will continue using the unit equivalent water height, as we did for capillary pressures.

The potential (Φ) is divided into three main potentials:

$$\Phi = \psi_z + \psi_m + \psi_o$$

where: ψ_z = gravity potential

ψ_m = moisture potential (or capillary potential, matrix potential) described above

ψ_o = osmotic potential

The osmotic potential is induced by the difference in salt concentration across a membrane, whether biological (plants) or mineral (clay).

Strictly speaking, an external potential (related to atmospheric pressure) and a kinetic energy term related to the fluid speed should be added to the previous potential energies. Since the flow speeds in porous media are generally very slow, however, this term is negligible.

Note that ψ_m changes sign and becomes a pressure potential (ψ_p) under the level of the aquifer piezometric surface.

The notion of total soil water potential is a very useful concept, since it provides an overall picture of the system, considering the soil-plant-atmosphere continuum as a single entity.

The possible water movements in the soil can therefore be identified. At a given time, the water profile of a soil is imposed by the equilibrium between the influxes (precipitation)/ outfluxes (evaporation) on the soil surface and the capillary supply by the water table (in the broad sense).

The flux direction will be determined by the potential difference between the two points considered, from the higher to the lower. If, to a first order, the kinetic energy is neglected, considering that the displacements are very slow, the total soil water potential is that defined above: $\Phi = \psi_z + \psi_m + \psi_o$. In addition, the soil surface is usually taken as height reference.

Figure 1-1.37 gives a very simple and highly schematic example of the distribution of the these potentials and therefore of the possible water movement, taking into account only the resultant of the moisture potential and the gravity potential. This problem is addressed implicitly when describing capillary drainage (§ 1-1.2.4B, p. 74) and more indirectly when examining the distribution of fluids in hydrocarbon reservoirs (§1-1.2.6, p. 102). In both situations, however, we are dealing with equilibrium profiles, which is not the case here.

We will first consider the examples of saturation profiles (Fig. 1-1.37 a). From an equilibrium profile (1) corresponding to the gravitational equilibrium of § 1-1.2.4B, near the soil surface, we observe variations due to the water influxes by precipitation (2) or water outfluxes due to evaporation (3).

These saturation variations induce correlative variations in moisture potential (Fig. 1-1.37 b). Since we chose the water equivalent height as unit of potential, the capillary equilibrium potential corresponds (by definition) to a straight line of gradient -1 cutting the y-axis at a negative value equal to the depth of the free (piezometric) water table. The influx or outflux profiles correspond to curves tending towards lower values for the outfluxes and higher values for the influxes (obviously, these are not absolute values but negative algebraic values).

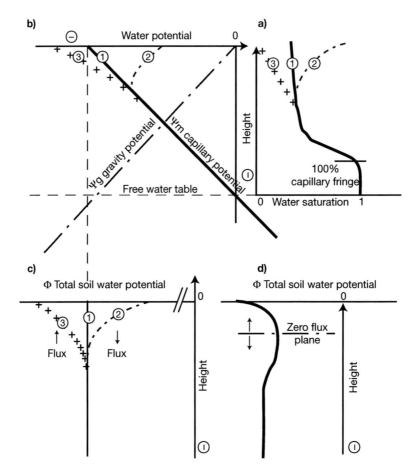

Figure 1-1.37 Highly diagrammatic example of distributions of potentials and saturations in a soil

a) Saturation profile and capillary pressure for a gravitational equilibrium state (1); water influx (2) or water outflux (3).

b) Potential corresponding to these states.

c) resulting total potential.

d) Example of zone with influx and outflux resulting in a zero flux plane.

The gravity potential corresponds to a straight line of gradient 1 going through the origin of the graph (soil surface).

In the simple hypothesis chosen, the water potential is equal to the sum of the previous ones. The resulting profile (Fig. 1-1.37 c) corresponds to a vertical line (equipotential) for the case of capillary equilibrium (by definition). In our examples of water influx and outflux, the potentials are locally greater or less than this equipotential, the water fluxes occurring either towards the top or the bottom.

Obviously, however, some saturation profiles may correspond to influxes and outfluxes successively on the same vertical. The resulting potential will be more complicated

(Fig. 1-1.37d) and may frequently have a zero flux plane, from which the water movements will occur both towards the top and the bottom of the section.

1-1.2.4 Capillary pressure curve measurement principle: Restored states, Centrifuge, Mercury Porosimetry

A) Desorber method
(also known as the restored state or semipermeable membrane method)

This method (Fig. 1-1.38) is based on the use of a semipermeable membrane, in this case a ceramic plate whose pores are so fine that once they are totally saturated with water their capillary resistance opposes the penetration of non-wetting fluid (gas or oil). This resistance may exceed one megapascal. The sample totally saturated with water is placed in a pressure cell, on the ceramic plate whose lower side is at atmospheric pressure. Since the ceramic is impermeable to the non-wetting fluid, the latter can be subjected to a pressure P, inside the cell, greater than that of the fluid in the sample, the water remaining at atmospheric pressure. The water expelled by the non-wetting fluid (drainage) flows through the ceramic; a graph of pressure in the non-wetting fluid (Pc) against water saturation can then be produced.

After this drainage phase, an imbibition phase can be carried out by measuring the quantity of water reabsorbed by the sample as the capillary pressure is released.

The restored state method produces homogeneous saturations inside the sample. This method probably gives the best results. It has one major practical disadvantage, however:

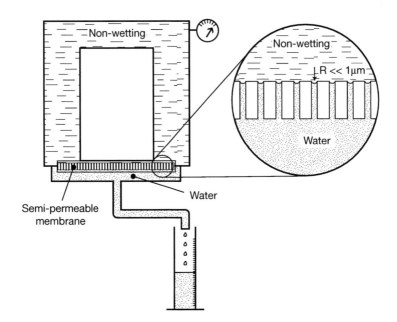

Figure 1-1.38 Restored state method for the measurement of capillary pressures

the long time required to reach equilibrium at each capillary pressure increment. Due to the fact that the pores of the semipermeable membrane are extremely narrow, its permeability is very low (10^{-5} to 10^{-7} D) and the differential pressures (i.e. the capillary pressures) are never very large.

In practice, the experiment is carried out:

- either in individual cells to measure the saturation variations of the unique sample by direct measurement of the liquid volumes produced or injected. The quality of this type of experiment is excellent;
- or in large cells with sufficient space for several dozen samples on the semipermeable membrane. After each pressure increment, the samples are removed from the cell (after a fast decompression step) to measure the saturations by weighing. This procedure is less accurate than the previous one, but is suitable for routine processing of a much larger number of samples. This method - the collective desorber method - is gradually being phased out.

Despite the fact that these experiments extend over a long period (at least several months), the desorber method is still widely used.

B) Centrifuge method

a) The principle of drainage by gravity

If a column of water-saturated porous medium, whose base is in contact with a free water level at atmospheric pressure, is subject to the gravitational field, water flows out of the top part of the porous block and is replaced by air. The pressure profile of this drainage can easily be calculated. At a height h above the reference water level (Fig. 1-1.39), the air pressure (P_{nw} pressure in the non-wetting fluid) is equal to atmospheric pressure (P_{at}). The water pressure is equal to the pressure at the reference level (P_{at}) reduced by the weight of the water column between h and this level, to respect the hydrostatic equilibrium, so $P_w = P_{at} - \rho_w \gamma_p h$, where ρ_w is the density of water and γ_p the gravitational acceleration.

By definition, the capillary pressure is equal to $P_{nw} - P_w = \rho_w \gamma_p h$.

Assuming that our porous sample is perfectly homogeneous and that we have a means of measuring the water saturation at various heights (for example by sampling the "sand pile" forming the block considered) we could plot the capillary pressure curve (Fig. 1-1.39) since we have $h = Pc/\rho_w \gamma_p$, (we will often refer to this standardised height - capillary pressure equivalence).

In the drainage by gravity experiment, we plotted a capillary pressure curve by measuring the saturation at various heights of a block subject to a constant force (the force of gravity in our example). The capillary pressure curve used to plot the graph corresponds to that of a sandstone of average permeability. In order to conduct our experiment a block several dozen metres high would be required. We can increase the acceleration considerably by using a centrifuge, however, and thereby reduce the height by the same proportion down to the few centimetres of the laboratory samples.

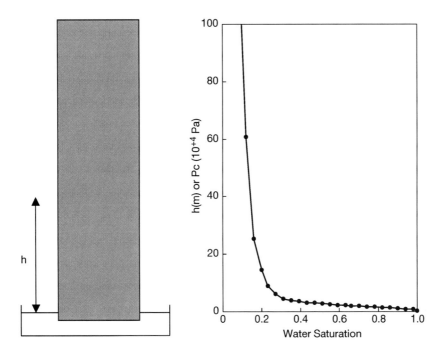

Figure 1-1.39 Schematic diagram of drainage by gravity

b) Centrifuge

The objects placed in a centrifuge undergo an acceleration γ of value corresponding to $\gamma = \omega^2 R$, where ω is the angular velocity and R the radius of rotation. A sample placed in a centrifuge cup (Fig. 1-1.40a) can be subjected to accelerations of up to 5.10^4 m/s^2 for traditional devices and much more for some special devices. Under these very high accelerations, the saturation profile determined on the example by gravity, undergoes very high anamorphosis. These profiles are shown on diagram 1-1.40b for a sample exhibiting the capillary pressure curve of Figure 1-1.39.

The cup has a transparent graduated tube to collect the water which drains out. A stroboscopic device is used to read the water level during centrifugation. The volume of water drained out and therefore the remaining saturation at each saturation phase can therefore be accurately determined. It is an average saturation for the entire sample, since the saturation profile varies within the sample. This variation is intrinsic to the phenomenon involved since the capillary pressure is a direct function of the water column height in the sample (the water saturation is theoretically equal to 1 on the lower side of the sample, diagram b). It is also related to the fact that the acceleration is not strictly constant in the sample (it depends on the radius of rotation). To a first approximation, we may consider that the average saturation is in the middle of the sample. The acceleration and therefore the capillary pressure can be calculated accordingly.

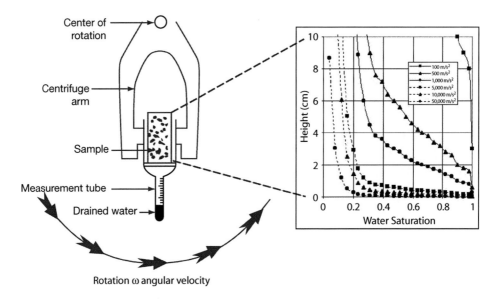

Figure 1-1.40 Centrifuge method for the measurement of capillary pressures
a) Schematic diagram of the centrifuge.
b) Sample saturation curve for various accelerations. The capillary pressure curve used is given on Figure 1-1-39.

This problem, the variability of the saturation and acceleration profiles, does not represent a real handicap for the centrifuge method since, if we assume that the sample is homogeneous, the true profiles can be recalculated. Numerous calculation programmes have been proposed. Since they do not all rely on the same modelling principles, the results obtained may differ slightly [Forbes, 1997].

The advantage with the centrifuge method is the relative speed (with oil/water, a good measurement lasts one day for each pressure stage). High capillary pressures and therefore low "irreducible" saturation values can also be reached.

We described the simplest case of first drainage, but other capillary equilibria can be produced, depending on the fluid surrounding the sample in the cup. For example, if a partially oil-saturated sample is centrifuged in a cup filled with water, forced imbibition will occur. This technique can be used to study cases of intermediate wettability (§ 1-2.2, p. 169)

C) Common methods in soil physics and soil mechanics

a) Measurement at ordinary pressures (up to 1.5 MPa)

Measurement in the laboratory

The desorber (restored state) method is generally used:

- For low pressures (suction), the water pressure is decreased to less than 1 atm by connecting the water phase of the (so-called semipermeable) suction plate to a reservoir placed lower than the sample.

– For the higher pressures, a cell capable of withstanding air pressures of up to 16 bars (pF = 4.2; wilting point), similar to that in the field of petroleum, is used. This is referred to as the "axis translation" technique since, for specialists in soil physics and soil mechanics, it corresponds to suction. This apparatus is known as the "Richard's cell".

In situ measurements

In situ measurements are becoming more frequent, especially in agronomy. Tensiometric probes are generally implemented and suctions of up to 1.5 MPa can be investigated with the most recent models. This probe (Fig.1-1.41) is based on the same principle as that of the desorber (Fig.1-1.38), except for the fact that the sample measured corresponds to the "totality" of the soil considered. A semipermeable ceramic candle, filled with water, is placed in the soil. Capillary equilibrium is reached across the semipermeable membrane between the water in the candle and the water in the soil, inducing a pressure drop (suction), which corresponds precisely to the parameter we are trying to measure.

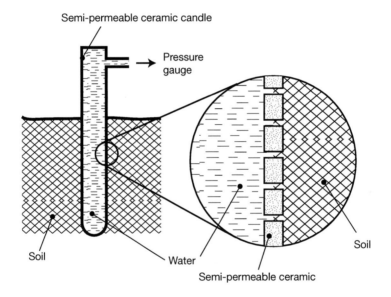

Figure 1-1.41 Principle of the tensiometric probe

Psychrometric probes based on measurement of relative humidity near the interstitial water are available for higher capillary pressures. These electronic devices, whose use is still not widespread, extend the measurement limit past 7 MPa [Delage *et al.* 1998].

b) High pressure measurement

To measure at very high capillary pressures, vapour phase equilibrium methods based on Kelvin's formula are employed. There is an equilibrium between humidity of the ambient air and suction:

$$(u_a - u_w) = (R_G T_K / \gamma_p M) Ln(h)$$

where: u_a = air pressure

u_w = water pressure

h = relative moisture, equal to the ratio of the partial pressure of water vapour in the atmosphere considered to the saturating vapour pressure. It depends on the temperature

R_G = molar gas constant

T_K = absolute temperature

γ_p = gravitational acceleration

M = molar mass of water

for information, at 20°C, $(R_G T_K / \gamma_p M)$ = 137.8 MPa

The most common method is to impose the moisture by saturated salt solutions (Tab. 1-1.9), in a closed, temperature-controlled container. The disadvantage with this technique is the long time taken to reach equilibrium. This technique is used in petroleum petrophysics to study very poorly permeable media.

Table 1-1.9   Relation between relative moisture and capillary force (Kelvin's formula); equilibrium relative moisture with some salt solutions, freezing point of water in capillary state

Suction (MPa)	Relative moisture (%)	Saturated salt solution (30°C)	pF	Equivalent pore thresholds (radius, μm)	Ionic force	Freezing point (°C)
0.01	99.993		2	15	$2.2\ 10^3$	
0.1	99.927		3	1.5	$2.2\ 10^2$	−0.08
1	99.277		4	0.15	$2.2\ 10^1$	−0.80
1.58	98.88		4.2	0.094	$3.38\ 10^1$	−1.26
2.83	98	$CuSO_4,5H_2O$	4.45	0.053		
7.18	95	$Na_2SO_3,7H_2O$	4.86	0.021		
10	92.7		5	0.015	2.2	−8.37
25	83.62	KCl	5.4	0.0060		
30.5	80		5.49	0.0049	6.0	
40.1	75.09	NaCl	5.6	0.0037		
43.8	73.14	$NaNO_2$	5.64	0.0034		
70	60		5.85	0.0021		
81.1	56.03	NaBr	5.91	0.0018		
100	48.4		6	0.0015		−81.0
126	40		6.1	0.0012		
141	36.5	$CaCl_2, 6H_2O$	6.15	0.0011		
221	20		6.35	0.00068		
305.4	11.28	LiCl	6.48	0.00049		
316	10		6.5	0.00048		

Other simple techniques may also be employed in laboratories with limited equipment, in which the soil to be measured is placed in contact with a porous material of known properties. Standardised filter paper can be used for example. The water saturation of the filter paper in equilibrium with the soil can easily be measured by weighing. This saturation corresponds to a pressure on the known capillary pressure curve of the filter paper. It also corresponds to the capillary pressure of the soil studied.

D) Mercury injection (Mercury porosimetry or Purcell test)

One very common method to measure the capillary pressure curve is based on the implementation of the mercury/vacuum system, according to the terminology generally used. Obviously, the "vacuum" corresponds to mercury vapour. These two fluids are employed in laboratories since they provide a fast means of obtaining a capillary pressure interpreted as such, but they can also be used to quantify the morphology of porous media.

Initially applied by chemists wanting to identify the porous medium of catalysts, this technique was quickly taken up by ceramics manufacturers, to study sintering phenomena. The method has been applied in petroleum technique under the initiative of Purcell [1949], hence the name of this test in this discipline. It is now widely used to characterise porous media of non metallic materials, both natural (rocks and soils) and artificial (ceramics, catalyst supports, etc.). This test is often referred to as the mercury injection porosimetry test.

a) Experimental method

Principle

After the sample has been washed to remove the hydrocarbons, dried, or freeze-dried if its texture is too sensitive to drying in an oven, as is the case with clay media (see § 1-1.1.2, p. 8), it is injected with mercury. Mercury injection starts once the sample has been placed under the best possible vacuum (e.g. 10 Pa). In practice therefore, the initial pressure in the porous medium is equal to the vapour pressure of mercury at the temperature of the experiment. This test must therefore be conducted in a temperature-controlled environment.

With the apparatus currently available, the mercury can be injected into the sample in small pressure increments; the volume injected is allowed to stabilise at each pressure stage.

If suitable equipment is available, it may sometimes be worthwhile conducting a variant of this experiment. In this case the mercury is injected in volume stages and the corresponding equilibrium pressures recorded. However, this type of experiment can no longer be interpreted strictly as a capillary pressure.

Modern devices cover a vast pressure range: from a few kilopascals (corresponding to the column of a few centimetres of mercury on top of the sample in the measurement cell) up to 400 MPa.

Samples containing minerals which could interact with the mercury (e.g. pyrite) must obviously be avoided.

Mercury being non-wetting, the capillary pressure is equal to the pressure applied on it, since the vacuum (mercury vapour) is the equivalent of the wetting phase. Mercury injection therefore corresponds to the drainage phase. The imbibition phase is recorded by gradually releasing the pressure in the mercury. It is recommended to measure the volume of mercury trapped in the sample at the end of the cycle. Divided by the total porous volume, this saturation S_{HgR} must be compared with the residual gas saturation S_{GR} at the end of the spontaneous gas/water imbibition process (§ 1-1.2.7, p. 114).

Since in practice the equivalent of the wetting phase does not move, no morphological aspect of the porous medium prevents the mercury from invading the entire space, especially since the maximum pressures applied allow access to extremely small spaces (dimensions less than 10^{-2} µm).

Precautions required to ensure correct representativeness

The main advantages of the method are its speed (a test carefully carried out on a volume of rock of about five cm^3 takes less than one day, excluding sample preparation), its exceptional repetitivity and its reliability. This test also simplifies the investigation, if required, of the least permeable media, mudstones included. It is one of the rare methods allowing analysis of samples of any shape, including those which have been broken up (e.g. cuttings).

With shaly samples, however, it is generally recommended to avoid these experiments in petroleum technique. The fact that these tests are not significant in the field of petroleum is not due to the method itself but to the preparation technique, in particular the dehydration of samples (this point is discussed in § 1-1.1.2, p. 8).

The advantages of porosimetry explain why it is implemented in numerous disciplines. Although the test is easy to execute, we must nevertheless remember that its representativeness depends on the application of clearly defined operating conditions. Those implemented by commercialised devices, however, tend occasionally to overlook some of these precautions (thorough degassing, thermostatting, compressibility corrections, etc.). Sometimes, the lack of sensitivity of the pressure sensors seriously limits the number of equilibrium stages, thereby cancelling out the progress brought by the new technologies which offer highly accurate control of the experiment. Inversely, some interpretation programmes provide a "specific area" calculation but this evaluation, which may be suitable for powder analyses, is meaningless in rock materials. One benefit of this test is the accurate control of the total porosity when the test is carried out according to professional standards. The porosity control is even better if the total sample volume is measured by immersion in mercury (see § 1-1.1.2, p. 9). The latter method offers the advantage of being consistent with the processing of the first experimental point (correction of surface conformance effects) which must take into account the surface irregularities of the sample. As a result, the subsequent saturation calculations are extremely accurate.

The main drawback of the method is the fact that it is destructive. The mercury-impregnated samples must be disposed of under conditions corresponding to their toxicity.

The essential operating conditions are:

– in-depth degassing of the sample. This preparation period often takes longer than the actual experiment: over ten hours under a vacuum of a few pascals for a sample of a few cm^3. The service laboratories may tend to overlook this phase;

– the experiment must be carried out under isothermal conditions;
– accurate control of the pressure at each step and of the injected volume stabilisation phase.

Concerning the data preprocessing, special attention must be paid to the following:

– correction of surface conformance effects which simulate macroporosity (Figure 1-1.42*);*
– the mercury column on the sample at the start of the test (low pressure) must be correctly taken into account;
– at the end of the test (high pressure), correction of the cell deformation, the mercury compressibility and, if necessary, the mineral phase compressibility. Although it is obvious concerning the deformability of the measurement chamber and the mercury compressibility (about 0.04 GPa^{-1}), correction of the mineral phase compressibility raises a difficult problem. The sample compression state is not accurately known, in fact, since according to the capillarity principle, the mercury/mercury vapour interface takes up the capillary pressure. The pressure in the mercury should only be transmitted to the solid on the surfaces in direct contact with the mercury (i.e. where the average radius of curvature is greater than that of the mercury/vapour interface). This correction is only important in case of low-porosity porous media such as crystalline rocks and quartzites (fractional porosity of about a few hundredths) and we recommend extreme caution in this respect.

To conclude on these methodological remarks, we must point out a problem which although not specific to mercury porosimetry may be extremely important when applying this method. The measurement technique must be adapted to the specificity of the materials

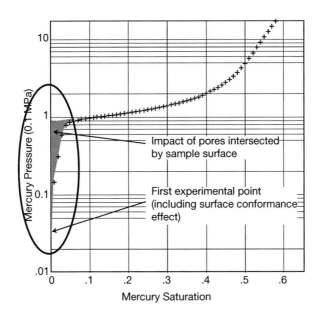

Figure 1-1.42 Zoom of the start of a mercury injection curve
showing the edge effect caused by the pores intersected by the sample surface

studied. Use of mercury porosimetry in petrophysics is financially less attractive compared with its use to characterise industrial materials (e.g. catalyst supports). The devices commercialised are therefore designed for this majority use. There are significant differences since the material investigated consists of fragments of millimetric granulometry (the time required for degassing, for example, is therefore much reduced) and the "large" throat radii are often not taken into consideration. These devices must therefore be tuned to petrophysics requirements.

Analysis and description of measurements

The Laplace equation, described in § 1-1.2.1B (p. 54), is as follows:

$$Pc = t_s \left(1/R_1 + -1/R_2 \right) = 2t_s/R_m$$

where R_1 and R_2 are the main radii of curvature of the interface, in the algebraic sense: positive or negative and R_m the average curvature of the capillary meniscus. If we liken the porous space containing the meniscus to a cylindrical capillary tube of radius R_c, we obtain Jurin's formula: $Pc = 2t_s \cos\theta/R_c$ where θ is the contact angle.

The characteristics of mercury are reputed to be relatively stable under standard experimental conditions:

– interfacial tension: $t_s = 480$ dynes/cm (mN/m)
– wettability angle: $\theta_{Hg} = 140°$

which gives:
$Pc(Hg) = 0.735/r$
with Pc in megapascals and r in micrometres.

We can therefore immediately transform the mercury injection curve Pc = function(volume of mercury injected) into a cumulative distribution of pore *threshold equivalent* radii.

– *equivalent*: since the porous medium is likened to the highly simplified model of a bundle of cylindrical capillary tubes;
– *thresholds*: since the pores are irregular. If we imagine a path through the porous space, it crosses a series of bottlenecks and bulges: the capillary pressure is only governed by the largest connection to a given pore (the "threshold").

The dimension of the bulges (in fact, pores in the usual meaning of the observation) cannot be measured by porosimetry. At best, the global volume of the pores whose "pore radius/throat radius" ratio is sufficiently high can be calculated, by recording the volume of residual mercury at the end of the mercury extrusion curve (equivalent to imbibition).

A finer analysis can be carried out by comparing the first mercury intrusion (1st drainage curve) with a second intrusion (2nd drainage curve) without extracting the residual mercury at the end of the final extrusion phase under vacuum (first imbibition curve). We also mentioned earlier a specific experiment (involving stages of injected volumes, a method similar to the "constant flow" capillary pressure methods) allowing a partial "pore radius/throat radius" interpretation.

Pore connectivity is another essential parameter of the porous medium which cannot be determined by porosimetry test.

b) Application of mercury porosimetry to estimation of capillary pressures

Conversion factor to convert Mercury/Vapour capillary pressure values into Water/Oil or Liquid/Gas values

Capillary pressure measured with mercury (Pc_{Hg}) can be converted into capillary pressure corresponding to other fluid pairs ($Pc_{reservoir}$) by multiplying by the ratio of interfacial tensions (see § 1-1.2.1.B, p. 55):

$$Pc_{reservoir} = Pc_{Hg} \, (t_{s\_reservoir} \cos \theta_{reservoir} / t_{sHg} \cos \theta_{Hg})$$

For practical purposes therefore, the conversion factors A ($Pc_{reservoir} = Pc_{Hg}/A$) are as follows:

- $Pc_{water-air}$ under standard condition (laboratory, pedology or hydrogeology): A between $5 < A < 6.6$;
- $Pc_{gas-water}$ under the most common oil reservoir conditions: between $18 < A < 25$;
- $Pc_{water-oil}$ under the most common oil reservoir conditions: $15 < A < 20$; this conversion is dangerous, however (see below).

The variations within the ranges given are partly due to interfacial tension variations (water salinity, oil composition, temperature, etc.) but above all to the fact that the contact angle θ is unknown. We have seen that this notion could only apply to perfect surface states. In practice, this angle is therefore no more than a conventional formalisation of wettability (cf. § 1-2.2, p. 167).

Gas/Water systems

For Gas/Water systems (gas reservoir, underground storage, hydrogeology or pedology), experience has shown that apart from shaly or unconsolidated rocks, there is a true equivalence between the results obtained by mercury porosimetry and by restored states. The use of mercury injection curves is therefore empirically acceptable for saturation calculations.

Oil/Water systems

For Oil/Water systems, conversion of capillary pressure measured using porosimetry is often unsuitable. The pressure stage (transition zone) is frequently underestimated and the pragmatic notion of "S_{wi}" cannot be measured. We are faced with the same problem, however, when trying to describe an oil reservoir by simplified "gas/water" restored states.

In oil reservoirs, only the use of a similar liquid/liquid system is representative. This is mainly due to the sensitivity of interfacial tension and wettability to the oil composition. However, as long as they are not interpreted too abruptly, acquisition of a large number of porosimetry tests may help define "petrophysical facies" (see "rock typing" § 2-1.4, p. 317) and in some cases this approach proves highly useful. This is typically the case when studying oil/water contacts disturbed by horizontal gradients of petrophysical variations (the "FWL" then remains horizontal) and not by the hydrodynamism of the aquifer. This point is discussed in § 1-1.2.6C below.

c) Parameterisation of the porosimetry curve and application of porosimetry to the morphological description of porous media

Parameterisation of the porosimetry curve

The high definition of porosimetry curves, when acquired according to the professional standards described above, can be used to determine capillary parameters which are difficult to measure with sufficient accuracy using the other techniques which offer very few measurement points. In particular (Fig. 1-1.43):

- the pressure plateau. This is a narrow capillary pressure band within which S_{Hg} varies significantly. This plateau corresponds to the most frequent throat radius (the modal radius) clearly identified by the porosimetric spectrum (see below);
- the displacement pressure (Fig. 1-1.43a) corresponding to the point where the tangent to the plateau intersects with the y-axis ($S_{Hg} = 0$) [Thomeer, 1960]. This is the pressure at which the non-wetting fluid "invades" the porous space;
- the "apex" corresponding to the apex of the hyperbola obtained by plotting the capillary pressure curve in bilogarithmic coordinates (Fig. 1-1.43d1). This point corresponds to the maximum of the variable "S_{Hg}/Pc" (Fig. 1-1.43d2) [Swanson, 1981]. This parameter is apparently rarely used;
- the coefficient n used to adjust a hyperbolic function in the Thomeer method, discussed in § 1-1.2.5.

The porosimetric spectrum

The mercury injection curve can be considered as the cumulative curve of the volumes of pores accessible by thresholds larger than that corresponding to the pressure considered. Note the equivalence between pressure and radius in a cylindrical capillary model:

$r = 0.735/Pc_{Hg}$; with Pc in megapascals and r in micrometres.

If the curve acquisition has been sufficiently fine (typically over fifty automatic stages), which is required to obtain a good representation of the capillary pressure process, independent of the dynamic effects, the derivative of this curve can be calculated to obtain a kind of histogram of equivalent pore threshold radii. Discussions are taking place concerning the derivation method [Lenormand 2003]:

- either directly with respect to the pressure increments;
- or with respect to the radii interpreted, before or after change of variables [from R to $\log_{10}(R)$]. This last procedure is the easiest to display; in addition, it is similar to the processing of granulometric spectra;
- to avoid artefacts, the curve, even if acquired at relatively high resolution must be smoothed once the various corrections have been taken into account and in particular the surface effects at low pressures, sometimes incorrectly interpreted in terms of macroporosity (see Fig.1-1.42).

From a practical point of view, since experience has shown that the equivalent access radius distributions are generally log-normal in natural media, the porosimetric spectrum is always plotted with the logarithm of the equivalent radius on the x-axis (Fig.1-1.43c2). The y-axis corresponds to the frequency of the radius considered. The unit chosen is generally the fraction of porous volume occupied by the pores whose access radius lies within the

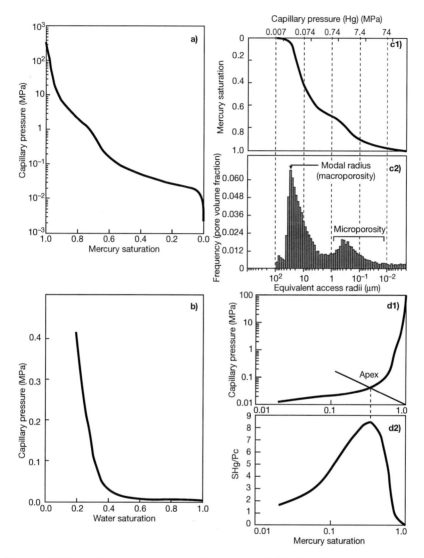

Figure 1-1.43 Porosimetry curve representation modes and definition of the main parameters. Example of Espeil limestone (see next Figure 1-1.44)

a) Conventional presentation of the mercury/vapour capillary pressure curve. The pressure is plotted on a logarithmic scale and the mercury saturation on a linear scale increasing from right to left for coherence with the graph of capillary pressure plotted against the wetting fluid.

b) Water/gas capillary pressure curve calculated with a parameter $A = 6.6$. The capillary pressure is plotted on a linear scale of maximum value 0.5 MPa which, for hydrocarbon reservoirs, corresponds to a very high capillary pressure.

c) Presentation of mercury porosimetry as a distribution of the equivalent access radii (porosimetric spectrum). The axes on the capillary pressure graph (c1) have been switched to clearly show the pressure/access radius equivalence. The graduation of the y-axis on the histogram (c2) corresponds to the fraction of porous volume occupied by the pores whose access radius lies within the elementary interval used for the derivation, in this case 1/15 of logarithmic decade.

d) Definition of "apex": d1) Capillary pressure plotted using bilogarithmic scale, similar to a hyperbola. d2) Curve $(S_{Hg})/(Pc)$ vs. S_{Hg}.

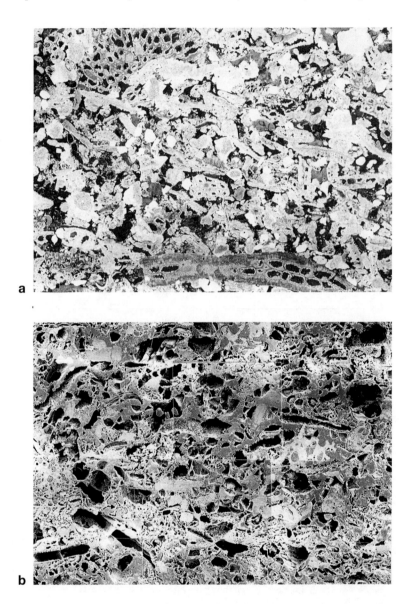

Figure 1-1.44 Photograph of a thin section (a) and epoxy pore cast (b) of Espeil limestone.

elementary radius interval used for derivation (1/15 of logarithmic decade for the example shown on the figure). During comparative studies, it is sometimes worthwhile choosing as unit the "absolute" value of the fraction of porous space concerned, simply multiplying the previous frequency by the fractional porosity value. Through this denormalisation, several samples can be compared directly. It was used on Figure 1-2.23, p. 157.

The porosimetric spectrum provides a very easy way to estimate:

- The modal radius of the "macroporosity" corresponding to the most frequent access radius ("plateau"). Comparison of this data, with an order of magnitude corresponding to the pore dimension acquired using a different technique (visualization § 2-2.2.2, p. 336; NMR § 1-3.3, p. 255, etc.), can provide a considerable amount of information. This ratio is sometimes tackled using the "displacement pressure" (defined above) instead of the "most frequent pore threshold radius", which is equivalent since these two parameters generally display excellent correlation.
- Whether the threshold distribution is unimodal, bimodal or multimodal, which the other methods are unable to detect through lack of sensitivity (a restored state test rarely exceeds 5 or 6 pressure stages) or due to the experimental principle ("S_{wi}" notion when using a wetting liquid). Any accessible macro and micro families (see porosity terms index) can be defined and quantified in this way. For a given set of rocks, the limiting value between micro and macro accessible porosity can therefore be defined. This value generally lies between 0.1 and 0.5 μm radius. As a result, the microporosity can be quickly and systematically identified, when large amounts of data need to be processed.
- Using these observations, we can quickly calculate the saturation value corresponding to the family of largest access, or the product of this value by the total porosity (similar to macroporosity) or the complement of this value (estimation of the microporosity). This explanatory parameter is important to understand the correlations with the other physical quantities (e.g. permeability).

We must constantly bear in mind, however, that porosimetry by mercury injection does not measure pore sizes but throat sizes and that the conclusions drawn by observing the porosimetric spectrum cannot be applied to the distribution of pore sizes as observed on thin section. Figure 1-1.45 is a simple example of two oolitic limestones showing in thin section significant interoolitic macroporosity. In mercury porosimetry, one is clearly bimodal with a "macroaccessible" porosity peak. The other has only one mode, quite spread out but essentially "microaccessible". There is a simple explanation: in the second case, the connections between the pores are highly cemented. This situation is very frequently observed in grainstones.

When studying the porous space of a reservoir, it is always recommended to compare the multimodal spectra with the morphological pore families that can be identified by observation at a comparable scale (e.g. thin section, scanning electron microscopy).

Example of application to the description of the porous space of sandstone of increasing clay content

The case of carbonates is often illustrated in the literature and we give numerous examples. We will therefore describe an example of clastic reservoir (Figure 1-1.46). Each mercury injection curve interpreted is accompanied, in addition to the usual petrographic and mineralogical analyses (not presented here), by observations under scanning electron microscope (SEM).

The series of plates of samples from the same reservoir illustrates increasing obstruction of the intergranular pores.

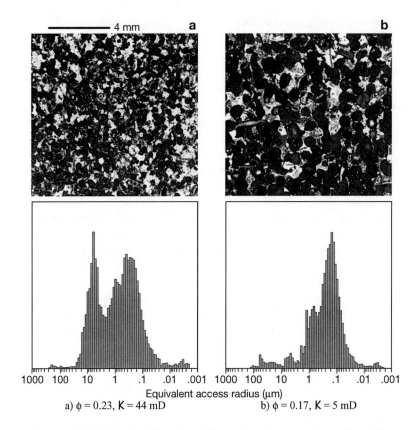

a) $\phi = 0.23$, $K = 44$ mD   b) $\phi = 0.17$, $K = 5$ mD

Figure 1-1.45   Example of limitation of the notion of access radius multimodality

The two oolitic limestones have two pore families: the intergranular macropores, clearly visible in thin section (photograph) and intraoolitic micropores (see porosity terms index). In mercury porosimetry (histogram of access radius), one (a) is clearly bimodal and the other (b) is unimodal (microporous). This is due to the fact that the throats to the intergranular macropores are very small.

– The first stage of this series corresponds to a clean sandstone (B1) whose histogram nevertheless shows low microporosity corresponding mainly to initial weathering of the potassium feldspars.

– Sandstone B2 shows the appearance of a few vermiculite aggregates typical of kaolinite/dickite. The illite/interstratified (I/S) phase first results in pore lining or grain coating (see § 1-1.1.3, p. 13) leading more to an increase in surface roughness than to individualisation of a clearly separate microporous phase.

– Then the clay phase increases (B3): the phenomenon develops with the total porosity remaining more or less constant. Since the volume content of the dry clay phase is low, some of the intergranular macroporosity is simply transferred to the microporosity. There is only a moderate impact on permeability since the equivalent macropore threshold radii are relatively unaffected.

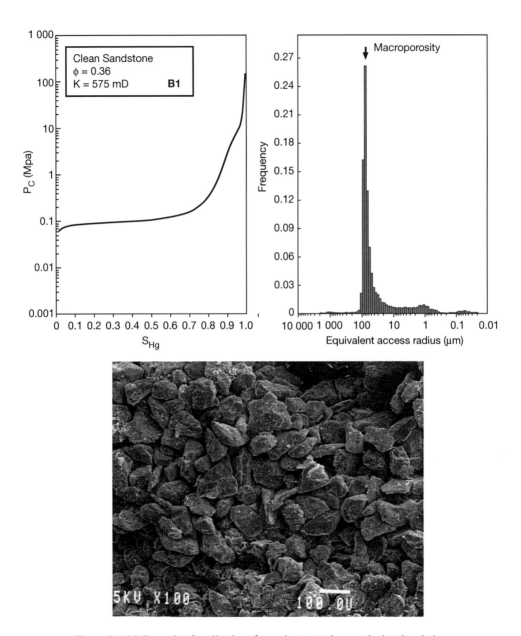

Figure 1-1.46 Example of application of porosimetry to the quantitative description of the porous space of sandstone of increasing clay content; (four plates)

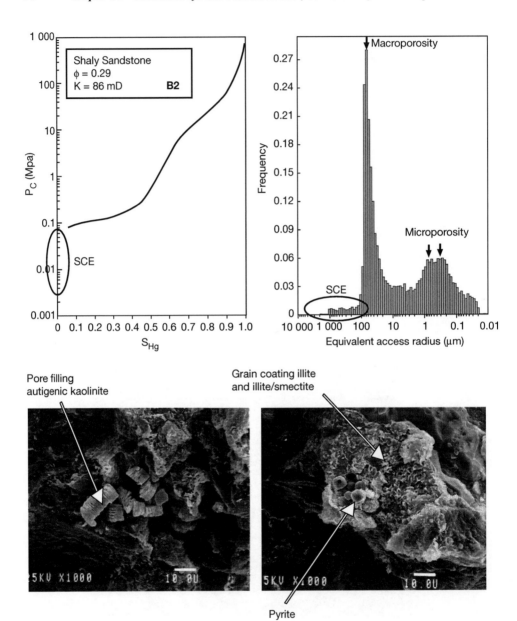

Figure 1-1.46 Continued, plate 2

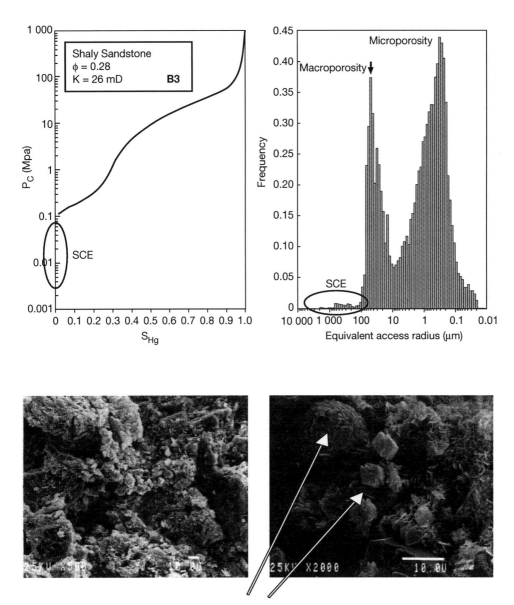

Pyrite and siderite in too small quantity to impact ϕ, K
but noticeable on density (2830 kg/m³)

Figure 1-1.46 Continued, plate 3

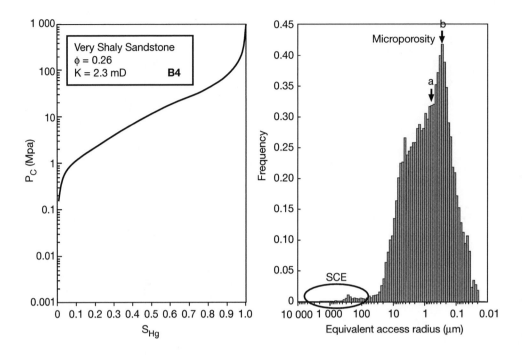

Silica-clay cement with pyrite spheroids.
Kaolinite/dickite cristals are masked by
illite/interstratified filling

Figure 1-1.46 Continued, plate 4

– In the last phase (B4), cementation increases, the clay phase is accompanied by silica-clay cement: the macroporosity observed in thin section decreases in relative volume but the pore threshold dimensions are so reduced that the "macroaccessible" porosity tends to disappear from the distribution histogram, merging with the microporosity. The total porosity value starts to be affected, but the most important impact is on the permeability, which is more severely reduced.

Behaviour of the capillary pressure curve at high pressures, effect of roughness

Using modern experimental equipment, mercury porosimetry can be conducted up to very high pressures (400 MPa, in routine tests) corresponding to equivalent capillary radii of a few nanometres. We mentioned earlier that compressibility corrections could be difficult to make at such high pressures. Nevertheless, mercury porosity at very high pressure offers a means of studying the roughness of pore walls.

This section of the curve corresponds to the phase where the mercury, after invading the pores themselves, moulds more and more closely to the surface roughness of the pores. The representation of this phenomenon could be simplified by replacing the model of cylindrical capillary tubes by a model with square capillary tubes.

In practice, there is a progressive boundary between pore surface state and microporosity. The examples shown on Figure 1-1.47 are highly explanatory as regards the clay phase: as long as it is present in relatively small quantities and disposed in pore-lining, it can be considered as being part of the surface state of intergranular porosity. As soon as this phase becomes more abundant, however, clear microporosity appears.

In conclusion, analysis of the porous medium must start with a critical comparison between the mercury injection curve, especially if it is multimodal, and observation of the sample at correct scales: the macroscopic description of the sample itself and its analysis on thin section provide a first scale of the organisation of the medium. SEM observation then provides access at a scale whose order of magnitude is similar to that investigated by mercury porosimetry. In particular, this approach is essential in case of multimodal behaviour and/or to describe the pore surface state.

Lastly, we must mention that, following the pioneering work of Fatt [1956] recent quantitative models of porous media are available, combining an interpretation of mercury

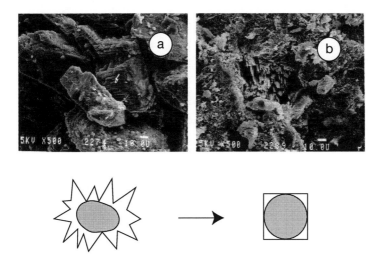

Figure 1-1.47 Example of changing roughness due to the corrosion of potassium feldspars in samples shown on Figure 1-1.46. a) sandstone B1; b) sandstone B2. The scale bar is equal to 10 micrometres. c) modelling the effect of roughness by capillary tubes of square cross-section

injection more advanced than the simple equivalent of cylindrical capillary tubes, with quantifications extracted from high resolution imaging. These models more and more accurately predict the multiphase dynamic behaviour (relative permeability, § 1-2.3, p. 179) of porous media in situations of widely varying wettability [McDougall and Sorbie, 1997; Laroche and Vizika, 2005].

1-1.2.5 Processing of capillary pressure data

Once they have been acquired on samples in the laboratory, the capillary pressure measurements must be processed and converted for analysis, in order to evaluate a saturation at a given height above zero capillary pressure (calibration of petrophysical interpretation of log analyses, reservoir modelling and calculation of initial accumulation) or to specify the static "petrophysical facies" or "rock types" (rock typing, § 2-1.4, p. 317). The "drainage" branches of the curves are used for these applications.

In dynamic digital modelling the "imbibition" branches, much more rarely acquired, must be used to simulate the most frequent cases of reservoir exploitation (water flooding or increase of the water level).

The main purpose of the conversions to be carried out on these experimental data is as follows:

- adapt the presentation of the parameters to suit the desired use;
- convert an often low number of experimental points into a continuous curve which can easily be used in a computer model;
- group into a limited number of representative curves the results obtained on numerous samples from the same layer.

In order to simplify the description, we will discuss the transformations of curves by successive phases, schematised in Table 1-1.10. Note that the transformations related to the change from plug scale to core scale (or scale of the interpretation of the log analysis) then from core scale to the scale of the digital model grids, are mentioned in § 2-1.3.

A) Capillary pressure curve transformation phases

a) Preliminary phase

This phase is designed to reduce the artefacts and ensure that the physical process is representative. It is more closely related to the processing of experimental values and to quality control. A certain amount of information regarding these problems can be found in the paragraph concerning each measurement technique.

Note that several different types of experiment are implemented to acquire the capillary pressures and that each one has its own advantages and disadvantages, which must be born in mind during the interpretation. *Whatever the case, it is recommended never to mix data from different types of experiment in the same processing batch.*

Table 1-1.10   Diagrammatic representation of the various capillary pressure data processing phases [according to B. Layan, private correspondence]

Acquisition and processing in the laboratory			
"Restored state" Surface conformance effects. **Pc; Sw$_c$**	*Centrifuge* Adjusting and calculation **Pc$_{cal}$; Sw$_{calc}$**	*Mercury Injection (Porosimetry)* Surface conformance Compressibility, ... **Pc; S$_{Hg}$**	*Other methods* Desorption, etc Interpretation (Kelvin, ...) **Pc$_{equiv}$; Sw**
Transformation of Y-axis values: Pressure *"Reservoir or in situ" transformation* Pc$_{res}$ = $Pc_{lab}([\sigma\cos\theta]_{res}/[\sigma\cos\theta]_{lab})$ *Leverett Function* $J = a(Pc/\sigma\cos\theta)\sqrt{K/\phi}$ *Height above Free Water Level* ht = $Pc_{res}/\Delta\rho_{res}\,g$		Transformation of X-axis values: Saturation *Specific "Porosimetry"* $S_w = 1 - S_{Hg}$ *S_{wi} Normalisation* $S_{wR} = (S_w - S_{wi})/(1 - S_{wi})$	
Curve by curve processing (adjustment)			
Power Law (Pc) $Pc = Pd\left(S_w - S_{wi}/1 - S_{wi}\right)^{\frac{-1}{n}}$ **Variant: replacing Pc by Ht**		Parametrisation for reservoir modelling	
		Pd = f(ϕ, K, ρs, Vcl,...) Pd > 0 S$_{wi}$ = f(ϕ, K, ρs, Vcl,...) $0 \leq S_{wi} < 1$ n = f(ϕ, K, ρs, Vcl,...) $0 < n \leq 1$	
Power Law (J) (rarely used curve by curve) $J = Jd/[(S_w - S_{wi})/(1 - S_{wi})]^{n_j}$		Jd = f(K, ρs, Vcl,...) S$_{wi}$ = f(K, ρs, Vcl,...) $0 \leq S_{wi} < 1$ n$_j$ = f(K, ρs, Vcl,...)	
Hyperbola (Thomeer) $(1 - S_w)/(1 - S_{wi}) = e^{\frac{-G}{\log(Pc/Pd)}}$		Pd = f(ϕ, K, ρs, Vcl,...) Pd > 0 S$_{wi}$ = f(ϕ, K, ρs, Vcl,...) $0 \leq S_{wi} < 1$ G = f(ϕ, K, ρs, Vcl,...)	
Processing of a set of curves			
Generalised WWJ method $S_w = 10^{(C_1 Pc^{C_2} + A_1 \log(K) + 100 A_2 \Phi)}/100$ **Variant: replacing Pc by Ht**		*Power Law (J)* $J = A/[(S_w - S_{wi}J)/(1 - S_{wi}J]^{nJ}$	

b) Transformations of the values on the curve coordinate axes

Transformations on the pressure axis (y-axis)

Three types of transformation are frequently carried out, for different reasons.

To adapt the curve to the required interpretation, the following transformations can be carried out:

- Transformation of *'in situ'* (or *"reservoir"*) coordinates to transform the pressures acquired into *"in situ"* equivalent pressures using the ratio of interfacial tensions t_s and the wettability expressed in $\cos\theta$.

$$Pc_{res} = Pc_{lab}([t_s \cos\theta]_{res}/[t_s \cos\theta]_{lab})$$

This point has already been mentioned in the chapter concerning mercury porosimetry.

The *"lab"* or *"res"* indices associated with a variable indicate "under laboratory conditions" and "under reservoir conditions", respectively

- transformation of *height (ht)* above zero capillary pressure plane (Free Water Level). This transformation will be extremely useful in the next paragraph (§ 1-1.2.6) which deals with the localisation of fluids in oil reservoirs. A few more detailed explanations could be found there.

$$ht = Pc_{res}/\Delta\rho_{res}\,\gamma$$

When interpreting these results with respect to deviated well recordings, it is important to think in terms of "true vertical depth" (TVD, Fig.2-1.19, p. 310) and not in terms of "driller depth".

Transformations on the saturation axis (x-axis)

Apart from the obvious transformation of saturations resulting from the porosimetry measurement in equivalent water content: $S_w = 1 - S_{Hg}$, normalisation by S_{wi}, the initial water saturation at maximum Pc, is sometimes carried out: $S_{wR} = (S_w - S_{wi})/(1 - S_{wi})$.

This transformation consists in only taking into account the fraction of the porous space in which the water is mobile. By using the variable S_{wR}, the "normalised" or "reduced" saturation, it is often easier to group curves from different samples into a single bundle (Fig. 1-1.48). The difficulty lies in the definition of S_{wi}, a more empirical and conventional than strictly physical value. This difficulty is discussed in several paragraphs, in particular § 1-2.3.3.

c) The Leverett transformation

It is a transformation [Leverett, 1941] of values on the pressure axis using the variable

$$J = a(Pc/t_s \cos\theta)\sqrt{K/\phi}$$

Variable J is dimensionless.

This transformation is worthy of special attention, since it is frequently used in reservoir geology to group together several capillary pressure curves with similar behaviour.

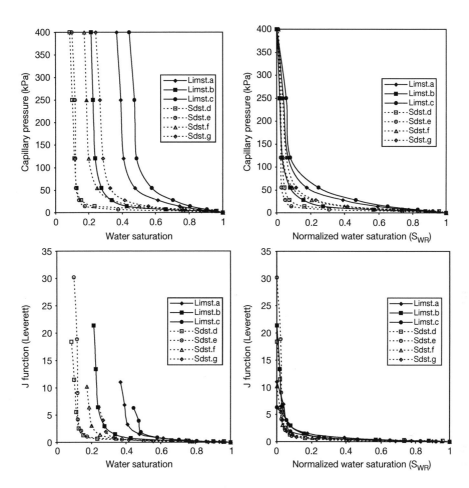

Figure 1-1.48 Example of capillary pressure curve after normalisation
of the water saturations (top right) and use of the Leverett transform (bottom graphs)

Note firstly that this transformation consists in normalisation, by the average hydraulic radius of the sample, of the equivalent capillary access radius for the pressure considered. We have seen (§ 1-1.2.1, p. 54) that the capillary pressure could be converted into an equivalent radius: $R_{cap\_Pc} = (2t_s \cos \theta)/Pc$. We also know (§ 1-2.1.3, p. 138) that the permeability K of a porous medium consisting of straight and parallel cylindrical capillary tubes is given by $K = \phi R^2/8$. Let R_h be this radius (very partial, but not incorrect, definition of the "hydraulic radius", see § 1-2.1.3C). We obtain $R_h = \sqrt{8}\sqrt{K/\phi}$ and therefore

$$R_h/R_{cap\_Pc} = \left(\sqrt{8}/2\right)\left(Pc/t_s \cos\theta\right)\sqrt{K/\phi}$$ where K is the single-phase permeability

which includes the expression of the J-transform. We can therefore calculate the value of the coefficient $a = \left(\sqrt{8\alpha}/2\right)$, where α is the adjustment coefficient related to the units used.

For example, $\alpha = 3.16$ if Pc is expressed in bar, t_s in dynes/cm (in mN/m), K in mD and ϕ is fractional.

Coefficient *a* is often missing since as a rule it is not relevant when dealing with J functions (as long as it is deleted from the entire database of the study considered). Some laboratories simply use α (equation with units) without taking account of the coefficient $\sqrt{8}/2$. In practice there are several variants therefore, and confusion between them may have led to frequent errors for reservoirs on which successive studies have been conducted. Consequently, it is essential to always check that the coefficient α is specified

Graphically, the use of the Leverett transform is equivalent to an anamorphosis parallel to the y-axis (Fig. 1-1.48). We see empirically that, for different samples from a set of rocks in a given layer, the J curves are often grouped if not on the same transformed curve, then at least tightly together. This grouping is mainly observed when the S_{wR} values are used on the x-axis.

When true, this specificity proves useful since it can be applied to "synthesise" the behaviour of a simple reservoir, a lithological level or a petrophysical facies ("rock type" § 2-1.4, p. 317). Note however that calculating this function, used to predict a saturation, involves the porosity and permeability variables but that these three physical quantities do not obey the same homogenisation rules in case of change of scale.

B) Curve adjustment

This operation is carried out to reduce a series of acquisition points to an analytical formula, easier to handle.

a) Adjustment by a power type function

This adjustment can be applied to capillary pressure curves, generally normalised (see above), but also to the Leverett-J functions, often normalised by S_{wi}. P_d is the displacement pressure, corresponding to the pressure at which the non-wetting fluid starts to invade the porous space (J_d is the Leverett transform corresponding to this pressure). It is illustrated on Figure 1-1.29 in particular.

$$Pc = Pd(S_{wR})^{\frac{-1}{n}}$$

$$J = Jd/[S_{wR}]^{n_j}$$

The advantage of a power function is that, if the saturation values are previously normalised by S_{wi}, it can easily be adjusted by linear regression after a logarithmic change of variables. If advanced statistical tools are available, it may be more relevant to use a non-linear regression, initialised by the parameters of the linear regression.

After the operation, the adjustment supplies three parameters on each sample:
- an equivalent of "S_{wi}";
- an equivalent of the displacement pressure, or its Leverett transform;
- a coefficient "n", function of the curvature of the curve.

In a set of samples representative of a reservoir, a subreservoir or a "rock type", the aim will then be to correlate these three values with parameters whose spatial variability is assumed to be known.

The remark made in the previous paragraph, concerning the caution required when changing scale, still applies.

b) Adjustment by a hyperbola type function: Thomeer transform

This method is based on the empirical observation [Thomeer, 1960] that, for a given sample, the mercury porosimetry results roughly form a hyperbola when the mercury saturation values and the corresponding capillary pressure values are presented in log/log (unlike the previous method based on an assumption of linear regression in log/log). The horizontal asymptote corresponds to the displacement pressure value. This type of representation is shown on Figure 1-1.43d (p. 85). Parameter G, characteristic of the hyperbola ("curvature"), can be used when comparing samples from the same reservoir.

This method can be extended to process the results of other capillary pressure types, by using the saturation normalised by S_{wi} and an equivalent of the displacement pressure. We obtain an expression of the normalised saturation, of the form:

$$S_{wR} = e^{\frac{-G}{\log(Pc/Pd)}}$$

c) Adjustment by a set of curves, Wright & Wooddy method

Another way of processing the results when conducting a synthesis study on reservoirs is to pre-identify groups of samples (by reservoir, subreservoir, lithology, rock type, etc.) and adjust a function on these groups.

Two methods are generally used:

- The first is to adjust a power law on the bundle of J-functions, generally normalised by S_{wi}.
- The second is to use a method which, although strictly empirical, is often relatively well predictive if the preliminary groups were well chosen.

This method has been described by Wright and Wooddy [1955]. For convenience, we will refer to it as the WW method. The authors observed a classification of curves according to permeability (Fig. 1-1.49a).

Hence the idea of looking for correlation between S_w and the logarithm of the permeability for a given capillary pressure Pc. Note that this is the generalisation of a well-known property of the irreducible saturation, summarised by the empirical formula: $S_{wi} = A \log K + B$. The relations between this classification of curves according to permeability and the Leverett transform are also quite strong.

The principle of the calculation process is as follows:

- Construct the following relation for about ten Pc stages: $S_{w@Pc(n)} = a_{(n)} \log(K) + b_{(n)}$.
- Check that the straight lines are roughly parallel and calculate a mean value of $a_{(n)}$ (Fig. 1-1.49b).

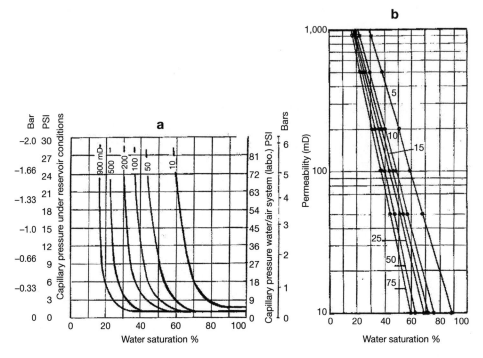

Figure 1-1.49 Principle of the WW method, [Monicard, 1980]

- Readjust the correlations $S_{w@Pc(n)} = a_{mean} \log(K) + b_{(n)}$ by imposing the multiplier a_{mean}.
- Then construct the correlation $b_{(n)} = \alpha Pc_{(n)} + \beta$.
- We thus obtain the relation $S_w = a_{mean} \log(K) + \alpha Pc_{(n)} + \beta$.

The pressures may have been converted previously into heights above zero capillary pressure (FWL).

The method can be improved by using more complete relations:

$S_{w@Pc(n)} = a_{(n)} \log(K) + b_{(n)} + c_{(n)}\phi$ firstly, and/or non-linear relations secondly.

Quite tedious (and dissuasive) when the calculations had to be carried out manually, this method has now been automated in core data interpretation software. The advantage of this method is that it produces good predictions if the sampling was carried out correctly. In contrast, since the coefficients have no simple physical meaning, it is impossible to check whether or not they are realistic during adjustment. The precautions required when changing scale are the same as those with the other methods.

C) Summary and conclusions

In order to choose between the numerous techniques available for normalisation and capillary pressure processing, a few basic principles must be borne in mind:

- The laboratory processing operations must be performed and validated. All methods suffer, to a greater or lesser extent, from artefacts which need correction.

– Remember that the petrophysical parameters involved in the relations do not all obey the same scaling processes.
– If the results corresponding to different types of measurement (centrifuge, porosimetry, etc.) are available, they must only be merged (after suitable normalisation) if there are insufficient measurements. Obviously, preference must be given to the method involving the fluid pair closest to the true conditions. For example, mercury injection is nearly always too optimistic for an "oil/water" reservoir. Although closest to the *in situ* process, the "Restored State" method is the one most likely to be affected by the least detectable artefacts (very long experiments). Especially at high values of Pc, care must be taken with the "collective" gas/water restored states (numerous samples processed simultaneously in the same cell).
– The first objective of this processing is to relate an estimation of the saturation to petrophysical quantities whose spatial distribution is either known or evaluated elsewhere. The approach taken depends on the application (log analysis or geological model) and on:
  • The availability of the support parameters (K, ρ_s, … are often found in "petrophysical log" but are not available in geological modelling).
  • The operator's "habits": the Leverett function is usually preferred in geological modelling. At any rate, it favours models involving the conventional notion of "S_{wi}". Saturation mapping is carried out in two phases: first the "S_{wi}" map then, if necessary, a superimposed map of a transition zone in S_{wR}, the normalised ("reduced") saturation.
– In spatial modelling, the Leverett-J function is often the most popular, although it does not remove the difficulties related to a change of scale. Saturation normalisation (by "S_{wi}") may be required to obtain a bundle of curves with as little dispersion as possible. The bundle(s) of J functions will be adjusted using a power type function. The aim is to correlate the coefficients of these functions with the parameters $\log(K)$, ϕ, V_{cl} (clay content), etc. $\mathrm{Log}(K)$ is often the best estimator. In contrast, if ϕ, V_{cl} and possibly ρ_s, can play a predictive role, they will be privileged for use in Production Geology.

Note that curve by curve adjustment of the J function is of little use. It is better to normalise the saturations curve by curve, then use the J functions with these reduced saturations, imposing $S_{wiJ} = 0$.

– In log analysis (or when producing a petrophysical log), the WW type method is the most efficient. The change of scale is easier in this configuration and the predictive performance is improved.
– Other types of adjustment, not described here, are sometimes applied (e.g. an exponential function: $S_w = \lambda Pc^{A\Phi + B}$). They may be more efficient but it is generally impossible to assign a physical notion to the adjustment parameters. They are of most use, however, to avoid having permeability as a support variable.

Table 1-1.11 summarises the most common practices.

Summing up, we observe that concerning the capillary pressure curves, as for all the other petrophysical properties, the various adjustment and correlation methods give satisfactory results as long as the porous media only contain a single type of porosity

Table 1-1.11   Summary of the most common reservoir characterisation practices
used to process capillary pressure curves

	Curve by curve processing	**Processing a set of curves**
Generalised WW method	Not applicable	After transformation into Pc_{res} in the general case or preferably after transformation into height
'Power Law' method	After transformation into Pc_{res} Use of the S_w values normalised by S_{wi} "*a priori*"	Use of the Leverett transform and the S_w values normalised by S_{wi} "*a priori*", in the most general case
'Generalised Thomeer' method	Often used after transformation into Pc_{res} and into S_w normalised by S_{wi} "*a priori*"	Rarely used, special cases

(unimodal media). In this case, this is due to the almost one-to-one relations between the various properties, through the basic normative parameter, porosity. But if several types of porous space cohabit in the medium considered, all these relations become inoperative since the global porosity measured is the sum of various "porosity types" present in unknown relative proportions. Concerning the capillary pressure curves, an attempt was made to find a solution by using the notion of "irreducible saturation". Implicitly, this corresponds to identifying and quantifying a type of "microaccessible" porous medium which would retain the capillary water. The porous space is normalised to the "macroaccessible" fraction for which relations can be derived between the various petrophysical properties (therefore including the capillary pressure curve). This is the justification for normalised saturation S_{wR}, applied so extensively we almost forget that the notion of irreducible saturation largely corresponds to a vague empirical concept (see § 1-2.3.3A, p. 186).

1-1.2.6 Location of fluids in oil and gas reservoirs

A) Accumulation of hydrocarbon in the reservoirs (secondary migration)

a) Overview on secondary migration

The hydrocarbons, produced in generally shaly, low-permeability source rocks, are expelled from these rocks during a process which is still poorly understood and in which compaction could play a major role. This phase is known as primary migration. We will restrict ourselves here to secondary migration, i.e. transport of hydrocarbons in permeable layers and trapping in the reservoirs. Capillarity plays a central role in these phenomena.

To describe this secondary migration, we can make the assumption that the permeable layers acting as drain or reservoir are strongly water-wet. These rocks, in fact, remained saturated with water for a considerable period of time before being placed in contact with the hydrocarbons. They are therefore water-wet, at least during the first phase of hydrocarbon accumulation. This phase will be described here (see § 1-2.2, p. 173, regarding the possible acquisition of oil wettability).

Hydrocarbons are driven to move by their "buoyancy" which is related to the difference in density ($\Delta\rho$) between hydrocarbons (ρ_o, ρ_g) and water (ρ_w). This force is used to overcome the capillary barrier (capillary pressure) opposing the penetration of non-wetting fluid. At the top of a hydrocarbon column of height h, the buoyancy induces a pressure difference with respect to water ($\Delta P = Po - Pw$) which is easy to calculate using the fluid equilibrium formula: $\Delta P = \Delta\rho \cdot \gamma \cdot h$, where γ is the gravitational acceleration (see paragraph § B. a below for further details).

Movement of the hydrocarbons therefore implies coalescence into accumulations whose height is sufficient to overcome the capillarity (see below § B. a for orders of magnitude). These accumulations will move upwards along the upper limits of the layers, or more precisely the permeability-capillarity contrasts. These contrasts may represent relative barriers which will be crossed when the height of the oil column has increased due to influx of additional hydrocarbons. These contrasts may also correspond to absolute barriers if the permeabilities are extremely low (compacted shale). This is the case of cap rocks which may lead to the formation of a reservoir if the geometry of the impermeable rock allows a trap to develop.

If the geometric structure forms a trap, the hydrocarbons accumulate (from the top towards the bottom of the trap). If the hydrocarbon influx is sufficient the geometric spill point (Fig.1-150a), corresponding to the lowest point of the trap, may be reached. The reservoir is full; the distance between the highest point of the trap and this spill point is called the trap "closure".

b) Capillary trapping (residual saturation) and traces of secondary migration

The notion of residual saturation caused by capillary trapping of a fraction of non-wetting fluid is defined in § 1-1.2.2.D, p. 59. At the end of the secondary migration, when there is no further influx of hydrocarbons, this residual saturation should last in the transfer zones, showing the path of the secondary migration.

Over the course of geological time, one might expect that the progressive disappearance of residual fluids, especially gases, is due to solubilisation (if there is water movement in the layer) or diffusion phenomena. As regards oil, an evolution into bituminous formations (tar-mats) is more likely.

What is the true situation? It is extremely difficult to provide an objective answer. It seems almost impossible to detect such traces of residual hydrocarbons, in fact, since they would be limited to relatively thin layers of rock in contact with permeability contrasts. Located by definition outside reservoirs, there is very little chance of samples being taken from this type of zone during coring. In addition, it is very difficult to detect by log analysis. We therefore see how difficult it is to provide an answer.

The notion of residual hydrocarbon also arises when, under the effect of tectonic movements (e.g. tilting), the reservoir closure decreases (relative elevation of the spill point) and the trap partly empties. In this case, any residual traces are easier to locate. It is likely, however, that there is a link between this residual fraction and the tar-mats sometimes present in large quantities below the current oil/water contacts. It is also highly probable that this phenomenon contributes to the uncertainty regarding the exact location of the oil/water contact surfaces often observed in practice.

B) Capillary equilibria and location of fluids in a reservoir

a) Evolution of capillary pressure in a reservoir according to depth

We have briefly outlined the accumulation of hydrocarbons in reservoirs. We will now examine the distribution of fluids in a reservoir at equilibrium. We will start from the following situation: we observe the coexistence in the reservoir of two fluids in continuous masses, one strongly wetting, water, the other non-wetting, oil (or gas).

Each fluid individually respects the law of hydrostatics, the pressure depending on the depth D and the density ρ:

$$P_w = \rho_w \gamma D + C_1 \text{ et } P_o = \rho_o \gamma D + C_2,$$

where C1 and C2 are adjustment constants.

Expressed in the coordinate system D against P (Fig. 1-1.50a), the two straight lines of different gradients intersect at a depth D_e such that $P_w = P_o$. The level corresponding to this depth plays an important role in the expression of capillary equilibrium. It is called the Free Water Level. We will use the abbreviation FWL. By knowing D_e we can determine the values of the adjustment constants C_1 and C_2 since at this depth.

$P_o = P_w$ and therefore $C_1 - C_2 = \Delta\rho\gamma D_e$ (where $\Delta\rho = \rho_w - \rho_o$).

Introducing a new depth variable, the height h above FWL.

$(h = D_e - D)$, we immediately obtain:

$\Delta P = P_o - P_w = \Delta\rho\gamma h.$

By definition, however, the capillary pressure Pc is equal to $P_o - P_w$. We therefore obtain the well-known equation.

$Pc = \Delta\rho\gamma h$, which provides a direct relation between the saturation in a reservoir and the capillary pressure curve.

b) Saturation distribution in a homogeneous reservoir

For a reservoir which is assumed to be perfectly homogeneous, the above formula immediately gives the saturation distribution from the capillary pressure curve $Pc = f(S_w)$, by making a change of variable on the y-axis. h can be substituted for Pc, since $h = Pc/\Delta\rho\gamma$.

Obviously, this logic is only quantitatively valid if the capillary pressure curve was measured with a pair fluids identical to those of the reservoir (or at least if their interfacial tension and wettability characteristics are similar). Otherwise a weighing factor must be applied, the ratio of the products $\sigma. \cos\theta$ (see above § 1-1.2.5)

We therefore obtain the saturation distribution shown on Figure 1-1.50b. We observe the three zones mentioned earlier (§ 1-1.2.2.C, p. 56), i.e. from top to bottom (decreasing capillary pressure):

– The saturation zone at maximum Pc ("irreducible" according to the former terminology) where the wetting phase (water) confined in the finest menisci is non-movable (the relative water permeability is zero). This zone where anhydrous oil is produced corresponds to the reservoir in the common meaning of the term. The main practical

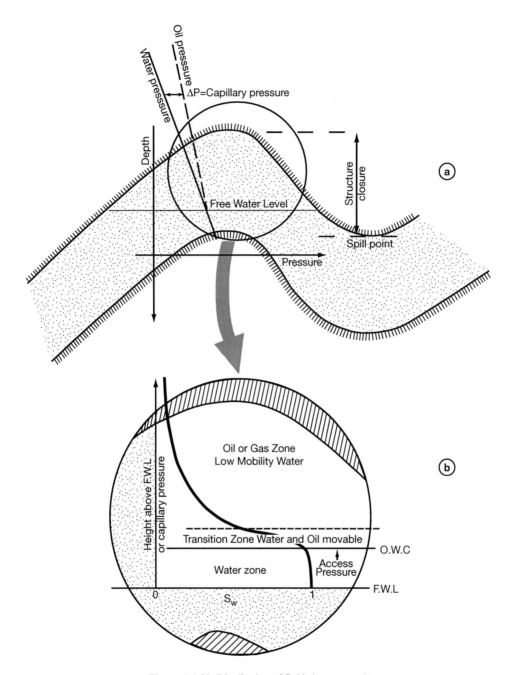

Figure 1-1.50 Distribution of fluids in a reservoir

a) Pressure profile in water and hydrocarbon, definition of the Free Water Level (FWL).
b) Standard saturation profile.

problem arising with this zone is how to determine the value of S_{wi}, a matter of great concern for reservoir specialists. We will discuss this subject briefly in § 1-2.3.3, p. 186.

– The transition zone where the two phases are movable (relative water and oil permeabilities non zero). The bottom of this zone corresponds to the appearance of hydrocarbons in the reservoir. This is the original (before start of production) oil-water contact (OOWC) or gas-water contact (OGWC) which could be quite different from the free water level (FWL).

– The capillary foot (or capillary drag). This is a very special zone, totally water saturated, but in which the (virtual) capillary pressure is non zero. The pore throat radius is too narrow for the non-wetting fluid to enter. In practical reservoir geology applications, this zone is of considerable interest to provide an insight into original oil/water contact (OOWC) anomalies.

c) Some simple examples of numerical values

We saw in § 1-1.2.5 that we can estimate the scale conversion factor to graduate the capillary pressure axis in heights above FWL, to obtain the diagram of a saturated reservoir. To simplify matters, we will use mercury porosimetry curves as capillary pressure value ($t_s = 460$ mN/m, $\theta = 140°$). Although this may not be the best way to investigate the notion of S_{wi}, it is sufficient for our purposes, where the main objective is to examine the capillary feet and the original oil/water contacts (OOWC). We will study a standard water/oil ($t_s = 30$ mN/m, $\theta = 0°$, $\Delta\rho = 200$ kg/m^3) and water/gas ($t_s = 50$ mN/m, $\theta = 0°$, $\Delta\rho = 700$ kg/m^3) case (see Table 1-1.7, p. 55).

On Figure 1-1.51, the maximum height scale (500 m) corresponds to the major petroleum zones (remember that the height of the hydrocarbon column may exceed one kilometre in some fields (e.g. the Middle East)).

At this scale, we mainly observe that:

– The water/hydrocarbon contacts (OWC, GWC) correspond to the free water level ($y = 0$ on Figure 1-1.51) for all the "permeable" reservoirs (single-phase permeability greater than several tens of millidarcy). There is no difference between OOWC and FWL when we consider the traditional sandstone reservoirs or the permeable carbonate reservoirs. This explains the sometimes marked lack of interest for this problem. In contrast, the "capillary foot" zone may take on considerable importance when dealing with very low permeability reservoirs, especially micrite type limestone reservoirs but also poorly porous sandstone reservoirs.

– Obviously, due to the larger density difference, the problems of FWL/GWC shift are less significant in gas reservoirs.

– As soon as we start to move up in the structure, the water saturation gradient as a function of the height is low, even with the simple calculation mode used (directly related to the porosimetry access radii).

We must remember that these remarks apply to a high oil column (important structuration **and** abundance of hydrocarbons). For "poor" petroleum provinces, the problems may be quite different, for example for stratigraphic traps (see § C. a and b below).

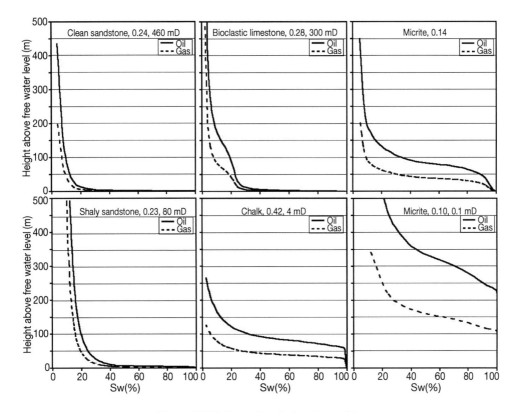

Figure 1-1.51 Examples of saturation profiles

C) Applications to some practical aspects of reservoir geology

The above description of fluid distribution corresponds to the idealised case of a perfectly homogeneous reservoir, both in terms of the petrophysical properties of the rock and those of the saturating fluids. In nature, the situation is never that simple. The first cause of complexity is due to the fact that natural reservoirs are heterogeneous. In this paragraph, we will discuss a few cases of reservoir geology where the application of these simple capillary equilibrium notions helps us to understand the phenomena involved.

a) Water/oil contact fluctuations

Determination of the water/oil contact and the free water level

As soon as we move away from the case of permeable reservoirs with zero transition zone, determining the exact value of these two levels may be less trivial than it would first appear.

Determining the original oil/water contact corresponds to the appearance of the first oil indices, which are not always easy to detect using log analysis in cases of low porosity and saturation. There are two ways to determine the free water level: either by identifying the oil/water contact in a drain with zero capillarity (in practice, an observation well left idle for a long period of time), or by measuring the pressure gradient in oil through mini-tests carried out at various depths (repetitive tests). Conducted according to professional standards both methods are expensive and consequently, when discussing real field cases, data uncertainty must always be taken into account.

Fluctuations due to petrophysical variations (fixed FWL)

General case

Imagine a large limestone reservoir changing, due to sedimentological evolution, from porous bioclastic grainstones to increasingly compact micritic formations exhibiting the petrophysical characteristics described on Figure 1-1.52. While the free water level (FWL) remains horizontal and stable throughout the structure, the original oil/water contact (OOWC) follows the value equivalent to the access pressure. In the assumption of Figure 1-1.52, the altitude of the water "level" increases from left (bioclastic grainstone) to right (compact micrite). To plot the curves shown on Figure 1-1.52 we adopted the standard oil characteristics used in the previous paragraph. The values are therefore representative. We observe that for the second saturation profile (pelletoidal grainstone) which still corresponds to a reservoir rock, there is already a shift of some ten metres. For the third and fourth profiles, the shifts reach several tens of metres.

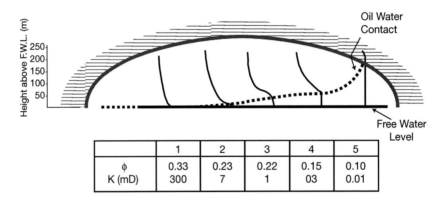

	1	2	3	4	5
φ	0.33	0.23	0.22	0.15	0.10
K (mD)	300	7	1	03	0.01

Figure 1-1.52 Diagram of irregular hydrocarbon/water contact
related to the lateral variation of the reservoir

Obviously, the example shown on Figure1-1.52 is highly diagrammatic. It is based mainly on an assumption of vertical homogeneousness. To return to a more realistic situation, we would have to introduce vertical variability in order to show "superimposed" water levels. This example nevertheless clearly shows that the notion of a flat and horizontal water level, so frequently encountered in permeable reservoirs, is completely meaningless as soon as we are faced with poorly permeable reservoirs, of micrite type for example.

Whenever we observe a fluctuation in water level, we must first consider the assumption of a petrophysical variation.

Case of fractured reservoirs

The saturation profiles 3 and 4 shown on Figure 1-1.52 correspond to poorly permeable reservoirs which will only yield good production if a network of open fractures creates high-permeability drains. These matrices, poorly permeable but still exhibiting high porosity, and therefore very oil-rich in the zones above the water "level", can supply the network of fractures. Some of the most productive oil fields in the world operate according to this scheme.

One important feature of open fracture networks must be pointed out. These fractures exhibit "conducting apertures" up to several hundred microns thick. As a result, they are no longer subject to capillarity and the oil/water contact corresponds to the free water level. We therefore observe in these reservoirs two different oil/water contacts, sometimes several tens of metres apart, corresponding to the two media of highly contrasted petrophysical properties, which coexist in these reservoirs (Fig.1-1.53). Since the open fractured medium represents only a minute fraction of the total porous medium, the oil it contains is impossible to detect using log analysis. Occasionally, some water saturated reservoir zones (matrix) yield excellent oil production if an open fracture intersects the well. However, this production only lasts as long as the open fracture network is "supplied" with oil... but this is outside the scope of our subject!

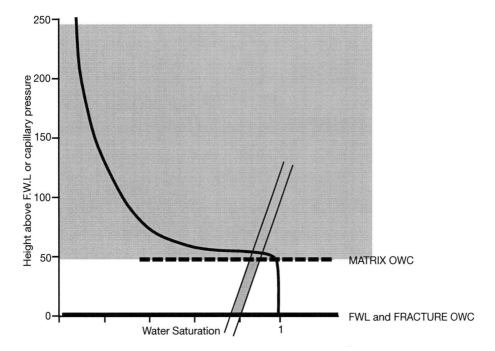

Figure 1-1.53 Position of the oil/water contact in fractured reservoirs

Extreme case of fluctuation of petrophysical origin: the stratigraphic trap

If we look at the fifth saturation profile on Figure 1-1.52, the access pressure of this very fine micrite corresponds to a height of more than 200 m above free water level. If the reservoir closure is less than this high value, the rock is totally water saturated and acts as a cap rock. If we imagine that the type 5 rock corresponds to a lateral facies variation upwards, in a monoclinal structure, we obtain a perfect example of stratigraphic trap. The stratigraphic trap corresponds to the extreme stage of petrophysical fluctuation of the water level.

We mentioned that this rock would act as capillary barrier (and therefore as cap rock) as long as the closure is not high enough to induce a capillary pressure greater than the access pressure. If the capillary pressure increases up to this threshold, then the "barrier" layer is invaded by oil and, at least theoretically, the oil previously trapped can escape across the "ex-barrier" and continue its secondary migration. We would therefore observe a geological equivalent of the penetration by the non-wetting fluid through the semi-permeable membrane in a restored state experiment (§ 1-1.2.4, p. 73).

Unlike the case of shaly cap rock, the notion of cap rock in "stratigraphic trap" is often related to the height of the oil column in the zone concerned. In shaly cap rocks, the access pressure is often more than one hundred bars (which corresponds to a 5 km oil column, according to our calculation assumption) and we may speak of absolute capillary barrier.

Fluctuations related to pressure variations

In the previous paragraph, we considered the case of a stable hydrostatic state and variable petrophysical properties. We will now investigate an opposite situation in which a pressure variation leads to a variation in OOWC (in a homogeneous petrophysical reservoir). Pressure variations are induced by two main causes.

Pressure variations related to hydrodynamism

The logic we outlined above is based on the rules of hydrostatics, i.e. it assumes pressure equilibrium, especially in the aquifer which acts as implicit reference. If the aquifer is active, in other words if there is a pressure gradient in the horizontal plane causing a flow, the free water level will be inclined in the same direction as the aquifer isobar lines. It will be "tilted" in the direction of flow (Fig. 1-1.54). This pressure variation is easy to calculate for a given aquifer flow. We will express this flow as filtration velocity (U). Remember (§ 1-2.1.1, p. 125) that the filtration velocity does not correspond to the displacement velocity of the water particles but to the total quantity of water which crossed a plane normal to the flow during the chosen reference period. The corresponding hydraulic gradient (in bar/cm) is equal to this velocity in cm/s divided by the permeability in darcy (definition of the darcy).

To use more "practical" units, we will say that a filtration velocity of 1 m/year corresponds to a gradient of about 300 millibar/km in a rock of permeability 1 D. In large captive water tables of high permeability, the filtration velocity is less than 1 m/year (Table 1-2.1, p. 126). We may therefore consider that this value is an upper limit of the situation found at shallow depths in terms of petroleum criteria.

The difference in depth (Δh) of the free water level induced by this hydrodynamic pressure variation (ΔP) is given by the hydrostatic formula: $\Delta h = \Delta P / \Delta \rho \gamma$. This value can be

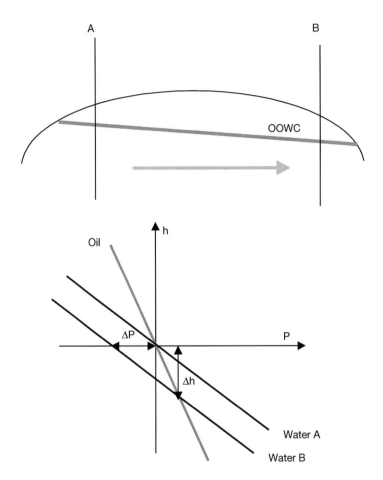

Figure 1-1.54 Hydrocarbon/water contact inclined due to hydrodynamism

checked graphically on diagram 1-1.54 where ΔP is clearly the capillary pressure at height Δh on the vertical B. In the extreme case of a filtration velocity of 1 m/year, the difference on the free water level value is therefore about 15 m/km, which corresponds to a very high gradient on the oil/water contact.

However, the filtration velocity values corresponding to the reservoir aquifers are much smaller. In a conventional oil field, therefore, the difference in water level related to "reasonable" hydrodynamism will be much less than 1 m per kilometre.

Hydrodynamism comes back from time to time as a possible explanation for water level "tilting". In our opinion this is often unjustified. The hydrodynamic explanation of "tilt" in the water level must be limited to very shallow reservoirs located on aquifers proven to be highly active. At the very least, we must avoid the absurdity of using hydrodynamism to explain a "tilt" associated with an aquifer with such little activity that water would have to be injected to maintain production.

Pressure variations related to modifications of the fluid types

We have seen that the main cause of the pressure difference between water and oil was the difference in their densities (Pc = $\Delta\rho\gamma h$). If a given reservoir is spread out geographically, however, density variations may be caused by a number of factors. These factors include (non exhaustive list – several factors may be found in the same reservoir):

- variation in oil type or gas content;
- variation in water salinity;
- variation in temperature under the effect of a geothermal variation.

Calculating the effect of these variations on the free water level and on the oil/water contact is more difficult than in the previous case since assumptions must be made, on a case by case basis, regarding the location of the pressure constants chosen as reference.

To give an example, Stenger [1999] used a temperature/salinity variation to explain a tilt of about 0.5 m/km in the FWL on the Ghawar field (Saudi Arabia). Note that this variation is only important since it extends over several hundred kilometres in this field, the largest in the world.

b) Anhydrous production in zones of high water saturation

In conventional permeable reservoirs, anhydrous oil production zones are generally associated with zones of lower water saturation This corresponds to the fact that the zones of high saturation can only be transition zones in which the water is movable. This rule does not apply in some reservoirs.

The extreme example of this problem was discussed when we mentioned earlier the case of fractured reservoirs producing under the water level corresponding to the matrix.

This phenomenon may be observed in a more "subtle" way, at matrix scale, in limestone reservoirs exhibiting highly contrasted double porosity (micro/macro – see Glossary Index). Figure 1-1.55 shows the epoxy pore cast (§ 2-2.1, p. 331) of an oolitic limestone exhibiting both significant intraoolitic microporosity, contributing in particular to the high porosity (0.34), and a well-developed intergranular macroporosity, inducing very high permeability (600 mD). The porosimetry curve, converted into saturation profile for a standard water/oil pair, indicates that at 25 m above free water level, the water saturation is about 0.5 (and 0.40 at 75 m). This high water saturation corresponds to almost non-movable water in the microporosity and the rock may yield abundant anhydrous production. This observation is particularly important for zones of low hydrocarbon column.

This represents a further example of the petrophysical features associated with the double matrix media encountered so frequently in carbonate rocks. This is why, during exploration phases, it is often recommended to carry out a systematic test of the porous limestone layers, irrespective of the saturation data obtained by log analysis.

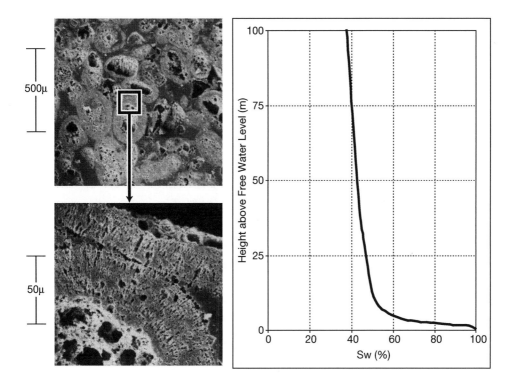

Figure 1-1.55 Example of reservoir which could exhibit anhydrous oil production
in a zone with high Sw: capillary pressure curve and epoxy pore cast observed under electron microscope

1-1.2.7 Capillary rise: Hirschwald coefficient, Apparent radius of capillary rise

A) Principle of capillary rise and definitions

Capillary rise is a special expression of the spontaneous imbibition defined in § 1-1.2.2.D
(p. 59). A wetting liquid rises up a porous network, under the effect of capillary forces alone,
expelling the air. A good description of this phenomenon is given in Dullien [1977] for
example.

a) Principle of the experiment

If we place the bottom of a test sample in contact with a volume of wetting liquid of constant
level (Fig. 1-1.56), the height of the capillary fringe (h) and the weight ΔW of fluid absorbed
by the sample can be measured against time.

The height of capillary rise and the weight of fluid absorbed increase in proportion to the
square root of the time until the capillary fringe reaches the top of the test sample. This
relation with $\sqrt{t}$ is a well-established fact, justified by combining Jurin – Plateau formula
(§ 1-1.2.1B, p. 54), which gives the capillary pressure (driving force), and Poiseuille's
formula (§ 1-2.1.3A, p. 138) which gives the flow rate in a capillary tube of radius R_c.

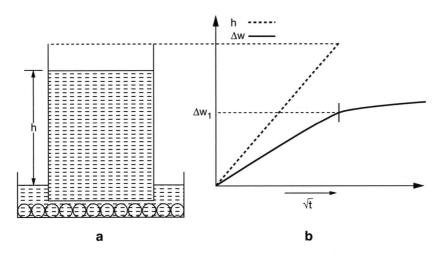

Figure 1-1.56 Capillary rise

a) Experimental diagram.
b) Variation of fringe height h and weight of wetting fluid absorbed ΔW against $\sqrt{t}$.

$Pc = 2t_s \cos \theta / R_c$; $Q = Pc(\pi R^4 / 8\eta h)$.

We obtain $h = \sqrt{(t_s \cos\theta / 2\eta)} \sqrt{R} \sqrt{t}$ since the differential expression of the flow rate in the cylindrical capillary is: $Q = \pi R^2 dh/dt$.

This equation is sometimes called the Washburn equation.

Obviously, this interpretation assumes that the forces of gravity are negligible compared with the capillary forces, i.e. that h is very small compared with the total potential height of capillary rise for the rock investigated. In most cases concerning consolidated rocks, for samples a few centimetres high, this condition is satisfied.

If the experiment is continued after the capillary fringe has reached the top of the test sample, we observe a sudden change in the kinetics. The rate of weight increase drops very sharply. At this stage, capillarity no longer plays a role: the weight increase is now due to dissolution and diffusion of the trapped air.

b) Hirschwald coefficient (or residual gas saturation)

Definition

When studying capillary properties, the most important parameter is the saturation state when the kinetics change. Numerous studies [e.g. Pickell *et al.* 1966] have demonstrated that this saturation state is practically independent of the experimental method. It is therefore an intrinsic characteristic of rock. These tests are extremely easy to carry out and consequently numerous results are available, especially in petroleum literature. They confirm the virtual independence of the result with respect to:

– the nature of the fluid pair (under perfect wettability condition);

- the pressure and temperature conditions (provided that the gas is saturated with vapour in the case of a liquid/gas pair);
- the flow rate, when imbibition is controlled (e.g. experiment on gas swept by water).

All the above experiments start with an initial dry material (free from wetting fluid). However, note that if the initial wetting fluid saturation (S_{li}) is not zero, then the trapped gas saturation (S_{tg}) will be affected by the initial saturation.

The most well-known expression used in petroleum techniques is Land's relation [1971]:

$$(1/S_{grR}) - (1/S_{giR}) = Constant,$$

with S_{grR} and S_{giR} corresponding respectively to the residual gas saturation and the initial gas saturation ($S_{gi} = 1 - S_{li}$) normalised (or "reduced" or "effective") by S_{wi}, i.e. $S_{grR} = (S_{gr})/(1 - S_{wi})$, as in the case of the S_{wR} discussed in § 1-1.2.5. In practice, below a value of S_{gi} of 0.5 to 0.6, a linear relation $S_{gr} = f(S_{gi})$ is a good adjustment. In all cases, the coefficients must be acquired experimentally.

Although petroleum laboratories mainly use the notion of residual gas saturation (S_{gr}) to define this intrinsic characteristic of rock, the saturation state at the end of spontaneous imbibition, non-petroleum laboratories prefer the term "Hirschwald Coefficient" (H) [Hirschwald, 1908, in German]. It is generally expressed as a percentage. It corresponds to the liquid saturation at end of capillary rise.

Thus H = $100(1 - S_{gr})$. Use of the term "Hirschwald coefficient" assumes an implicitly zero initial wetting fluid saturation (S_{li}).

It seems logical to assume that the mercury (non-wetting) saturation at the end of the extrusion phase in the mercury porosimetry experiment is an equivalent of S_{gr}. Pickell *et al.* [1966] experiments led to the same conclusion. Applications of this observation to petroleum geology are developed by Wardlaw and Cassan [1979]. We must nevertheless make the following reserve [e.g. Bousquié, in French, 1979]: this equivalence is only valid for the macro-connected porous networks. In micro-connected porosity (e.g. $R_c < 0.5$ µm), there is generally a large difference between the Hirschwald coefficient measured using mercury and using air. The same applies in some cases of macro-connected porous networks with smooth pore walls (formed from macrocrystals) such as sucrosic dolomite.

Variation of the Hirschwald (or S_{gr}) Coefficient with porosity

The raw compilation of the numerous data available concerning the H *vs.* φ relations might suggest that there is a considerable degree of confusion in these relations. Figure 1-1.57 gives an example of this apparent lack of relation, especially in the case of high porosity rocks. However, when porosimetry measurements indicating the relative importance of the macro-connected porosity ($R_c > 1$ µm, for Fig. 1-1.57) are available, we observe a fairly clear drop in the Hirschwald Coefficient as the proportion of macroporosity increases.

This is due to the fact that microporosity (associated with the micritic phase in limestones or the silty-clay phase in sandstones) does not trap any residual gas. Although a fairly general observation, this point must be considered carefully. It is observed by weighing the mass of liquid present in the sample at the end of the experiment and by comparing it with the total porous volume. However, if we measure the volume of gas

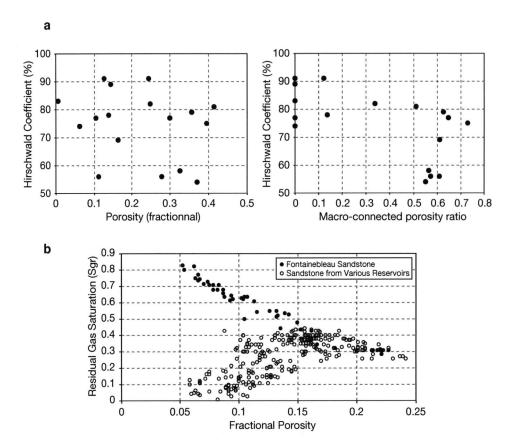

Figure 1-1.57 Hirschwald Coefficient (H) or Resisual Gas Saturation ($S_{gr} = 1 - H/100$) *vs.* porosity
a) Graph of Hirschwald Coefficient against total porosity and against macro-connected porosity/total poros-ity ratio. According to Bousquié's data [in French, 1979] on a set of limestone rocks.
b) Graph of S_{gr} vs. porosity According to Hamon *et al.* data [2001] on various sandstones.

expelled from the sample during imbibition using a bell jar type device under atmospheric conditions, we see that a significant part of the air may in some cases remain trapped in the sample, in a reduced volume, i.e. at a pressure above atmospheric [Egermann, personal correspondence, 2005]. This point is less surprising if we refer to the trapping mechanism described in § 1-1.2.2D (p. 63). At the start of capillary rise, the wetting phase content is very low and the capillary pressure very high. The air globules, *a priori* at atmospheric pressure may be trapped in capillary water at high "negative" pressure. During imbibition the pressure difference between these globules and the water may be maintained, at least partially, whereas the water pressure will tend towards atmospheric. We must nevertheless point out that experiments concerning the residual gas saturation, conducted under pressure in the petroleum laboratories, would appear to confirm the absence of trapping in the microporosity. As a result, considerable caution must be taken when interpreting gas trapping in microporous medium.

In conclusion, concerning the Hirschwald coefficient:

- In monomodal macroporous rocks (e.g. clean sandstones), H increases with porosity. This can be explained by the increased number of connections with the porosity and the correlative reduction of the pore radius/throat radius ratios.
- H is always very high in the microporosity.
- Consequently, in limestones or silty-clay sandstones, H tends to increase as the porosity decreases, this increase being related to the increasing microporosity.

In sedimentary rocks, the values most frequently encountered are in the region of 70% and extend from 95% to 50%. Values down to as little as 10% may exceptionally be found in very special sandstones or carbonates with poorly connected vugs. Lastly, leaving the field of sedimentary rocks, we must point out that in pumice stone, the Hirschwald coefficient is extremely low (virtually total trapping of air in the vugs, even though these vugs are interconnected by very narrow accesses).

Note that in soil science, the phenomenon of capillary trapping is rarely mentioned. The fact that air is not trapped, or only very slightly, by the very fine media probably explains the lack of knowledge of this process. There is in fact so little knowledge that the experimental soil saturation standards generally implement imbibition type techniques. Investigators may therefore observe, on the rare occasions when trapping occurs, apparently inconsistent behaviour.

c) Apparent radius of capillary rise

Definition

We have seen that the straight capillary model provided a good explanation of the kinetics of capillary fringe rise and could be used to establish the Washburn equation:

$h = \sqrt{(t_s \cos\theta / 2\eta)} \sqrt{R} \sqrt{t}$. This formula has been validated experimentally for a large number of fluids. Note that for a given test sample, the only parameter which depends on the material is R. This is one of the intrinsic parameters of the porous medium, so rare that this point is worth mentioning.

The apparent radius of capillary rise (R_{apc}) is defined as the radius of the straight cylindrical capillary tube in which the capillary rise of a meniscus under the effect of the capillary forces alone would occur at the same speed as in the material considered. It is easy to measure from the capillary rise slope (B), $h = B \sqrt{t}$ (Fig. 1-1.56).

$$R_{apc} = B^2/(t_s \cos\theta/2\eta)$$

The practical unit for measurement of B in the laboratory is the centimetre per $\sqrt{hour}$. To convert B^2 into SI units, we must therefore multiply by $27.7 \ 10^{-9}$ (rounded). For the water/air pair, the expression $t_s \cos\theta/2\eta$ corresponds to about 30 SI units. By rounding, we obtain the estimation:

$$R_{apc} = B^2 \text{ with } R_{apc} \text{ in nanometres and B in cm}/\sqrt{hour}$$

Value of the apparent radius of capillary rise and relation with permeability

The most surprising point, in a first analysis, is the very low value of these radii. For the limestone sample set shown on Figure 1-1.58, the maximum values do not exceed 100 nm in order of magnitude, whereas some of these limestones have pore throat equivalent radii (mercury porosimetry) of 10 μm of order of magnitude. This ratio, which may exceed 10^3, between the two types of equivalent radii is explained by the fact that all the untrapped bulges of the porous space reduce the driving capillary force whilst considerably increasing the volumes of wetting liquid displaced by capillary rise.

Note that the relations R_{apc} *vs.* ϕ or R_{apc} *vs.* K (Fig. 1-1.58) seem relatively complex since the antagonistic roles of macroporosity and microporosity observed for H and K (§ 1-2.1.4B, p. 156) are combined in this case.

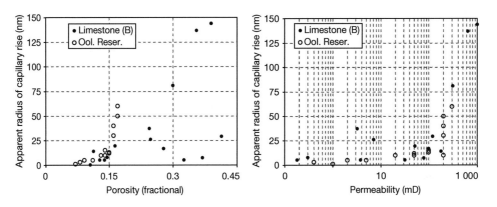

Figure 1-1.58 Relation between porosity, permeability and apparent radius of capillary rise
Limestone (B): Sample of limestone rocks, Bousquié's data [1979].
Oolitic reservoir.

B) Advantage of capillary rise for fast characterisation of reservoir rocks

The apparently complex relations described above between the capillary rise parameters and porosity-permeability represent an incentive to reconsider the advantage of this experiment for statistical characterisation of reservoir rocks, whose principle is described in § 2-1.4 under "rock typing". It is important to note that:

– The experiments which can be used to determine H (and $1 - S_{gr}$) and R_{apc} are very simple and therefore inexpensive. By using organic solvents, we can even avoid the most serious wettability problems. H and R_{apc} can therefore be systematically measured on the cores, like ϕ and K_{air}.
– By definition, H is a physical equivalent of residual gas saturation and, less directly, an analogue of residual oil saturation. These two parameters play a central role in reservoir characterisation.
– R_{apc} can be placed in direct relation with the wetting fluid relative permeability in a sample in residual saturation state. It is therefore an analogue of the relative permeability at the end point S_{or} (see § 1-2.3.3, p. 188).

Despite all these advantages, the parameters H and R_{apc} have not yet been widely used in petrophysical analysis. Recent publications [e.g. Hamon *et al.*, 2001] indicate that this situation is changing.

C) Application to pedology and conservation of building stones

In soils and surface formations, the process of capillary rise enables plants to obtain water and mineral salts without reaching the water table. The root systems therefore remain under aerobic, non-asphyxiating conditions. In periods of drought, however, the competition between capillary rise in the soils and evaporation results in the transit and accumulation of mineral salts towards the surface. This phenomenon is exaggerated by excessive irrigation in hot, dry regions. It leads to sterilisation of agricultural soils due to salinisation.

In Building and Public Works, these phenomena are also known for their sometimes damaging effects: capillary rise and deterioration induced by permanent humidity in the structures, together with accumulation of salts in the evaporation fringe. The crystallisation pressure causes the materials to crumble. This is frequently observed on old buildings (Fig 1-1.59, see also photograph in Jeannette *et al.* [in French, 1992]).

Figure 1-1.59 Example of damage partly caused by capillary rise (**a**) To prevent this damage, it was usual in traditional building to lay a bed of impervious rock, e.g. millstone grit (**b**) or quartzite (**c**). Paris suburbs

Crystalline growths are not the only sources of damage in building stones however. The expansion of water during freezing is also a serious phenomenon which disfigures and sometimes even ruins buildings.

These phenomena develop when the materials used are sufficiently porous/permeable for water to flow. This said, there is less damage when the volume of air trapped by spontaneous imbibition is sufficient to act as a kind of expansion chamber. Measurement of the porosity trapped by capillary rise (equivalent of residual gas saturation, S_{gr}, in petroleum technique) then becomes an indirect but easily accessible criterion of resistance to frost splitting. Geomorphology is another discipline dealing with the study of these properties during the investigation of surface formations generated during glaciation (slope breccias, etc.)

Several laboratory methods were normalized in the late 1970's for the study of the protection of building stones, monuments and civil engineering structures.

They include the two types of measurement, carried out in the air/water system. One is based on the dynamics of capillary rise and the other focuses on the end point:

We may mention as an example the recommendations issued by Unesco/RILEM [1978]:

- a criterion of capillary rise rate corresponding to measurement of the quantity of water absorbed by a sample which has one side in contact with a free water surface maintained at constant level, the lateral sides of the test sample are waterproofed and the top side is left in the open air $W = A\sqrt{t}$, where W is the mass of water absorbed per unit area immersed (kg/m^2) and t is in seconds. We obtain an equivalent of the apparent radius of capillary rise;

- a criterion of saturation coefficient after imbibition by complete immersion of the sample at atmospheric pressure until the mass becomes constant. Since this criterion is "rarely established", however, by convention the measurement is taken after 48 hours immersion. Obviously, this is an equivalent of the Hirschwald coefficient.

In France, there is an indirect test of frost susceptibility by spontaneous imbibition, standardised by AFNOR – France (standard NF EN 1925): The procedure is as follows (Hirschwald coefficient): the samples (diameter 40 mm and length 160 mm) previously dried at 60°C are immersed up to $^1/_4$ of their height for 1 hour, then up to $^1/_2$ of their height for 23 hours and then totally immersed for the next 24 hours. This progressive immersion method may be contrasted with immediate total immersion, possibly extended for several days if significant water absorption is observed. Field geologists and geomorphologists have carried out this test for many decades. One of them [Lautridou, personal correspondence, 1980] mentions a change from 80% to 85% when immediate total immersion is replaced by the AFNOR – France standardised test, for a quarry limestone (Fleury/Orne – France).

Overall, frost susceptibility can be defined as follows: the more porous a material is, the more susceptible it is to frost. But, at a given porosity, however, the lower its water absorption coefficient (or Hirschwald coefficient) (in other words the higher the residual gas saturation), the greater its resistance to frost splitting.

In addition, more advanced tests take into account the competition between evaporation and capillary rise through special experimental arrangements. These sophisticated tests can be used to determine the properties of the least porous rocks (e.g. granite). These tests estimate the resistance of materials to frost splitting, taking into account the processes actually involved in non-immersed materials (elevated masonry for example): capillary recharge compensating for water evaporation and migration at the start of the freezing phase (cryosuction).

From a physical point of view, a distinction must be made between the cocurrent imbibition tests, where the fluids flow in the same direction, and the countercurrent tests, where the fluids flow in opposite directions. The latter case is typical of immediate total immersion. Note that the French standardised test (Hirschwald coefficient) is hybrid but mostly countercurrent if the material is poorly porous/permeable. These conditions may be encountered in materials used in Building and Public Works. The flows in most natural processes (pedology, hydrogeology and petroleum reservoirs), however, are cocurrent.

1-1.2.8 Overview on three-phase capillary equilibria

Cases of three-phase saturation (water, oil, gas) may be observed, for example in the "gas-caps" (gas zone over an oil reservoir). The three-phase domain is complex and difficult to control under laboratory conditions. We will restrict ourselves to simple cases here where - by assumption - the initial system is water wet.

Traditionally, gas is considered as the "most non-wetting" fluid. Consequently, the capillary pressure is generally considered as a function of S_g with a total liquid phase $S_o + S_w$.

More recent studies involve the notion of spreading coefficient:

$$C_s = t_{gw} - (t_{go} + t_{ow})$$

where t_{gw} = gas/water interfacial tension
t_{go} = gas/oil interfacial tension
t_{ow} = oil/water interfacial tension

If C_s is negative, the oil does not spread and relatively high oil saturations may remain in the gas zone.

If C_s is positive, the oil spreads over the water surface.

When C_s is positive, spreading is controlled by another coefficient:

$$\alpha = [t_{ow} (\rho_o - \rho_g)]/[t_{go} (\rho_w - \rho_o)]$$

if α is positive, there is a critical height above which the oil saturation is virtually zero.

In practice, t_{gw} is often greater than $(t_{go} + t_{ow})$ and the traditional method is then valid. However, in this complex field, it is always preferable to contact highly specialised teams, before starting serious evaluations.

CHAPTER 1-2

Fluid Recovery and Modelling: Dynamic Properties

Characterising the flows in the geological layers and modelling the production of hydrocarbon reservoirs or aquifers involve the "dynamic properties": the **intrinsic permeability** when the porous medium is totally saturated with a single fluid and the effective permeability leading to the concept of **relative permeability** when several non-miscible fluids coexist. To correctly describe capillary phenomena during a multiphase flow, however, the **wettability** state of the reservoir rock must be known. These three points are discussed in this second chapter.

1-2.1 INTRINSIC PERMEABILITY

1-2.1.1 Definitions and Darcy's law

Permeability characterises the ability of a porous medium to allow fluids to flow. There may be a certain degree of confusion with this definition due to the fact that hydrogeologists consider the flow of a single fluid, water, and therefore use the concept of Hydraulic Conductivity (k), in line with Darcy's historical description [1856, in French]; concerning the reference to Darcy, it is important to point out to the interested reader that only five pages of section D in the appendix of Darcy's book deal with this subject. A copy of these few pages is included in [Marle, 2006]. As soon as various fluids (i.e. with different viscosities) are involved, however, the notion of permeability (K), an intrinsic characteristic of the porous medium considered, must be introduced.

A) Hydraulic conductivity: k

a) Darcy's experiment. Definitions of hydraulic conductivity (k) and filtration velocity

Darcy's historical experiment has retained all its educational value. A tube (Fig. 1-2.1) is partially filled with a sand column of variable height (l) over a filter of cross-section S. Two

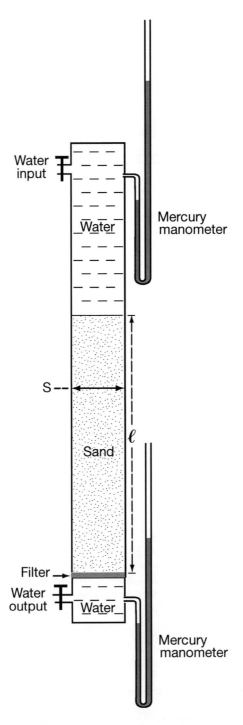

Figure 1-2.1 Darcy's experiment

mercury manometers, at the ends of the column, indicate the input and output pressures. These pressures are converted into water piezometric heads (h_1, h_2). The upper part is connected to the mains water supply. Taps at the input and output can be used to adjust the piezometric heads. The results of the second experiment (in which the sand height is constant and the input and output pressures are variable) are shown on the graph (Fig. 1-2.2a). It shows that there is a straight line relation between the flow and the difference in piezometric heads, corroborating Darcy's equation:

$$Q = kS(h_1 - h_2)/l$$

where Q is the volume flow rate, S the cross-section in a plane perpendicular to the direction of flow, l the length of the block and k a proportionality coefficient known as the **hydraulic conductivity**. To simplify, the flow rate can be expressed per unit area. This new parameter $U = Q/S$ has the same dimensions as velocity. It is known as the Filtration Velocity (and sometimes Specific Discharge or Darcy Flux). Similarly, by defining the hydraulic gradient $G = (h_1 - h_2)/l$ we obtain the general form of Darcy's equation:

$$U = kG$$

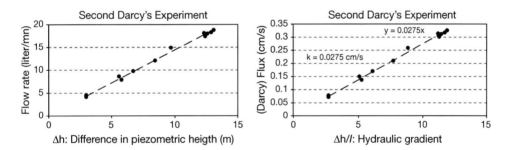

Figure 1-2.2 Results of Darcy's second experiment; a) Raw data; b) Same data expressed as filtration velocity (= specific discharge, = Darcy flux) and hydraulic gradient

Since the hydraulic gradient **G** is dimensionless, the hydraulic conductivity **k** has the same dimensions as velocity. On Figure 1-2.2b, which shows the experimental data of Figure 1-2.2a, but with axis values corresponding to the general form above, we measure the hydraulic conductivity of the Darcy block directly (slope of the regression line). It is approximately 0.0275 cm/s. We will say more about the units during the comparison with intrinsic permeability.

b) Remark on filtration velocity and fluid displacement velocity (seepage velocity)

The filtration velocity is not really a velocity in the common sense, it simply has the same physical dimension. The average velocity of the fluid through the porous medium, which is the velocity of interest in the common sense, is sometimes called the **seepage velocity**. This velocity is measured in hydrology, using dye markers, for example. *It is equal to the filtration velocity divided by the fractional porosity of the medium concerned.* The filtration velocity corresponds to the flow rate divided by the total area of the medium, solid plus pore if we may say so, whereas seepage velocity is limited to the porous space. Obviously, the seepage velocity is always greater than the filtration velocity. Table 1-2.1 gives some orders of magnitude of seepage velocities found in hydrology. The velocities of deep (several km) aquifers encountered in petroleum sedimentary basins are much smaller.

Table 1-2.1   Orders of magnitude of seepage velocities and corresponding filtration velocities

Aquifer type	Porosity	Seepage velocity	Filtration velocity = seepage velocity.porosity	Permeability (order of magnitude)	Hydraulic gradient mB/km
Large aquifers of medium depth; e.g. "Sables Verts" (Greensand) of the Paris Basin, Sahara "Continental Intercalaire" aquifer	0.3-0.4	3 m/year	1 m/year	5 D	60
Alluvial deposits of fast flowing rivers (e.g. the Rhine water table)	0.3-0.4	1 km/year	300 m/year	500 D	180
Karst network or low-porosity and fractured environments (sedimentary or volcanic rocks)	$\ll 10^{-3}$	10 m/hour i.e. about 100 km/ year	–	–	–

B) Intrinsic permeability (K)

a) Generalisation of Darcy's equation and definition of intrinsic permeability

The hydraulic conductivity of a block obviously depends on the block itself and the fluid. Darcy's law can be generalised for all fluids in non-turbulent flow by including the viscosity (η) and thereby defining a property intrinsic to the porous medium: the permeability (K). In its simplest form, Darcy's equation is written:

$$Q = (K/\eta)\ S(\Delta P/\Delta l)$$

Where ΔP is the pressure difference at the ends of the block of length Δl.

This expression of Darcy's law corresponds to the isotropic case, i.e. when K is independent of the direction considered. In practice, permeability is very frequently highly dependent on the direction. We will discuss this anisotropic case in detail in § 2-1.2.2, p. 283.

In addition, note that this equation implicitly assumes the existence of a single homogeneous fluid phase in the porous space (to define a viscosity). Consequently, this permeability is sometimes called the "single-phase" or "absolute" permeability as opposed to the "effective permeabilities", when several non-miscible fluid phases coexist in the porous space (the term "relative permeability" – dimensionless – is defined by the ratio between this effective permeability and a reference permeability (see § 1-2.3, p. 179).

The term single-phase permeability may be reserved for experimental acquisition. From one fluid to another, in fact, experiments may reveal some differences demonstrating that the notion of *intrinsic permeability* is simply an approximation, although acceptable in most cases

b) Remark on vertical flows

In ($\Delta P/\Delta l$) of the above equation, we recognise a form similar to the previous hydraulic gradient G. Note however that in Darcy's experiment (like in most hydrogeological applications), the hydraulic gradient corresponds to a piezometric head difference at the two ends of the block divided by the fluid displacement distance. Use of piezometric heads allows us to include the effect of gravity, making the direction of flow with respect to gravity unimportant. In Darcy's experiment for example, the flow is vertical but as long as we compare piezometric heads, measured with respect to a common reference, this point has no impact on the result.

If "absolute" pressures are used, however, as in the above equation, a possible hydrostatic pressure related to the presence of a column of fluid in the block must be taken into account. The simple equation is only valid for a horizontal flow. In the other cases, the pressure equivalent to the fluid column ($\rho\gamma h$) must be added to the pressure measured at the block output.

c) Permeability units

Since the dimension of K in Darcy's generalised equation is an area, the SI permeability unit is therefore the m^2. It is a gigantic unit and no natural environment comes anywhere near this permeability.

The unit generally used is the darcy (D). The darcy is related to the former CGS system since it corresponds to the permeability of a porous medium which permits a flow of 1 cm^3/s per cm^2 of area perpendicular to the direction of flow of a fluid with viscosity 1cP under a pressure gradient of 1 normal atmosphere per cm.

The definition includes the old pressure unit "normal atmosphere", equal to 0.101325 MPa, which explains why the equivalence to the SI unit is not a "round number". The darcy can therefore be converted into SI units using the following relation:

1 darcy = 0.986923 E-12 m^2

In practice, the equivalence 1 D = 10^{-12} m^2 is quite satisfactory.

The viscosity chosen in the definition (1 cP) corresponds to that of water at ambient temperature. We can therefore calculate the permeability of a medium of known hydraulic water conductivity (k). In the definition of the darcy, the flow rate per unit area (flux, filtration velocity) corresponds to 1 cm/s. Concerning the hydraulic gradient, by considering that one atmosphere is equal to 10 m piezometric head of water, we see that the pressure gradient in the definition of the darcy is 10^3 (10 m/1 cm). Expressed in m/s, the hydraulic conductivity (water) of a medium is therefore equivalent to 10^{-5} times its permeability expressed in darcy.

Thus: K (permeability) = k (water conductivity) . $(\rho\gamma/\eta)$

where ρ = density of water

γ = gravitational acceleration

η = viscosity of water

The use of two different concepts (conductivity, permeability) by two scientific communities with little contact (hydrogeology, petroleum) may lead to confusion, so you will forgive us if we emphasise one simple but fundamental point: *hydraulic conductivity and permeability are two different physical quantities* with different dimensions, one equivalent to velocity, the other to area, and only permeability is intrinsic to the porous space considered. *The equivalence 1 D "=" 10^{-5} m/s is only meaningful when the fluid concerned is water.*

In practice, when soil-mechanics specialists and hydrogeologists use the term "permeability", they often mean "hydraulic water conductivity". Hydrogeologists do not use this notion very frequently, however, preferring the concept of "transmissivity", which is the product of hydraulic conductivity by the thickness of the level considered. Transmissivity is one of the parameters obtained directly from well tests ("pumping tests"). This is also due to the fact that hydrogeologists virtually never measure permeability in the laboratory.

d) Orders of magnitude of permeabilities in geological materials.

The range of permeabilities encountered in geological materials is extremely vast. On Figure 1-2.3, which indicates orders of magnitude on a porosity-permeability diagram with logarithmic scales, we see that the permeability values extend over nearly 15 orders of magnitude, from the clays of the cap rocks at less than 10^{-9} D up to the coarse-grained alluvial deposits at more than 10^4 D.

Obviously in practice, a physical quantity with such extreme variations can only be represented on a diagram with logarithmic scale. Permeability distributions in the homogeneous geological layers are in fact expressed as log-normal law.

1-2.1.2 Intrinsic permeability measurement principle

The principle used to measure permeability simply consists in measuring the volume flow rate associated with a pressure difference. Care must nevertheless be taken to avoid certain disturbing effects. For example, those due to the flow itself (in case of non-laminar flow, Darcy's law would no longer be applicable) and those related to a modification of the porous space due to the measurement (plugging, swelling of clays). These special effects will be discussed in the section where they are likely to be most relevant. Concerning the actual methods, for practical reasons we will separate the "high" and "low" permeabilities. The limit between the two will be set at about 0.1 mD (i.e. 10^{-16} m^2), but obviously this is only an order of magnitude.

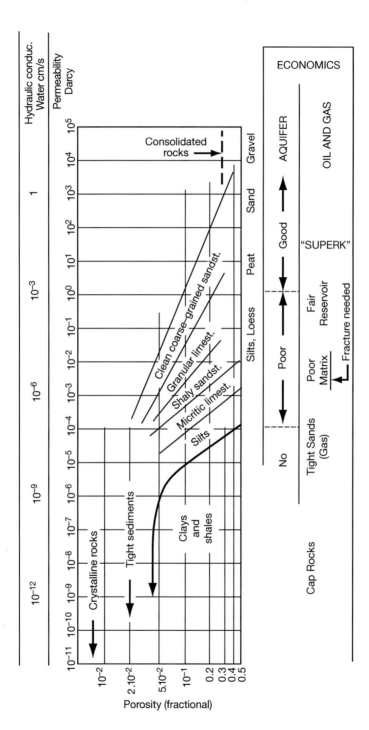

Figure 1-2.3 Range of permeabilities encountered in geological materials

A) Measurement of high permeabilities

Measurement using liquids and measurement using gases will be dealt with separately. In the latter case, estimation of the volume flow rate is slightly harder since it depends on the pressure.

a) Liquid permeability measurement

Measurement principle

This is the simplest method. Numerous experimental variants are possible, based around three practical points:

- Impermeabilisation of the lateral sides of the sample of porous medium. For routine measurements on cylindrical sample, it is generally carried out using a special mechanical apparatus such as the Hassler cell (Fig. 1-2.4a) in which a plastic sleeve is "stuck" against the sample by excess radial pressure. A coating can also be applied to impermeabilise the sides In this case, measurements can be taken on large, parallele-pipedic samples.
- Measurement of pressure difference. Obviously, the pressure can be measured using manometers at the ends of the sample (or better still, a differential manometer connected to the ends). For special experiments, however, the hydrostatic pressure of a column of liquid from a constant level tank can be used. Very simple but accurate apparatus (Fig.1-2.4b) can be produced in this way, based on Darcy's experiment.
- Measurement of the liquid flow rate. Various types of flowmeter can be used (ball, propeller, etc.), as well as a simple graduated container (Fig.1-2.4b).

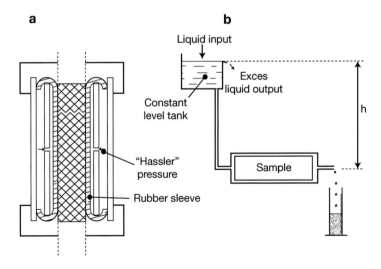

Figure 1-2.4 Examples of experimental apparatus to measure permeability:
a) Diagram of the Hassler core holder.
b) Diagram of simplified apparatus to measure liquid permeabilities.

Phenomena likely to affect the measurement of liquid permeabilities

- Turbulent flows. In point of fact this effect may be observed with all methods: if the fluid flow rate is too high, turbulence occurs in the pores and the energy dissipated will no longer be proportional to the flow rate. This phenomena is clearly illustrated on the graph of volume flow rate (Q) against head loss (ΔP). If the permeability is measured using several pressure values, care must be taken to remain within the linear section. The transition zone between laminar flow and permanent flow corresponds to the Forchheimer effect, described below.
- Plugging by transport of fine particles. This case applies more specifically to measurement with liquids which entrain very fine particles more easily than gases do. Blocked at the pore throats, these particles progressively plug the pores, thereby lowering the values of the measured permeability. So that this phenomenon does not disturb the measurement, during the experiment it is essential to make sure that the permeability is not modified over time as the flow continues, or when the flow direction is reversed.
- Swelling of clays depending on the saturating fluid. This is a major effect when studying shaly sandstones. Clay minerals may react in low salinity water or in some liquids (alcohols, glycols, etc). We then observe a swelling of the clay phase in the intergranular space, causing the rock to disintegrate in some cases. The reduction in pore dimensions produced by this phenomenon may have a considerable effect on permeability values.

b) Gas permeability measurement

Measurement principle

The measurement principle is the same as for liquid permeability, but in this case gas compressibility must be taken into account: only the mass flow rate will be constant along the sample during the experiment, whereas Darcy's equation concerns the volume flow rate.

If the approximation of perfect gas and constant viscosity over the pressure range used is acceptable, it is easy to integrate along the sample (dl) the expression

$Q = Q_n P_n/P = (K/\eta_a) S(dP/dl)$ where Q_n and P_n are the flow rates (volume) and pressures at point n of the sample. Air permeability is given by the formula:

$$K = 2\eta_a Q_a (L/S) [P_a/(P_{in}^2 - P_{out}^2)]$$

where Q_a is the flow rate measured at pressure P_a (atmospheric), η_a the viscosity, and P_{in}, P_{out} the input and output pressures.

"Constant head" experimental apparatus may be used (similar to that shown on Figure 1-2.4): a differential manometer and a gas flowmeter can be used to measure the head loss and flow rate, respectively.

One type of apparatus is extremely easy to use: the falling head permeameter (Fig. 1-2.5). A water column, placed in depression in a calibrated tube returns to equilibrium by sucking air from the exterior through the sample to be measured. The time for the column to fall between two given marks is proportional to the permeability.

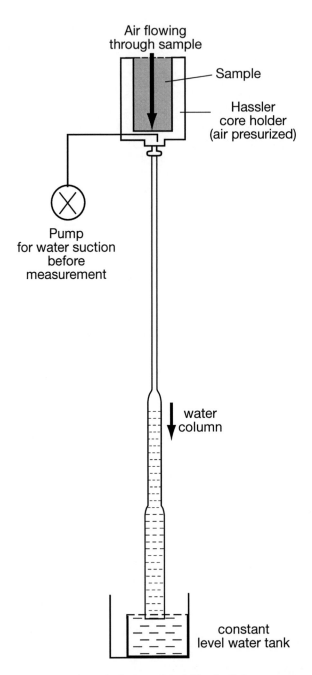

Figure 1-2.5 Schematic diagram of the falling head air permeameter

The Klinkenberg effect

Gas permeability measurements (K_g) may be disturbed by the so-called Klinkenberg effect [Klinkenberg, 1941]. It mainly concerns low permeabilities ($K < 10$ mD) measured under low average pressure. Traditionally, this phenomenon is explained by a change in the way gas molecules move through the pores whose diameter approaches the mean free path of the gas: molecule slippage along pore walls requires less energy than Darcy type flow. The permeabilities observed are therefore higher than those measured with a liquid (K_l).

This phenomenon is demonstrated by comparing the permeability values obtained by measuring at different average pressures (P_m). These values converge towards a point at "infinite pressure" corresponding to the liquid permeability value. This is illustrated on Figure 1-2.6 which gives the "historical" Klinkenberg values.

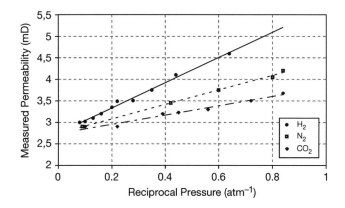

Figure 1-2.6 Effect of average gas pressure on the permeability value measured.
(Klinkenberg's experimental data [1941])

It can also be demonstrated by comparing liquid and air permeabilities at low pressure (atmospheric) for media which exhibit no interaction with the liquid (e.g. clean sandstones). Figure 1-2.7 shows experimental results on Fontainebleau sandstone [Jacquin, in French, 1964].

In the laboratory practice, the safest way of avoiding the consequences of the Klinkenberg effect, for poorly permeable samples, is to take measurements at various pressures, as shown on Figure 1-2.6.

To estimate the Klinkenberg effect on measurements taken at a single average pressure (with the falling head permeameter for example), charts can be produced (Fig. 1-2.7) by using the formula corresponding to straight lines of the same type as those shown on Figure 1-2.6:

$$K_g = K_l(1 + b/P_m)$$

where b is a coefficient which depends on the capillary diameters (and therefore on the permeability, in a semiempirical approach). Using values of b proposed by the American Petroleum Institute [1952], we plotted the curve of Figure 1-2.7 for the pressure P_m corresponding to atmospheric pressure (falling head permeameter). Jacquin's experimental

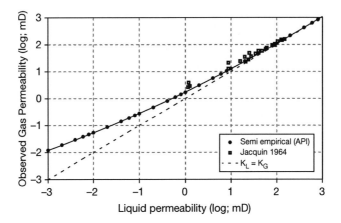

Figure 1-2.7 Comparison of the results of semiempirical correction and Jacquin's experimental values [1964]

values [1964] follow this trend fairly well. We can therefore propose a correction formula of the following type:

$$\log(K_l) = a[\log(K_g)]^2 + b[\log(K_g)] + c$$

with at atmospheric pressure: $a = -0.0616$, $b = 1.2652$, $c = -0.3021$ and K in mD.

Concerning the curve shown on Figure 1-2.7 and the corresponding regression above, we must emphasise the fact that it is a semiempirical formula purely intended to provide orders of magnitude of the Klinkenberg effect. It is only valid for values less than 100 mD.

The Forchheimer effect (quadratic effect)

Another effect disturbing the permeability measurement appears as the flow rate increases. Darcy's equation is only valid for laminar flow. When the fluid velocity increases, however, and before the flow becomes strongly turbulent, the head loss takes a quadratic expression formulated by Forchheimer [1901]:

$$\Delta P/\Delta L = \eta V/K + \beta \rho V^2$$

where   V = fluid velocity
        β = Forchheimer coefficient (zero if no effect)

Descriptive explanations emphasise the fact that the fluids travel in the natural porous space through a succession of bulges and bottlenecks. The fluids accelerate in the pore throats.

This quadratic effect has often been neglected in the laboratory.

In practice, this effect and the Klinkenberg effect are observed simultaneously. The experimental data must be interpreted by an iterative linear regression (three parameters: K, β and b – of the Klinkenberg formula –). The result, especially the Forchheimer β factor, is highly dependent on the quality of the data. Whatever the case, experience has shown that in order to obtain correct measurements, tests must be carried out with backpressure flow.

B) Measurement of low permeabilities

When the permeabilities are too low for the permanent flow to be fast enough to stabilise (measurement at constant flow or pressure difference), the measurement is taken by interpreting a transitory flow.

Several methods are available:

a) The "Drawdown" method (Fall off)

A tank (T1, Fig. 1-2.8) of known volume (V_{T1}) is filled with gas at a known pressure then, via a valve (V_{si}), it is connected to the sample, previously placed under vacuum. The pressure drop against time is recorded and interpreted. The derivative of this experimental curve can be used to evaluate K_1, b (Klinkenberg coefficient) and β (Forchheimer coefficient) since we have seen above that the influence of these effects is non-negligible with low permeabilities, which is the case here:

$$\Delta P/\Delta L = \eta V/(K_1(1 + b/Pm)) + \beta.\rho V^2$$

The flux through the sample is governed by:

$$Q \text{ (mass flow rate)} = V_{T1}.(dP/dt)$$

Which assumes that the gas used obeys Boyle's law: $P.V = Cte$

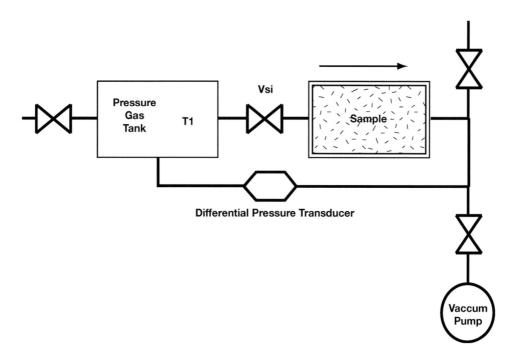

Figure 1-2.8 Schematic diagram of the "Drawdown" method

Numerical resolution is obtained using an iterative multivariable regression method initialised, for example, by a realistic value of the coefficient b.

The difficulties encountered with this method are related to the temperature sensitivity (thermal effect of decompression), the need to use a nearly "perfect gas" (helium), the significant Forchheimer effect at start of flow and the need to use a sufficiently sensitive differential pressure transducer.

However, it is easy to automate this method (automatic measurement for large series). Service companies generally use this technique.

b) The "Build Up" method

In this method (Fig. 1-2.9), gas is injected in the sample at constant flow rate, using a closed loop control system operating a regulating valve (V_{re}). The sample output is connected to a pressure tank of known volume (T1), of the same order of magnitude as the pore volume of the sample studied. The input/output pressure difference is recorded, as well as the pressure in the output tank.

The calculation is based on the above relation concerning compressible fluids:

$$K = \eta Q_v \, (L/S) \, [2P_v/(P_{in}^2 - P_{out}^2)]$$

where: η: gas viscosity

      L and S: sample length and cross-section

      Q_v: volume flow rate measured (calculated) at pressure P_v

      P_{in} and P_{out}: absolute pressures at sample input and output

the volume flow rate being calculated from the mass flow measured and the density of the gas used, knowing the average gas pressure in the sample. This average pressure is also used by the Klinkenberg correction to be taken into account.

This test is more accurate than the "drawdown" method (fewer artefacts, less significant thermal effect), but requires high-quality measurement transducers. Automation for measurement of large series is difficult.

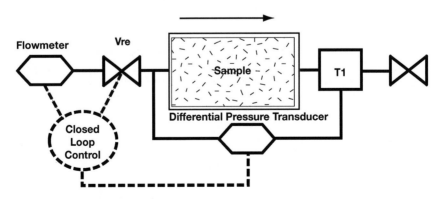

Figure 1-2.9 Schematic diagram of the "Build-Up" method

This technique, like the next one, requires a highly accurate closed loop system which has only been possible since the advent of advanced electronic transducers (sensitivity, stability, etc.). These pressure transducers and the closed loop control mass flowmeter are now sufficiently sensitive to remain within the range of laminar flows and generally avoid the Forchheimer corrections.

c) The "Pulse" method

The sample is connected (Fig. 1-2.10) to a large input pressure tank (T1) and to a smaller output tank (T2). The pressure is first balanced in the two tanks and in the sample. Then, after isolating the input tank, the pressure is decreased slightly in the sample and the small tank. The shut-off valve (V_{si}) is then opened and the drop in input/output pressure difference measured.

Numerical resolution is based on the derivative of the pressure change at each measurement step. The sample porosity and the fluid compressibility C_f must be known in order to apply the notion of dimensionless pressure by analogy with the interpretation of well tests [Bourdarot 1998]:

$$k_{adim} = K/\phi \eta C_f$$

The advantages and disadvantages are similar to those of the "Build Up" method. Another advantage with the "pulse" method is that it can be applied to measurements with liquids.

These three methods can be applied with no major additional experimental difficulties to samples subject to an effective stress close to that existing in the reservoir. This point is essential in view of the fact that permeability is highly sensitive to stress, in case of poorly permeable samples (see § 1-2.1.5).

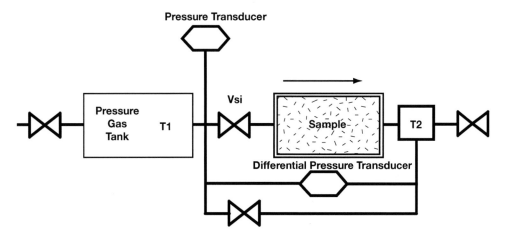

Figure 1-2.10 Schematic diagram of the "Pulse" method

1-2.1.3 Geometric parameters affecting permeability and simple models

For a better understanding of the permeability variations observed in natural materials, it is useful to understand the relative importance of certain parameters of the porous space on permeability. In our opinion, the simplest method is to study the permeability of a few elementary geometries. These models will also be extremely useful when estimating the permeability of materials for which true measurements are unavailable.

A) The straight capillary or parallel fracture model: relative effect of porosity and pore dimension

a) Straight parallel capillaries

As regards permeability, the simplest model of a porous space is that of a bundle of straight capillary tubes of the same radius. Using Poiseuille's law, we can calculate the flow rate in a capillary of radius R and length l:

$$Q = (\pi R^4/8\eta) \, (\Delta P/\Delta l)$$

It is easy to estimate the porosity of this type of model which would have n capillaries per unit area: $\phi = n\pi R^2$

Considering a block of unit dimension and for a unit pressure gradient, we can use simplified expressions of Poiseuille's law and Darcy's law:

$Q = n(\pi R^4/8\eta) = K/\eta$ hence $K = n(\pi R^4/8)$ and bringing in ϕ:

$$K = \phi R^2/8$$

Although this model is quite remote from the reality of geological materials, it may be worthwhile expressing the above formula in "usual units". If we use micrometres for the radii and percentages for the porosity we obtain, in mD: $K = 1.25\phi R^2$.

This easy calculation on a very simple model leads us to a conclusion which we will find on numerous occasions during our study of the ϕ-K relation: **The capillary radius value has a much greater effect on the permeability** (in this case raised to the power 2) **than the porosity** (power 1).

b) Flat parallel fracture model

The flat parallel open fracture model is considered symmetrically. In this case, the laminar flow formula must be used:

$$Q = (e^3/12\eta) \, (\Delta P/\Delta l) \text{ per unit fracture length}$$

where e is the fracture opening. In this case, the unit block porosity is $\phi = ne$ and hence:

$$K = \phi e^2/12$$

The result is therefore identical to the previous one: the fracture aperture thickness, raised to the power 2, plays the major role.

Taking the "usual units" already used above (micrometres for the thickness and percentages for the porosity) we obtain, in mD:

$$K = 0.833\phi e^2$$

The flat fracture model is more appropriate to a real case: that of fractured reservoirs. Considering a single fracture per metre, with an opening of 500 μm and inducing only 5.10^{-4} porosity, we observe that the permeability exceeds 10 D, obviously in the direction parallel to the fracture plane.

On this very simple example, we measure the effect of an open fracture on the total reservoir permeability.

B) The Purcell model to deduce the permeability from the porosimetry curve

The porosimetry curve (§1-1.2.4D, p. 79) supplies for each porosity fraction the value of the porosity access equivalent radius. Purcell [1949] suggested modelling the porous space of the rock as a set of capillaries of variable radii, but whose radius-porosity distribution is given by the porosimetry curve. The previous formula could then be applied for capillaries $K = 1.25\phi R^2$ (in usual units), integrating it on the porosimetry curve:

$$K = 1.25\int R^2 d\phi_{(R)}$$

Obviously, the natural porous media are very different from this model but Purcell had assumed that there must be a roughly constant ratio between the permeability of a sample and the result of the previous calculation. Purcell demonstrated on a large number of samples that this coefficient, known as the "lithological factor", was about 0.2. Numerous studies [e.g. Bousquié, 1979; in French] have corroborated this observation.

The main difficulty with Purcell's method lies in interpreting the porosimetry for large throat radii. In Purcell's law, the radius is raised to the power two, so the very large radii have a considerable effect. Unfortunately, these large radii are generally unreliable (e.g. surface anfractuosities of the sample) and many failures of the method can be attributed to using the start of the porosimetry curve "blindly": it is therefore essential to only take into account the porosimetry radii from a clearly defined access pressure (§ 1-1.2.2.C, p. 56) and to avoid the surface conformance effect (Fig. 1-1.42, p. 81).

Numerous improvements/complications have been suggested for Purcell's method. Considering the highly simplistic nature of these models and from a strictly practical point of view, the only ones to be taken into account are possibly those which systematise the deletion of the non-significant "large radii" which, most of the time, are artefacts due to sample surface conformance effect.

Eventually, attempts could be made to improve these models, using simple models of capillary tubes with star-shaped cross-section, instead of circular. This would improve the simulation of the rough surface states of the pores which, in Purcell's elementary model, are included with the microporosity as such (see Fig.1-1.47, p. 93).

Purcell's method and its derivatives offer a real potential, a potential which may not always be fully exploited:

– This method can be used to estimate rock permeability when only small fragments (e.g. cuttings) are available.
– For specialised studies, if both permeability and porosimetry data are available, it may be worthwhile comparing the experimental permeabilities and the "Purcell" permeabilities in order to calculate a lithological factor which may be a petrophysical indicator.

Using the standard lithological factor of 0.2, it is easy to remember the order of magnitude of the contribution of 1 porosity point (1%, i.e. 10^{-2}) to the total permeability depending on the access radius (usual units: micrometre, mD):

$$K_{(1\%)} = R^2/4$$

and to emphasise the leitmotiv of our comments concerning the ϕ-K relations: rocks formed uniquely of microaccessible porosity (R < 1 μm) never exceed a few millidarcies permeability whereas rocks with large macroaccessible porosity (R > 10 μm) may exhibit very high permeabilities, sometimes highly variable from one sample to another, since a small variation in the radius will induce significant permeability variations.

C) Carman-Kozeny equation and normalisation by the square of the grain diameter

a) Carman-Kozeny equation

The Carman-Kozeny equation is frequently mentioned since it can be used for permeability estimations, for example in modelling calculations, although introducing the tortuosity (τ) makes it harder to use in practice than one might think.

In a network of cylindrical capillaries, we obtained the relation $K = \phi R^2/8$. For a network of tortuous capillaries of any cross-section, we obtain a similar formula by introducing the hydraulic radius (R_h, equal to area/perimeter ratio of a capillary cross-section), the tortuosity of the current lines (τ) and an almost invariable shape parameter A:

$$K = (A/\tau_{(\phi)})\, \phi R_h^2$$

However, R_h is proportional to $\phi/\sqrt{s}$, where s is the specific area of the porous space. Since $\sqrt{s}$ is itself inversely proportional to the grain diameter (d), Carman and Kozeny [Carman, 1937] determined for the permeability of an intergranular space:

$$K = B\phi^3(d^2/\tau_{(\phi)})$$

where B is a constant for a given medium.

We see that the permeability of an intergranular space (isogranular) is proportional to the cube of the porosity, this point being fairly well corroborated for the Fontainebleau sandstones (§ 1-2.1.4A, p. 146). It is also proportional to the square of the grain diameter, and this point is discussed below.

We must nevertheless return to the notion of tortuosity which, in a capillary model, quantifies the mean developed length of a current line (l_d) joining the two ends of the model, with respect to the true length of the latter (l_m): $\tau_{(\phi)} = l_d/l_m$. According to the authors, the exact expression of $\tau_{(\phi)}$ may vary. Note also that one of the disadvantages with the notion of tortuosity is its dependence on porosity. For a given "porous architecture", the tortuosity is not a constant. We will return to this point in the paragraph concerning the formation factor (§ 1-3.1.1C, p. 202)

b) Normalisation by the square of the grain diameter (d^2)

The square of the grain diameter (d^2) is a useful parameter to normalise the permeability of the intergranular spaces. Figure 1-2.11 illustrates experimental data [Pellet, personal correspondence] obtained on sintered glass of different granulometries. If we use the dimensionless number K/d^2, the relations with porosity become independent of the granulometry. On Figure 1-2.11, there is no difference between the series of 50 μm granulometry and that of 250 μm. For porosity values greater than 15%, the slope of the relation K/d^2 vs. φ is close to 3, as predicted by the Carman-Kozeny equation.

This K/d^2 normalisation may be extremely useful in reservoir characterisation studies, since, for a given facies, it allows us to plot φ-K relations independent of the granulometry.

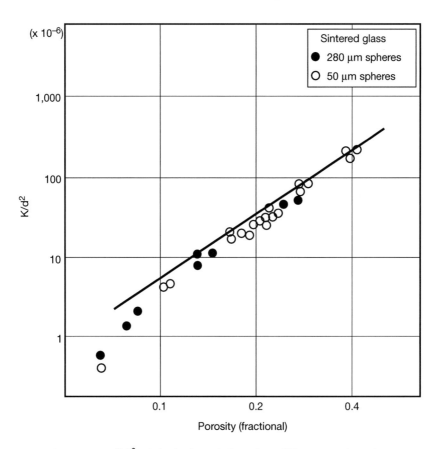

Figure 1-2.11 K/d^2 relation in sintered glass of two different granulometries. [Pellet, personal correspondence]

When considering relatively isogranular media (e.g. sand), an approximate estimation of the permeability may also be obtained by applying the K/d^2 normalisation to the ϕ-K formula derived from measurements in Fontainebleau sandstones (§ 1-2.1.4A, p. 147).

D) Notion of percolation threshold

The notion of percolation corresponds to a phenomenon which is both simple when it is described qualitatively and highly complicated in its theoretical and quantitative mathematical implications. Obviously, we will only discuss the first aspect of the subject, which may lead to interesting conclusions as regards permeability.

We will consider an assembly formed from connections (e.g. the circles on Figure 1-2.12) arranged on a network (e.g. a grid of 16×16 sites on Figure 1-2.12). If a random process is used to create (or delete) a number n of connections out of the N possible connections (256 on the figure), it is interesting to know the probability of the network providing continuity between its outer limits. On Figure 1-2.12a, the network on the left allows percolation between the lateral sides via the cluster of grey circles, but not the network on the right.

The curve showing the probability for a cluster to be "percolating" against the number of connections on the network (n/N) is extremely important. The mathematics required to calculate this function are often complicated. The function of Figure 1-2.12b was built empirically from a count out of 100 clusters obtained from a program given on the Internet [Gonsalves 1999]. On the x-axis, the number of connections has been converted into porosity (pore area, in black, divided by the total area of the network, to approximate a geological example (rock with vug porosity). Note that the maximum porosity (all connections made) corresponds to 0.785, which is the proportion of a square taken up by the inscribed circle. In order to represent the "vug porosity" values graphically, we give 4 examples of clusters for various probabilities (none being percolating).

As long as the connection densities are below a certain threshold (about 0.3 porosity, on our example), the connection probability is zero. Our "vuggy rock" is impermeable. This value is called the percolation threshold. Above this threshold, the probability increases very quickly. The medium will therefore become "exponentially" connected.

From this very brief description we must remember 2 points which will prove extremely useful when discussing ϕ-K relations in consolidated rocks.

– Below a certain porosity threshold (the "percolation threshold"), the permeability will be zero. In a porous medium formed from spherical pores, which is rare in rocks since limited to vug media, the percolation thresholds correspond to porosities of about 0.30. Despite its simplicity, the example shown on Figure 1-2.12 reproduces the values supplied by the theorists. For intergranular type media, we do not know about available theoretical results but the empirical results (such as those obtained on Fontainebleau sandstones (see below) lead to percolation thresholds of about 0.05 porosity. Only porous media formed from randomly-organised open cracks may exhibit extremely low percolation thresholds.
– For porosity values close to the percolation threshold, the network connectivity changes extremely quickly. The experimental results leading to power laws with very high exponents observed on narrow porosity bands therefore seem to be realistic.

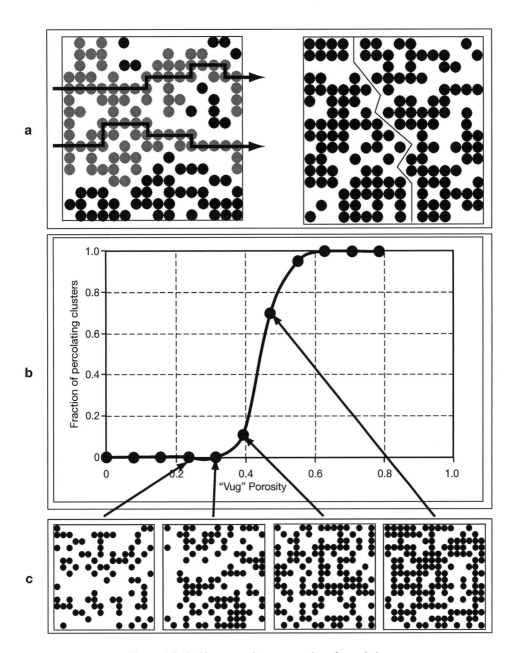

Figure 1-2.12 Diagrammatic representation of percolation

a) Example of networks with the same connection density, percolating (on the left) and non-percolating (on the right).

b) Fraction of percolating clusters plotted against the connection density (black circles) expressed as "vug porosity".

c) Examples of clusters for various connection densities (none is percolating).

1-2.1.4 Porosity/Permeability relations in rocks

The porosity/permeability relations in rocks are extremely useful in practice, mainly since permeability is a "no-logging parameter": despite all attempts (and the corresponding publications), there is no reliable log analysis method to continuously measure the permeability of the reservoir rocks in the borehole.

Discrete hydraulic methods are obviously available (e.g. mini-test), but they are costly. Reservoir geologists therefore would like to find a relation between porosity (easily measured using log analysis techniques) and permeability, so that it can be easily deduced. Most of the time, this is a risky operation.

The complexity of the ϕ-K relation is related to the complexity of the porous space itself. We will therefore start by describing the simple case of the "ideal" intergranular porous medium as encountered in the Fontainebleau sandstones. We will then study the carbonate rocks and the common sandstones.

A) Simple porous networks: Example of Fontainebleau Sandstone

a) Fontainebleau sandstones (Fig. 1-2.13)

The Fontainebleau sandstones (Paris region, France) are a rare example of simple natural porous media (intergranular porosity) exhibiting large porosity variations (from about 0.02 to 0.28) with no major change of grain granulometry. This is an ideal example on which to study the porosity/permeability relation.

Fontainebleau sandstone consists mainly of quartz grains subjected to a long period of erosion and good granulometry sorting before being deposited during the Stampian age, in coastal dunes. The original deposit consists of quartz sands with subspherical monocrystalline grains of diameter in the region of 250 µm [Jacquin 1964, in French]. These dunes underwent a complex and still poorly understood geochemical evolution leading, firstly, to the total dissolution of bioclastic limestone fragments probably abundant originally and, secondly, to a more or less pronounced siliceous cementation. Silica was deposited between the grains, as quartz in crystalline continuity with them (syntactic cement). This syntactic cementation explains the holomorphous crystal shape frequently taken by the grains. It also explains why the pore walls sometimes correspond to almost perfect crystalline planes (bottom photograph on Figure 1-2.13). The composition of Fontainebleau sandstones is therefore exceptionally simple: 99.8% silica (mainly quartz crystals).

It may be surprising to note that the variable cementation does not result in a significant variation in the apparent grain diameter. This is due to the fact that cementation occurs by progressive plugging of the intergranular space.

The exceptional simplicity of the solid phase matches the simplicity of the porous phase: it is exclusively intergranular. The microcracks at the grain contacts, observed in some cases studied below (Fig. 1-2.16), must be considered separately. Although negligible in terms of volume (porosity), these microcracks may have major consequences on the acoustic properties or on the permeability in the range of very low porosities.

500 µm

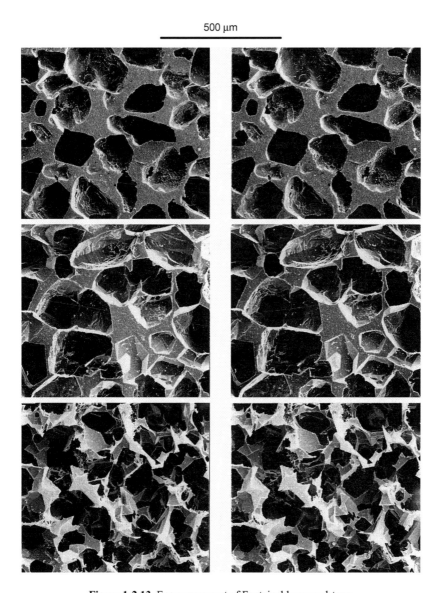

Figure 1-2.13 Epoxy pore cast of Fontainebleau sandstone,
Observation under scanning electron microscope, stereographic representation.
The mean porosities of the samples (from top to bottom) are: 0.28, 0.21 and 0.05. [Bourbié *et al.*, 1987]

b) The φ – K relation

Numerous studies have been conducted on the porosity/permeability relation. A first publication [Jacquin, 1964, in French], concerns a small number of samples (about 60) but includes accurate measurements of grain dimensions in thin section, thereby allowing normalisation by the parameter K/d^2. A study on 240 samples of diameter 40 mm and length between 40 mm and 80 mm [Bourbié and Zinszner, 1985] indicates a double trend for the $K - \phi$ relation. For high porosities (between 0.08 and 0.25), all the experimental points lie fairly well on a curve of type $K = f(\phi^3)$. Note that the power 3 corresponds to that of the Carman-Kozeny equation. For low porosities ($\phi < 0.08$), we may observe large exponents suggesting a percolation threshold (§ 1-2.1.3D).

Example of the Milly la Forêt "normal" sandstones

In this book, we give results of a much larger sampling (but of the same type as the 1985 samples and including them). To simplify the analysis, it is best to identify the origins of the various groups of samples studied. The results shown on Figure 1-2.14 concern about 340 samples from a very restricted geographical area (Milly la Forêt). Due to their geological "unity", the quality of the porosity/permeability relation is quite exceptional. The subdivisions (MZ2, etc.) correspond to different blocks measuring several decimetres in size, obtained from various points in a limited number of quarries. Note that the permeabilities of block MZ10 are slightly above the average: the granulometry is probably slightly coarser.

Air permeability values are measured in "room condition" (falling head permeameter, § 1-2.1.2A, p. 131). The experimental permeabilities below 100 mD have been corrected for the Klinkenberg effect according to the semiempirical formula (§ 1-2.1.2A)

On Figures 1-2.14 the permeability axis uses a logarithmic scale (corresponding to the log-normal distribution of permeabilities). We describe both types of axis used for porosity. On the left hand figure, we use a linear axis (normal distribution of porosities) and on the right hand figure a logarithmic axis to show the power laws.

In semi-logarithmic representation, which is generally used for reservoir characterisation, note the exceptional quality of this ϕ-K relation. To our knowledge, this is the only example of its type.

In logarithmic representation, we can see the slopes of the power laws: for the porous samples ($\phi > 0.09$), a slope of about 3.25 gives a fairly good picture of the $K = f(\phi^n)$ relation.

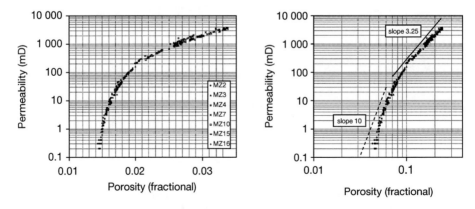

Figure 1-2.14 Porosity-Permeability relation in the Milly la Forêt "normal" Fontainebleau sandstones. The porosities are presented in the two usual ways: linear scale (semi-log graph on the left) and logarithmic scale (on the right). The subdivisions (MZ2, etc.) correspond to different blocks

The value of the exponent is similar to that observed in the earlier studies. For low porosity samples (0.04 to 0.06), the Klinkenberg correction increases the value of this exponent even more, reaching 10. Such high exponents can only be explained by a percolation threshold (about 0.05 porosity).

To profit from the exceptional quality of the results on the Milly la Forêt sampling, we calculate the polynomial regression best fitting the experimental values, in order to obtain a basic datum for estimation of the ϕ-K laws of intergranular porosity media.

A polynomial regression of order 3, on logarithmic values of ϕ and K, of type

$$\log K = a(\log \phi)^3 + b(\log \phi)^2 + c(\log \phi) + d,$$

a	b	c	d
11.17	−40.29	51.6	−20.22

with the above values (corresponding to the case of **porosities expressed as percentages**) gives excellent results for porosities between 4% and 25%.

"Microcrack" facies

Some poorly porous samples have "microcracks". They consist of strongly pronounced grain joints that can be observed on thin section but even more clearly on epoxy pore cast (Figure 1-2.16). These "microcrack" facies have been identified due to the very strong acoustic anomalies generated by these cracks [Bourbié and Zinszner, 1985]. These facies must be considered separately when studying the ϕ-K relation. Figure 1-2.15 shows some sixty values corresponding to this type of sample (4 different series). As previously, the semiempirical Klinkenberg correction was carried out on the permeability values. By comparison, the values for the Milly samples are expressed as a sliding geometric mean (for clarity purposes).

Note that the permeabilities of the "microcracked" samples are significantly larger than the Milly samples of equivalent porosity. This seems normal since the "crack porosity" becomes more efficient in terms of permeability than the "intergranular porosity". Note also

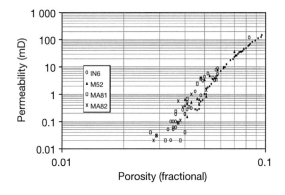

Figure 1-2.15 Porosity-Permeability relation in "microcrack" Fontainebleau sandstones. The "standard" values correspond to the sliding geometric means of Milly samples

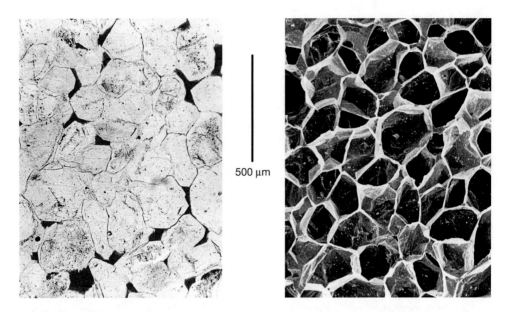

Figure 1-2.16 Fontainebleau sandstone with "microcrack" facies ($\phi = 0.06$). Photograph of thin section (red epoxy injected, § 2-2.1.1, p. 325) on the left and of epoxy pore cast, on the right (§ 2-2.1.3, p. 331). "Microcracks", which correspond to grain contacts, can be clearly seen on the epoxy pore cast

that, for very poorly porous samples ($\phi < 0.04$), the drop in permeability is less significant than might have been expected by extrapolating the Milly results: the percolation threshold effect is less marked, once again due to the special properties of "crack porosity".

This shows that, even in what seems to be the simplest case, adding a second phenomenon may significantly disturb the analysis. The microcrack facies must be identified and isolated in order to restore the "simplicity" of the Fontainebleau sandstone ϕ-K law and to demonstrate the unquestionable percolation threshold at about 0.05 porosity.

Lastly, note that this microcrack porosity is extremely sensitive to the effect of differential pressure (see § 1-2.1.5, p. 164 and § 2-1.1.2, p. 267). It is highly likely that application of the moderate differential pressure would be sufficient to close these microcracks and therefore eliminate their petrophysical effect. This is quite spectacular concerning elastic wave propagation (§ 1-3.2.4, p. 233). No data are available on the permeability of Fontainebleau sandstones with "microcrack" facies measured under pressure, but it is highly likely that application of a differential pressure would remove this microcrack effect.

c) Conclusion

Fontainebleau sandstones represent an ideal example on which to study the ϕ-K relation. They can be used to determine a practical standard for the intergranular porous space of isogranular packings. The polynomial regression on Milly values (mean grain diameter $d = 250$ µm) could be extended to different granulometries by using the K/d^2 normalisation.

The percolation threshold at about 0.05 porosity is also clearly determined; it will prove extremely useful when discussing ϕ-K relations in double-porosity limestones.

B) Porosity/Permeability relations in carbonate rocks

The situation with carbonates is strikingly different from that observed in Fontainebleau sandstones. Figure 1-2.17 shows the ϕ-K relation (air permeability) for a set of about 1 500 limestones and dolomites samples (diameter 4 cm) corresponding to a large variety of petrogaphic texture. Note that in line with standard practice, and in spite of the fact that it is poorly adapted to the power laws, we have adopted the semilogarithmic representation which makes the graphs much easier to read on the porosity axis. The permeability dispersion is very high since, on the porosity interval most frequently encountered in reservoir rocks (0.1 to 0.3), the values extend over nearly four orders of magnitude. Put so bluntly, it is clear that there is no ϕ-K relation! Considering the microtexture of the rocks, some general trends may nevertheless be observed.

a) Dolomites

On Figures 1-2.17, the points corresponding to dolomites and dolomitic limestones are separated from the limestones. The dolomite/dolomitic limestone/limestone separation was made using the criterion of matrix density (§ 1-1.1.5, p. 26) choosing 2 770 kg/m^3 as the lower limit of the dolomite and 2 710 kg/m^3 as the upper limit of the limestones. We will only consider the case of the dolomites. Far fewer points are available (about 50) than for the limestones. We can nevertheless make a few important comments. Although dolomites are present throughout the ϕ-K space (Fig.1-2.17), they are mostly represented in the region of high permeabilities. The reason is quite simple: there is no microporosity in true dolomites whereas it plays a major role in limestones (this point will be strongly emphasised in the next paragraph).

Sucrosic dolomite, vuggy dolomite

Two main contrasting features can be observed in dolomites.

- The dolomite has a granular structure (sucrosic dolomite) and, in the ϕ-K space, these dolomites are similar to sandstones due to absence of microporosity.
- The dolomite has a vuggy structure, in which case the porosity is concentrated in sometimes very poorly interconnected vugs and permeability is relatively low.

Dolomitic drains, "Super-K"

We can see on Figure 1-2.17 that the sample with the highest permeability (0.3, 70 D) is a dolomite. This sample comes from a dolomitised zone inside an oolitic barrier (France, core sample).

This type of dolomite is interesting due to the central role it plays in some petroleum reservoirs by creating drains which are thin (less than 1 m) but which exhibit very high permeability (up to 100 D). Some excellent examples of these "super-Ks" can be found in the Ghawar field (Saudi Arabia). A detailed description is given in Meyer *et al.* [2000]. A simplified example of these super-Ks is illustrated on Figure 1-2.18. It corresponds to the

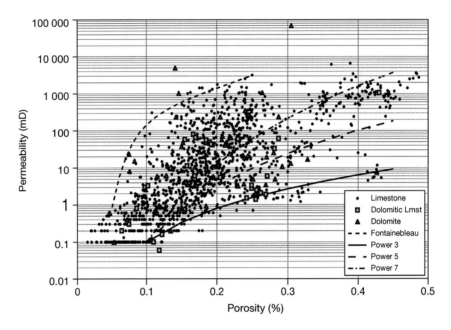

Figure 1-2.17 Porosity-Permeability relation in carbonate rocks
(air permeability, Klinkenberg effect not corrected, about 1500 samples of diameter 4 cm)

case of a very good producing well (over 10 000 barrels per day). The production log analysis (flowmeter: FLT1) shows that most of the production comes from a single very thin layer. At this layer, the drilling hole diameter, as shown by the caliper (CAL1), is larger than the nominal value (caving in a highly brittle rock) and the porosity log analyses (NPLE and RHOBE) are disturbed by this caving phenomenon. We may consider, however, that the layer porosity is very high (0.30 or more).

This reservoir has been cored and recovery is almost perfect (more than 95%) but the super-K cannot be observed on the core porosity-permeability log (Log(Kh) and CORE-PORE). The highly brittle rock forming the super K was flushed away during the coring operation. This is an example of the principle according to which, even during an excellent coring operation, some extremely important details of the reservoir may be overlooked.

Figure 1-2.19 shows photographs of thin sections from the French example discussed above (the samples do not belong to the same database as that used for Figure 1-2.17 and are therefore not represented). The samples were prepared using several resins (§ 2-2.2.2, p. 336), but to understand the ϕ-K relation, the undifferentiated total porosity (blue, yellow, red) is sufficient. Both samples are totally dolomitised ($\rho_{ma} > 2\ 800\ kg/m^3$) and the oolites clearly visible on the photographs are only "ghosts". Photograph **a)** ($\phi = 0.18$; K = 4 D) shows a very vuggy system which is highly connected and responsible for the high permeability. The sample of photograph **b)** ($\phi = 0.06$; K = 0.1 mD), although located at a depth of less than 50 cm below sample *a*, corresponds to a totally different facies. It has only very poorly connected vugs, hence the very low permeability. On photograph *b*, we can see extensive white areas corresponding to large dolomite crystals. These large crystals take the

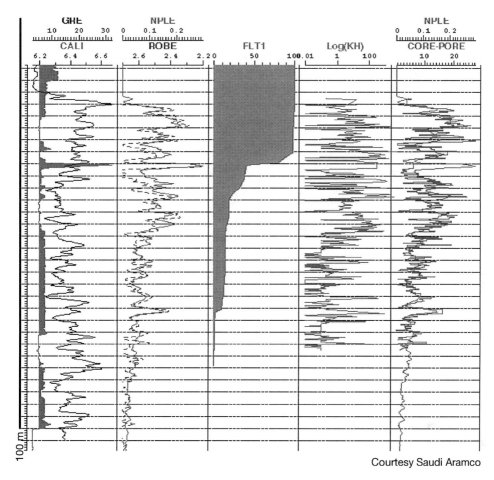

Courtesy Saudi Aramco

Figure 1-2.18 Example of well log analysis in a Super-K dolomitic reservoir

GRE: γ-ray in API unit; CAL1 (grey curve) caliper graduated in inches.
NPLE (continuous curve): Neutron porosity (fractional porosity) ROBE (dashes): bulk density.
FLT1: flowmeter (as a percentage of the total production).
Log(KH): air permeability on plugs with horizontal axis (logarithmic scale in mD).
CORE-PORE: Porosity on the same core samples (scale in percentage), the neutron porosity curve is duplicated for comparison (excellent agreement).

geometrical positions of the extensive red areas on photograph *a*. They are the result of cementation after the vacuolarisation phase of photograph *a*. This recrystallisation has been extremely erratic, which explains the petrophysical variations between samples located several tens of centimetres apart in the geological series.

b) Limestones

The values reproduced on Figure 1-2.20 correspond to some 650 samples from the database of Figure 1-2.17, for which a limited amount of information concerning the microstructure is available.

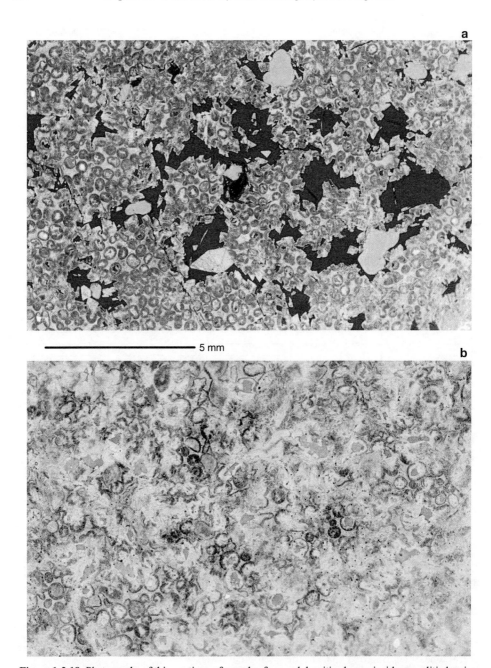

Figure 1-2.19 Photographs of thin sections of samples from a dolomitised zone inside an oolitic barrier
(France, core sample). The samples were prepared using 3 resins (§ 2-2.2.2, p. 336).
Pores of access radius less than 0.3 μm are shown in blue. The red areas correspond to porosity displaceable
by spontaneous imbibition, the yellow areas to trapped porosity

Photo **a)** (top): $\phi = 0.18$; $K = 4$ D.
Photo **b)** (bottom): $\phi = 0.06$; $K = 0.1$mD, note that all the vugs correspond to trapped porosity.

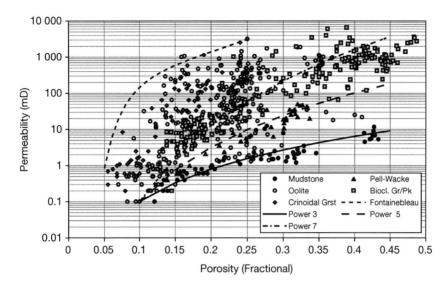

Figure 1-2.20 Porosity-Permeability relation in limestone rocks of permeability greater than 0.1 mD. The values are extracted from the database of Figure 1-2.17

The "Power 3, 5, 7" labels correspond to the relations $K = f(\phi^n)$ with powers 3, 5, 7.

Example of oolitic limestones

Oolitic limestones clearly illustrate the variability of the ϕ-K relations. These rocks, whose microtexture is clearly defined for the petrographer (oolitic grainstone), are dispersed throughout the ϕ-K space (Fig. 1-2.21). On the 0.25 porosity line, the permeability values extend from 1.5 mD to 2 D.

There is a simple explanation for this diversity: in oolitic limestones, the distribution of the pore access radii is always bimodal (apart from the rare cases of exclusive microporosity). The microporosity inside oolites, always present but in variable proportions, plays virtually no role in the permeability. The proportion of macroconnected porosity between the oolites is highly variable. It determines the permeability value.

Figure 1-2.21 shows a few examples of macroconnected/microconnected porosity ratios, from photographs of epoxy pore casts (§ 2-2.1.3, p. 330). Limestones *d* and *e* reveal only intraoolitic microporosity, their permeability is therefore very low. The other limestones exhibit macroconnected and microconnected porosities in variable proportions.

Limestone *a* exhibits exceptional intergranular macroporosity, which explains its high permeability in spite of the fact that its porosity is relatively low. Limestone *f* exhibits only moderate intergranular macroporosity and significant intraoolitic microporosity. What makes this limestone so unusual, however, is the very high moldic porosity inside the oolites. This abundant (in terms of pore radius) macroporosity can only be accessed through the intraoolitic microporosity. It is therefore known as microconnected porosity. This explains the relatively low permeability of this exceptionally porous oolitic limestone.

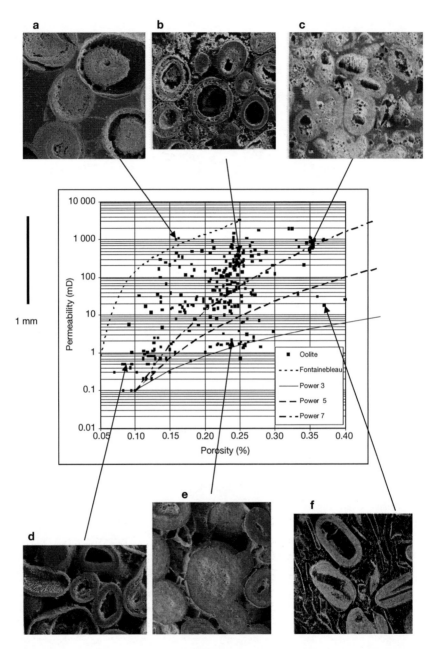

Figure 1-2.21 Porosity-Permeability relation in oolitic limestones (values extracted from Figure 1-2.17). Epoxy pore casts (§ 2-2.1.3, p. 330) photographed using scanning electron microscope (SEM). Same scale for all six photographs

Limestones *a*, *b*, *d*: core samples.

Limestone *c*: "oolite miliaire" outcrop, Normandy (France).

Limestone *e*: outcrop, Chaumont (France); Limestone *f*: outcrop, Brauvilliers (France).

Micrites

Micrites (Fig. 1-2.22) are rocks mainly formed from **micr**ocrystalline calcite. The φ-K relation, however, is quite different from those of the oolitic limestones. In the φ-K space, the micrites are grouped along a line corresponding approximately to a power 3 law going through φ = 0.1; K = 0.1 mD. This relative simplicity of the φ-K relation in micrites is explained by the fact that they only have a single type of porosity. The porosimetry spectra (Fig. 1-2.23) are clearly unimodal and contrast with the other limestones.

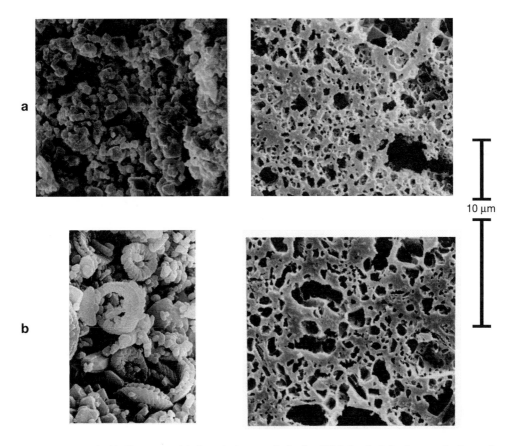

Figure 1-2.22 Micritic limestones (Mudstone) photographed using SEM. On the left, photograph of natural rock (fracture); on the right, epoxy pore cast

a) Core sample (Middle East); φ = 0.29; K = 1.5 mD.
b) White Chalk (outcrop) from the Paris Basin; φ = 0.44; K = 6 mD.

Main trends and limiting values of φ-K relations in limestone rocks

So far, we have emphasised the extreme diversity of φ-K relations. However we can make a few observations which will prove useful when studying limestone reservoirs:

– Limits of the φ-K space.

The various φ-K relations described in this section respect the "matrix value" criterion (§ 2-1.3, p. 311). They come from plugs and the minimum homogenisation volume is millimetric. We can see on Figure 1-2.20 that these values define a φ-K space whose limits are quite well defined

• By the Fontainebleau sandstone line towards the high permeabilities
• By the mudstone line (power 3 law) towards the low permeabilities

Although these limits are very broad, they are nevertheless practical. When studying reservoirs, special attention must be paid to values outside this area. They generally indicate measurement errors or faulty samples (e.g. fissured plug). In the other cases, however, a special study could prove well worthwhile.

– Main trends

Some main trends in the φ-K relation may also be observed according to the petrographic texture of the limestones (Fig. 1-2.20).

• The very poorly permeable limestones (K < 0.1 mD) have not been shown on Figure 1-2.20. The porosity of limestones whose permeability is greater than this low value is generally more than 10%, apart from the important exception of crinoidal limestones which have very little microporosity, hence the high permeabilities. Some oolitic grainstones lie within the same area of the φ-K space, for the same reason: proportionally very low microporosity.

• For the other types of limestone, we observe a point of convergence at about φ = 0.1; K = 0.1 mD. If power 3, 5, 7 law graphs are plotted from this point, we observe that the mudstones are grouped on the power 3 law (see above), the wackestone-pelstones around the power 5 line and the bioclastic grain-packstones around the power 7 line. Although these are obviously very general observations (except for the mudstones), they may prove useful when looking for orders of magnitude, during modelling for example.

c) Conclusion on the φ-K relations in carbonates

We have described the vast diversity of φ-K relations in limestones and have returned on several occasions to the unique cause of this diversity (and of this absence of φ-K relation in the strict sense). In limestone rocks (apart from mudstones), at least two types of porosity affect the petrophysical characteristics, in quite different ways.

• The microporosity (microconnected porosity) always present in the allochems and matrix (mud).
• The macroconnected porosity sometimes found between the allochems (intergranular), and also – rarely – in the dissolution vugs if they are present in sufficient numbers to be interconnected.

Only macroconnected porosity plays a significant role in permeability. We would therefore need to plot the $\phi_{macroconnected}$ *vs.* K relations to obtain a rough estimation. This approach would, however, be both costly in terms of porosimetric measurement and of little practical application, since the φ-K relations are of most use when there are no samples (drilling).

The porosimetric diversity of carbonates is represented on Figure 1-2.23 where the porosimetry/permeability relation can be checked qualitatively (Purcell model, § 1-2.1.3B, p. 139). Although a basic point, we must reiterate the fact that the most important value required to understand ϕ-$\mathbf{K}$ relations is the equivalent pore access radius. The dimension of the pore itself has virtually no impact on permeability.

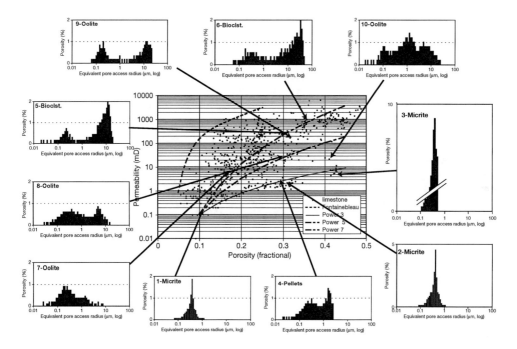

Figure 1-2.23 Overview of porosimetric diversity in limestone rocks

To draw the spectra, the occurrence frequencies (y-axis) have been represented as bars of 1/20 of a logarithmic decade thickness and height proportional to the "absolute" value of the porosity concerned expressed in "points" of 10^{-2} fractional porosity. The accessible porosity between two pore access radius values is therefore equal to the area of the portion of histogram. The scale is the same for all spectra, but the y-axis of micrite 3 (chalk) had to be cut off at the top due to the extremely high modal value.

Note the clearly bimodal distributions for the bioclastic and oolitic facies with the "microporosity" accessible through access radii of less than 1 µm. Micrites exhibit only microporosity.

One extreme example is that of the limestones (and especially the dolomites) with large moldic pores (Fig. 1-2.24) connected only by a very fine matrix, thereby inducing only low permeability. This remark partly explains the disappointment often felt when interpreting NMR log analyses in terms of permeability. The geometric data deduced from NMR is much closer to the pore radius than the access radius (§1-3.3.2, p. 255; fig 1-3.33)

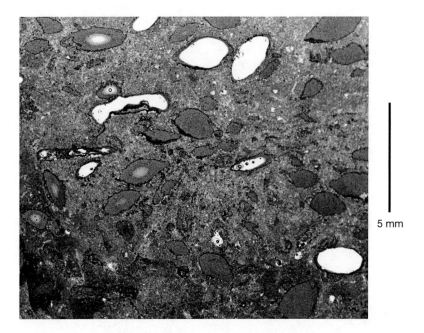

5 mm

Figure 1-2.24 Example of dolomitic limestone (ρ_{ma} = 2750 kg/m³) with large,
poorly-connected moldic pores. The mean sample characteristics (ϕ = 0.18; K = 7 mD)
are not necessarily representative of the area photographed (heterogeneity). Despite saturation under
vacuum, the red epoxy resin did not have time to invade all the vugs due to the low permeability

C) Porosity-Permeability relations in sandstones (comparison with carbonate rocks)

The ϕ-K relation in Fontainebleau sandstones is an exception in the natural environment. In
the general case of clastic rocks, petrological variations of various origins (granulometry,
cementation, etc.) disturb in particular the ϕ-K relations, such that even when limited to a
reservoir or a given formation, the ϕ-K relation in sandstones, as in limestone rocks, is
generally only a cloud of points. After providing an overview, we will briefly examine the
main petrological parameters affecting this relation.

a) Overview concerning the ϕ-K relations in sandstone reservoirs

Some examples of these relations are shown on Figure 1-2.25. This figure was produced
using the results of a compilation of Nelson & Kibler [2003]. This outstanding work is
available on the Internet as a USGS "Open File". The data concern exclusively
measurements on core samples from oil or gas reservoirs. Most are "routine" measurements
taken on 1 inch (25.4 mm) diameter samples under ambient conditions.

The results of Figure 1-2.25 must be compared with those of Figure 1-2.20 concerning
limestone rocks. Note, however, that different sampling principles were used for the
databases corresponding to these two figures. Since the figure concerning the limestone
rocks includes numerous outcrop samples, it may be slightly more representative of the

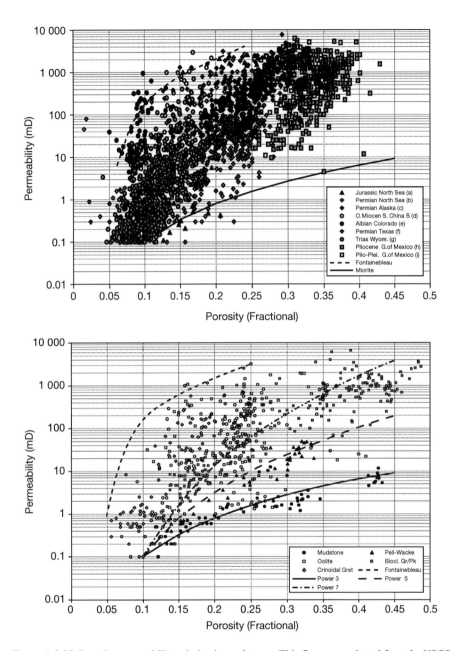

Figure 1-2.25 Porosity-permeability relation in sandstones. This figure was plotted from the USGS compilation [Nelson and Kibler, 2003]. The various families of points correspond to the data of:
a) [Aase *et al.*, 1996], b) [Amthor and Okkerman, 1998], c) [Atkinson *et al.* 1990], d) [Bloch, 1991], e) [Keighin *et al.,* 1989], f) [Langford *et al.*, 1990], g) [Lindquist, 1988], h) [Marzano, 1988], i) [Reedy and Pepper, 1996]. To allow comparison, Figure 1-2.20 (limestones) has been reproduced and values less than 0.1 mD have not been shown. The total number of points is over 2 500

diversity of porous facies. To allow easier comparison with Figure 1-2.20, only permeabilities ≥ 0.1 mD have been taken into account.

Note firstly that the porosity distributions are clearly different (Fig. 1-2.26). We observe a shift of about 0.07 porosity between the sandstone and limestone sets. Obviously, the samplings considered are not fully representative of all these rocks, but we think that this difference in porosity distribution corresponds to the true situation. Due to the possible abundance of microporosity in matrix (mud) and allochems, as well as to the intergranular macroporosity, consolidated limestone rocks may on average be much more porous that the sandstones. A sandstone of 0.35 porosity can only be unconsolidated whereas a limestone of more than 0.40 porosity may form an exceptional ashlar (e.g. "Banc Royal" of the Paris Lutetian quarries).

In contrast, note the similarities between the permeability distributions. The differences observed towards the low permeabilities are relatively insignificant since the low permeabilities are probably biased in the sampling of limestone facies.

Figure 1-2.25 shows that the ϕ-K relations are highly dispersed in sandstones. In a first analysis, the dispersion seems just as high as in limestones (Fig. 1-2.20). Note, however, that

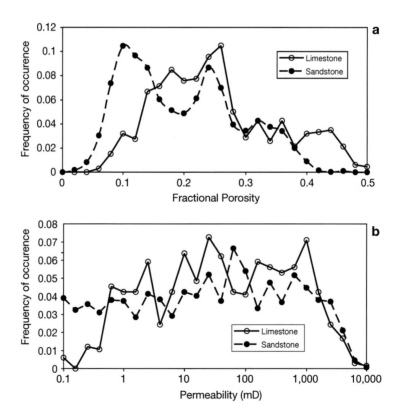

Figure 1-2.26 Distribution of porosity (a) and permeability
(b) values in the sandstone and limestone samples shown respectively on Figures 1-2.25 and 1-2.20

different methods were used to differentiate between the sets of points on the two figures. For limestones, selection was carried out using petrological criteria, thereby *a priori* improving the φ-K relation. For sandstones, selection was based purely on geographic and stratigraphic criteria (by reservoir). It is likely that if this selection method had been used for Figure 1-2.20, there would no longer have been any visible correlation.

b) Influence of petrological characteristics on φ-K relations

Figure 1-2.25 shows the dispersion of φ-K relations, even for samples grouped according to stratigraphic origin. This situation can be improved somewhat by introducing additional criteria offering better sorting of the samples by porous network type. In reservoir characterisation, only petrological criteria are readily available. Some basic examples of the effect of petrological characteristics on permeability are given on figure 1-2.27.

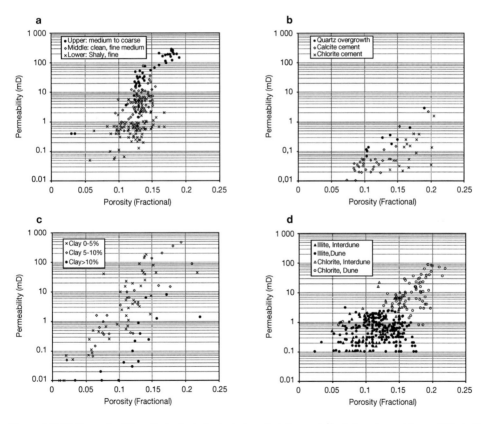

Figure 1-2.27 Example of the effect of petrological characteristics on φ-K relations in sandstones. Effect of:
a) granulometry [Taylor and Soule, 1993].
b) cement type [Langford *et al.*, 1990].
c) clay content [Muller and Coalson, 1989].
d) clay type [Thomson and Stancliffe, 1990].

Granulometry

Granulometry is one of the most important petrological factors affecting the ϕ-K relation of sandstones. We emphasised above (§ 1-2.1.3C, p. 140) the major effect of the grain diameter (*d*) on the permeability of a packing of isogranular spheres and the practical benefit of working whenever possible on the dimensionless value K/d^2 rather than on the permeability as such. The diagram of Figure 1-2.27a, plotted using data of Taylor and Soule [1993], illustrates what may frequently be observed. We must nevertheless beware of making hasty generalisations. The notion of granulometry also includes the sorting of grain dimensions. It would be more appropriate to speak of granulometric spectrum. Sorting also has a major effect on permeability; it is easy to appreciate that small grains located inside the intergranular space created by large grains considerably reduce the permeability. In sandstone, grain sorting and average diameter are not independent in a given sedimentological type. This dependence may lead to compensation effects on the ϕ-K relation, making the granulometric influence less evident. An example of this situation, concerning unconsolidated sands, is given below.

Cement type, clay content and clay type

The cement type and the clay content also have major consequences on permeability, depending on the extent to which they plug the intergranular spaces. Examples of situations often encountered in the sandstone reservoirs are shown on Figure 1-2.27b [Langford *et al.*, 1990] and *c* [Muller and Coalson, 1989]. The effect of the clay mineralogical type is another well-known phenomenon. The fibrous illite invading the intergranular space reduces the permeability much more than the chlorite covering the pore walls. Diagrams of these clay structures are given on Figure 1-1.7, p. 17.

c) Practical limits of the notion of ϕ-K relation, on an example of unconsolidated sands

Numerous studies, often confidential, are conducted on the unconsolidated sands frequently found in the deep off-shore deposits and representing a major financial stake. One particular publication listing numerous laboratory results [Levallois, 2000] can be used to measure the practical limits of the notion of ϕ-K relation. Figure 1-2.28 shows graphs of the values measured in one of the reservoirs studied. The number of samples concerned is quite low but the selection is representative of all the results.

We observe that there is no ϕ-K relation in these "clean" sands, i.e. of very low clay content. In contrast, there is a surprisingly good inverse relation between the porosity and the average grain diameter. This *a priori* paradoxical observation can be explained by examining the similar situation concerning the "sorting" parameter (equal to the square root of the ratio of the values in the third and first quartiles of the granulometric curve). In these deep sea turbiditic facies, there is a close relation between granulometry and sorting: fine sands are sorted best. The opposite situation would probably be observed in beach or dune sands. Consequently, the lack of correlation between permeability and average grain dimension is no longer surprising.

This example illustrates the highly frequent case where strong correlations between various petrological parameters influencing the permeability in opposite ways (granulometry, sorting, clay content, etc.) confuse the ϕ-K relation to such an extent that it

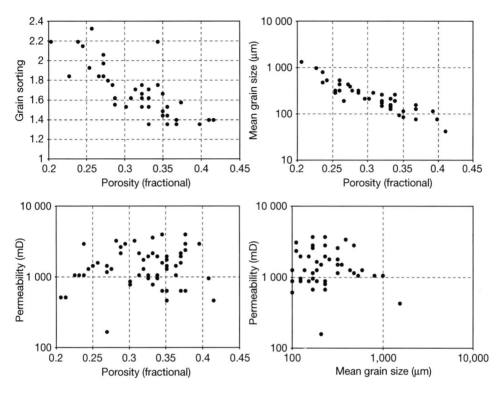

Figure 1-2.28 Porosity, Granulometry, Sorting and Permeability relation in clean unconsolidated sands of a deep off-shore oil reservoir. Data of Levallois [2000]

becomes meaningless. In reservoir characterisation, for each case we must try to adapt the study to the specific petrological features encountered.

1-2.1.5 Effects of stress and temperature on the intrinsic permeability

Before describing the effect of stress variation on permeability, we will mention a point discussed earlier on several occasions: when studying the effect of external conditions on a petrophysical property, a distinction must be made between what exactly should be attributed to the variations of the porous medium itself and what should be attributed to the variations induced in the saturating fluids. Permeability provides a clear illustration of this principle. Fluid viscosity is the most important factor when measuring permeability since it governs the volume flow rate. The viscosity of some liquids is highly sensitive to temperature. One well-known example is that of glycerol, whose viscosity varies by a factor of 100 between 10°C and 60°C (this large variation is put to good advantage in many laboratory experiments). If we vary the temperature during a permeability measurement experiment, we may observe a substantial variation in the flow rate. The correlative viscosity variation must therefore be taken into account to conclude that there has been no permeability variation (or only an extremely small variation).

Similarly, pressure also has a fairly large effect on the viscosity of some fluids, especially oils, the viscosity often doubling between 0 MPa and 50 MPa. When interpreting the permeability variation under the effect of a pore pressure variation, overlooking this phenomenon may significantly bias the result.

A) Effects of temperature on intrinsic permeability

As reminder, after the previous comments, note that temperature has only a negligible effect on permeability. Obviously, this remark only applies for variations observable under normal conditions in the surface layer of the Earth's crust (variation of a few dozen degrees, for example).

This observation can be related to the value of the linear expansion coefficients of common minerals, which is of the order of 10^{-5}. If we apply these coefficients to estimating the variation in pore threshold radius, we obtain an order of magnitude of a few thousandths for a hundred degrees Celsius. We can therefore appreciate that the simultaneous variation in permeability is extremely low.

The expansion may be strongly anisotropic and the expansion coefficient of calcite is negative in some directions in the crystal. Temperature variations in limestone rocks with no porosity (marble) may therefore produce substantial modifications at the intercrystalline joints. This phenomenon helps explain the rapid weathering of some marbles exposed to solar radiation. Large permeability variations may be observed in this type of rock, whose initial permeability is very low, much less than one microdarcy.

Although strictly speaking far removed from the petrophysical context, we may mention another case of permeability variation related to a change in temperature: fracturing phenomena observed near water-injection wells. During the exploitation of many oil fields, the pore pressure is maintained in the reservoir by injecting large quantities of water, whose temperature is generally well below the reservoir temperature (geothermal gradient). This variation in the rock temperature may generate fractures which significantly modify the permeability around the well, at the metric and decametric scale. Obviously, this fracturing caused by a thermal shock must not be confused with the "hydraulic fracturing" induced by excess pressure in the injected fluid.

B) Effects of stress on intrinsic permeability

a) Limitations of the description

Numerous studies have already been conducted on this subject and we will therefore restrict ourselves to a very brief summary. Interested readers will find detailed information in Tiab and Donaldson [1996] and interesting references in Ferfera *et al.* [1997]. Like the effect of stress on porosity, to which it is closely related, this subject is at the crossroads between Rock Mechanics and Laboratory Petrophysics, which does not make it any easier to deal with!

An initial difficulty is due to the fact that the notion of differential pressure is not sufficient to characterise the mechanical state of the material (far from it). To investigate the subject in detail, we must therefore consider the differential pressure as well as the deviatoric stress. This makes the experiments and presentation of the results more difficult.

For a pragmatic approach, we will make two remarks:

- We will restrict ourselves to the effects related to stresses low enough to avoid exceeding the pore elastic deformation limit. Beyond this limit, the significant and irreversible deformation of the porous space (see pore collapse § 1-1.1.6A, p. 42) leads to complex phenomena and to often antagonistic consequences on permeability (increase, decrease). There is no guarantee that these mechanical conditions will often be observed in practice.
- When studying the variations in a petrophysical property, the relative importance of these variations must be considered with respect to the dispersion of values of this property in nature. Obviously, a 20% variation in permeability will result in an equivalent variation in the flow rate, but since when characterising a layer, no matter how homogeneous it may be, the uncertainty on the mean permeability value is much greater than this value, the importance of such a variation should perhaps be kept in perspective. This is an important debate which would probably provoke conflicting arguments.

b) Two main types of behaviour: "pore porosity" and "crack porosity"

Figure 1-2.29 gives a highly diagrammatic summary of the contrasting behaviour of the two types of porous structure (§ 2-1.1.2A, p. 267) under the effect of stress. Note that the measurements shown on the figure correspond to a routine study conducted on sandstone reservoirs [Keighin *et al.,* 1989], not a study on the effect of stress. There is a striking contrast between the most porous samples ($\phi > 0.15$) for which the variation in permeability between 2 MPa and 20 MPa seems small in logarithmic scale, and the less porous samples for which the variation may be greater than one order of magnitude.

With these low porosity samples, the measurement pressure is generally 40 MPa. The pressure used in the laboratory has been calculated so that it is as close as possible to the differential pressure actually present in the reservoir and the least porous samples come from deeper reservoirs. We are convinced, however, that this is not the main cause of the difference in behaviour. As with compressibility (§ 1-1.1.6, p. 46) or elastic wave velocity (§ 1-3.2.4, p. 233), there is a power law type relation between the permeability and stress and the largest variations are already observed at a differential pressure of 20 MPa.

This example, drawn from reservoir characterisation, illustrates the contrasting behaviour of the two main types of porous structure (pore and crack) described in more detail in § 2-1.1.2A. We can easily understand that this contrast is greatest in the case of intrinsic permeability. We have seen earlier (§ 1-2.1.3, p. 138) that the permeability induced by a crack varies in proportion to the cube of its thickness. Since the thickness of a crack may vary extensively under the effect of stress, the permeability in turn may vary considerably. In contrast, with "pore porosity" where the major permeability variation parameter is the pore access radius, since this radius varies much less than the thickness of the microcracks (arching in the solid phase), the permeability will be far less dependent on the stress.

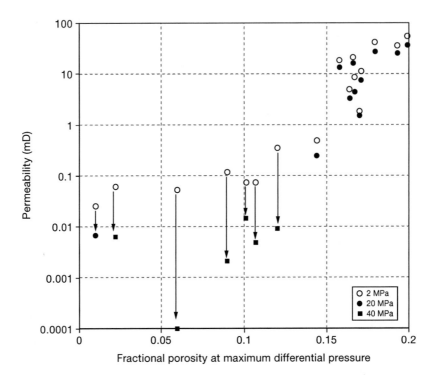

Figure 1-2.29 Effect of differential pressure variation on air permeability (Klinkenberg). Measurements by Keighin *et al.* [1989] collated in [Nelson and Kibler, 2003]. In the φ-K diagram, the porosity values shown are those measured at maximum differential pressure (20 MPa or 40 MPa depending on the samples)

c) In conclusion: suggested experimental approach

In conclusion, when characterising a geological formation, we would suggest adopting different experimental approaches depending on the type of porous medium, identified using porosity/permeability data acquired under ambient conditions.

Rocks with possibly predominant "microcrack porosity": Permeability < 0.1 mD, Porosity < 0.05

The most common examples of this type of medium are the crystalline and metamorphic rocks, the marbles and, as regards oil exploration, the tight sands. These are low porosity/permeability sandstones but may correspond to financially very attractive gas reservoirs, especially if the reservoir pore pressure is very high (great depth). The permeability variation under the effect of stress may reach several orders of magnitude, to such an extent that there is no longer anything in common between the petrophysics of the sample under ambient conditions and the same sample in confinement. Obviously, the measurements must be taken under stress for permeability in particular, but this is also true for the other petrophysical properties.

We must always bear in mind that this type of porous network is highly prone to anisotropic effects (§ 2-1.2.2, p. 282). Anisotropy in these networks is caused by a number of factors, whether the distribution of microcrack directions is itself anisotropic, or the cause of network modification (stress), or both. This point must always be checked, for example by taking measurements in two or three orthogonal directions.

Rocks with probably negligible "microcrack porosity": Permeability > 10 mD, Porosity > 0.15

These are porous sedimentary rocks with almost exclusively "pore type" porosity. The effect of stress on permeability may lead to permeability variations of a few percent (very porous and permeable rocks) up to several dozen percent (low or average permeability rocks). Although important when considered as "absolute values", these variations are not sufficient to change the "nature of the porous space" and must be kept in perspective in view of the wide dispersion of permeability values, as was suggested earlier. A large number of measurements taken under ambient conditions will be more useful when characterising the reservoir than a smaller number of measurements in confinement, since only very few of these will be of any use to demonstrate the basic assumption.

Special cases: intermediate cases and highly porous, low-permeability rocks

In practice, with cases in between the two previous ones ($0.05 < \phi < 0.1$; $0.1 < K < 10$ mD) it may be difficult to decide. Generally, a petrographic analysis will remove any uncertainty. At the start of the study, however, it is certainly more cautious to check the effect of stress on a limited number of samples.

Note also the special case of very porous, low-permeability rocks ($\phi > 0.4$; $K < 5$mD). They may be easily-identifiable sedimentary rocks with very fine grain, such as chalks or diatomites. Due to their high porosity, these rocks are extremely compressible and the pore deformation elastic limit may correspond to low differential pressures. Nevertheless, this does not make them particularly sensitive as regards permeability, obviously as long as we do not exceed the elastic limit. We may however observe cases of vuggy porous media with crack connections. These cases include carbonate rocks, although very rarely, but above all lava (some types of pumice correspond to this definition). Stress may have a large effect on permeability and it should be checked. Generally, in case of porous media for which few data are available, e.g. lava, care should obviously be taken regarding the conclusions drawn.

In petroleum practice, the effects of stress on permeability become extremely important in deep, low-permeability reservoirs, subject during production to a considerable drop in pore pressure (and therefore to a significant increase in the effective stress).

1-2.2 WETTABILITY OF RESERVOIR ROCKS

1-2.2.1 Definition and measurement of intermediate wettability

A) Introduction: specificity of the situation in porous environment

In Section 1-1.2 we described capillary pressures in the case of "perfect wettability", i.e. where one of the fluids displays a clearly marked affinity for the solid, homogeneously across the entire wall of the porous space. The simplicity of this case lies in the fact that only the geometry of the porous space needs to be considered to understand capillary phenomena (the only important parameters of the fluids are their surface tensions t_s and, possibly, the wettability angles θ). By using the "normalisation" parameter $t_s \cos \theta$, we were able, for a given rock, to process equally well the cases water/oil, water/gas or mercury/vapour.

In nature, this favourable situation is only found in very specific cases, such as that of a rock which has never been in contact with anything but water (or brine) and which is invaded by liquid or gaseous hydrocarbons (secondary migration § 1-1.2.6, p. 102). This is also true near the surface, in vadose zone, where the porous space contains both water and air. But even in this case, it is not uncommon to observe states (at least transient) of non-water wettability (see Fig. 1-1.27, p. 53), under the effect of biological phenomena such as microorganisms covering the porous walls.

Petroleum is a mixture of a large number of chemical compounds, including molecules exhibiting a special affinity for solids. This affinity may be more or less marked for certain minerals. We can therefore see that when a porous rock is invaded by oil, deposition of these molecules may have a significant impact on the nature of the relations between the solid and the fluids, in other words, by definition, the wettability. The example we give for the oil reservoirs may be almost identical to the situation in surface geology (hydrogeology, pedology), for example in the event of deliberate or accidental injection of chemical products.

Consequently, the capillary behaviour described in Section 1-1.2 must be reconsidered. If these phenomena led to a clear wettability reversal, i.e. if the rock became as oil-wet as it was previously water-wet, then we could simply apply the principles of Section 1-1.2, exchanging the fluids. This type of wettability reversal can be carried out in the laboratory by depositing silane molecules, for example, on the entire porous wall. In nature, however, this is very rarely the case. The modification undergone by the solid/fluid interface is heterogeneous and, together with the fact that the porous wall is sometimes very rough, it induces phenomena which are extremely difficult to quantify accurately at microscopic scale. This explains why the definition of wettability when studying geological materials can only be based on a "phenomenological" approach.

To make it easy for non-specialists to understand the wettability phenomena observed in rocks, a few elementary but useful remarks must be made. The first, and most important, is that *the diagnosis* (and quantification) *of petrophysical wettability is not the result of a "rigorous physical analysis"* based, for example, on the interactions of electrostatic forces, *but the result of a particular experiment*. Several types of experiment (see below) giving different results can be performed. It is therefore not surprising that the wettability state of a rock varies depending on the test conducted.

The second remark is that, unlike porosity or single-phase permeability, *wettability is not an intrinsic characteristic* of the rock, *but a variable state of the interface between the mineral solid and the fluids* contained in the porosity. When defining this state, the fluids and the history of their relation with the solid are as important as the solid itself.

Lastly, it should be added that, in everyday language, oil wettability may not be the reciprocal of water wettability. "Oil wettability" is often the observation deduced from a particular experiment discussed below. Although the same may be true of "water wettability", the result is often actually more conclusive, being based on a number of observations. It would often be more appropriate to speak of a trend towards oil wettability.

The debate between "innate" and "acquired" is a recurrent feature in some fields of biology. The same applies in petrophysics with the problem of wettability since it is never easy to distinguish clearly between the causes intrinsically related to the nature of the porous space: geometry, mineralogy (innate) and those related to the nature of the fluids and the particular history of their coexistence (acquired). The debate is of much more practical importance than it seems, since only the "innate" parameters can be practically extrapolated by using geology methods.

For detailed information on wettability, readers can refer to the bibliographic work of Anderson [1986-87].

B) Phenomenological approach: wettability as the result of an experiment

a) Terminology remark on "capillary pressure, drainage and imbibition"

Before describing the two main methods used to measure wettability, some of the terminology must be clarified:

In Section 1-1.2, in line with convention, capillary pressure has been defined as being the difference in pressure across the interface between the non-wetting fluid and the wetting fluid, in other words the excess pressure required to "force the capillary barrier". This definition is meaningless if we want to define which is the wetting fluid. Nevertheless, this term is always used in the experiments. Capillary pressure (sometimes called effective or differential pressure) is equal to the algebraic difference between the pressure in oil (or gas) and in water. This pressure can be positive (basic case of water-wet situation) but it can be negative if the pressure in water is actually higher than in the second fluid.

Similarly, the wettability assumptions concerning drainage and imbibition must be clarified. An objective definition of drainage is as follows: operation leading to a reduction in water saturation, imbibition is the opposite phenomenon (increase in water saturation), the adjective "spontaneous" refers to the return to a minimum value of (effective) capillary pressure and "forced" when the absolute (not algebraic) value of (effective) capillary pressure increases under the effect of external forces (e.g. centrifuge).

b) The USBM method

The USBM (US Bureau of Mines) method [Donaldson *et al.*, 1969] is based on the observation that in order to increase the saturation in non-wetting fluid of a porous space, energy must be put into the system (conversely, an increase in wetting fluid releases energy). This observation is clear in case of strong wettability. On a drainage capillary pressure curve (Fig. 1-1.29, p. 57), the area under the curve corresponds to the integral of the product of a volume (saturation) by a pressure (capillary), in other words to work.

This observation can be generalised by measuring the energy required (area between the curve and the x-axis) to increase the saturation of each fluid successively and estimate that the ratio of these areas is an indication of wettability.

We will consider the following sequence of experiments (Fig.1-2.30) conducted, for example, in a centrifuge:

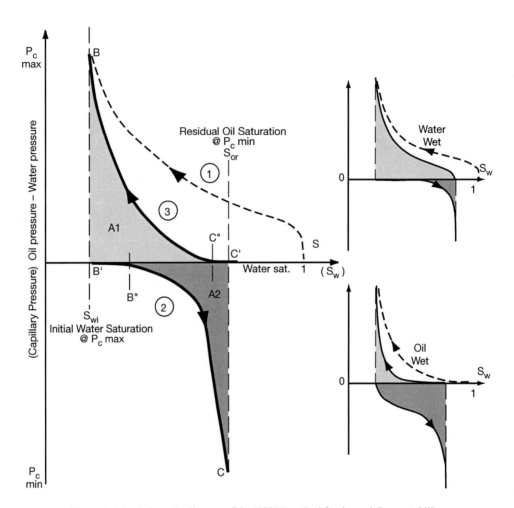

Figure 1-2.30 Schematic diagram of the USBM method for determining wettability

A sample originally totally saturated with water (point S, Fig. 1-2.30) is immersed in oil in a centrifuge cup where it is subjected to increasing accelerations, up to very high values. The sample undergoes a "first drainage" (phase 1 on Figure 1-2.30; see also Figure 1-1.34, p. 65) bringing it into a state of irreducible water saturation (point B on Figure 1-2.30). Stopping the centrifuge does not change this state, even though there is no longer any differential pressure since the sample is completely surrounded by oil (point B' on Figure 1-2.30).

The measurement as such starts at this stage. The sample is transferred into a cup filled with water and centrifuged again. We can apply the same calculation methods as with a "traditional" drainage ($\Delta P = \Delta \rho \gamma h$), but since the water phase is now the continuous phase outside the sample the sign of $\Delta \rho$ must be inverted. The pressure in the water is greater than that in the oil and we are moving along branch No. 2 on Figure 1-2.30 ("forced" imbibition under the effect of "negative" capillary pressures). Note that the "negative" capillary pressures are only observed from point B" on Figure 1-2.30 since the saturation interval B' – B" corresponds to a spontaneous imbibition phase.

Under very high accelerations, from the saturation corresponding to point C, the oil mobility becomes zero and we reach the residual oil saturation state, a generalisation of the notion of residual saturation discussed in paragraph § 1-2.3.3, p. 188.

Stopping the centrifuge (point C') does not change the saturation (see above). The sample once again placed in a cup filled with oil undergoes a second forced drainage phase (curve 3) which could possibly be preceded by spontaneous drainage (from point C' to point C").

The area (A1) between this second drainage and the x-axis corresponds to the energy required for the oil to penetrate into the system. Equally, the area A2 between curve 2 and the x-axis corresponds to the energy required for the water to penetrate. If the medium is water-wet, this energy will be very low (ideally A2 = 0) and conversely for A1 if it is oil-wet. In the USBM method, wettability is characterised by the logarithm of the area ratio $[W_{USBM} = \log(A1/A2)]$. If $W_{USBM} > 0$, the medium is preferentially water-wet, if $W_{USBM} < 0$, the medium is preferentially oil-wet. For example, a medium with USBM index of about 1 is strongly water-wet.

c) Amott's method

Amott's method [Amott, 1959; Cuiec, 1975] is based on estimating the relative importance of the fluid fractions displaceable during spontaneous and forced operations (imbibition/drainage).

After first being totally saturated with water, a sample is brought into a state of "irreducible" water saturation by a long period of oil flooding (or possibly by centrifuging) until no more water is produced.

- In a first step, this sample is placed in water where it undergoes spontaneous imbibition, displacing a volume V_a of oil (Figure 1-2.31a). The sample then undergoes forced imbibition (waterflooding or centrifuging in "water" environment) displacing an additional volume of oil V_b. V_a is the volume of oil displaced spontaneously by the water and $V_a + V_b$ the total volume of displaceable oil (spontaneous and forced). The index Iw is defined: $Iw = V_a/(V_a + V_b)$. If the medium is perfectly water-wet, we obtain the situation of § 1-1.2.2D, p. 59, and after the spontaneous imbibition phase, the residual oil is totally trapped by the capillary forces, forced displacement by water will produce no oil, $V_b = 0$ and $Iw = 1$.
- The second step consists in carrying out the reverse operation with oil: the sample in "irreducible/residual" oil saturation state (note once again the terminology ambiguity as soon as the wettability state is no longer fixed) is immersed in oil (Figure 1-2.31c), drainage spontaneously displaces the volume of water V_c. Forced drainage (flooding or centrifuging) produces an additional volume V_d. We can therefore calculate an index $Io = V_c/(V_c + V_d)$. We will once again use the example of perfect water wettability from § 1-1.2. Spontaneous drainage will not displace any fluid, so $V_c = 0$ and $Io = 0$.

We can define a unique wettability parameter $W_{IA} = Iw - Io$ of value between –1 and 1. A value of 1 corresponds to the case of perfect water wettability, as mentioned previously.

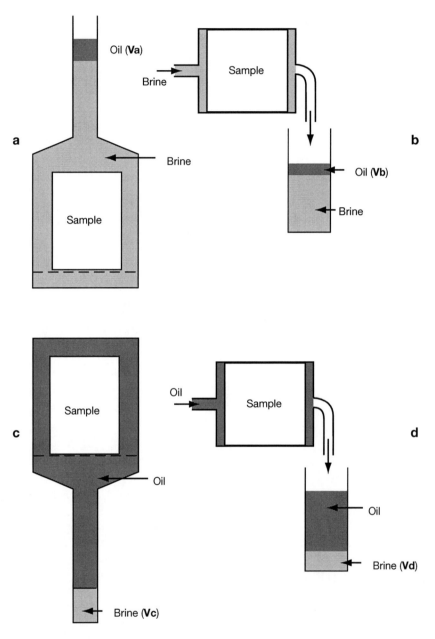

Figure 1-2.31 Schematic diagram of Amott's method for determining wettability

Inversely, a value of −1 corresponds to a case of perfect oil wettability, a situation which is relatively rare in nature. The low values, corresponding to the states for which both fluids are displaced spontaneously in similar proportions, represent the intermediate wettability states.

C) Microscopic approach: direct observation of fluid positions

By definition, wettability is the relative affinity of a fluid for a solid surface. For example, the photographs of Figure 1-1.27, p. 53, clearly show that the glass surface is not mercury-wet. Cryomicroscopy techniques (§ 2-2.2.1, p. 334) can be used to make similar observations in porous media, up to the scale of the micrometre. We can therefore consider mapping wettability at pore scale [Robin, 2001]. At least theoretically therefore, we can calculate the percentage of the surface of the porous space wettable to a particular liquid and also observe the relations between wettability and the type of this surface according to mineralogical and/or roughness criteria.

Note, however, that the observation concerns the relative location of the fluids and that wettability is deduced from this location. This observation has major consequences if we consider that the wettability state can be modified by transferring molecules from a liquid to the solid surface without there necessarily being any fluid movement and therefore any variation in the location of the various phases, during the ageing of the "restored" samples (see below) for example.

This direct observation method must therefore be restricted to the cases where fluids in chemical equilibria have been displaced shortly before the observation. For example, assuming that pressurised cores could be sampled in a pristine reservoir, the method would be poorly adapted to drawing conclusions on wettability. In contrast, still taking the same assumption of core protection, the method would be quite suitable for investigating a reservoir which had already produced oil, i.e. where the remobilisation of fluids led to an "update" of the relative phase positions.

D) Remark on drag hysteresis

We mentioned drag hysteresis when describing the capillary pressure curves in case of strong wettability: the phenomenon of "drag hysteresis" which characterises the shift observed between the second drainage curve and the imbibition curve (not to be confused with the phenomenon of "trap hysteresis" between the first and second drainage). Although not specific to intermediate wettability, it is likely that this phenomenon may become even more important in this case. The simplest explanation is that the wettability angle varies with the direction of displacement of the non-wetting fluid. This rather general phenomenon seems to be significantly amplified when the solid surface is very rough [Morrow 1970], which is frequently the case in geological porous media.

1-2.2.2 Wettability of reservoir rocks

Wettability plays a very important role in the oil recovery processes in reservoirs since it has a direct impact on the proportion of oil recovered and on the kinetics of this recovery. We will discuss this point in the "End Points" paragraph (§ 1-2.3.3, p. 186). The obvious practical importance of wettability has led to numerous research studies. Nonetheless, it is still difficult to obtain an accurate picture of wettability in reservoirs since numerous difficulties are encountered when trying to upscale results from laboratory to reservoir. As we have already pointed out, this is due to the fact that wettability is a state which may turn out to be highly unsteady.

Figure 1-2.32 shows examples of reservoir rock wettability evolution, over the course of geological time and depending on the laboratory sampling and processing conditions.

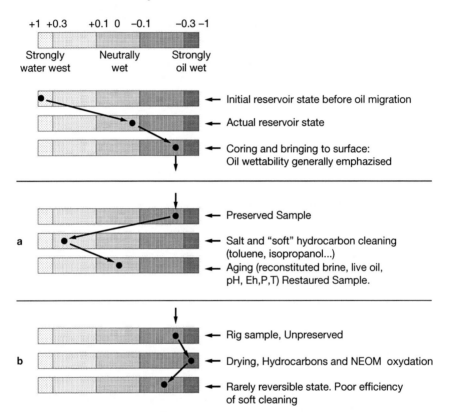

Figure 1-2.32 Diagrammatic example of reservoir rock wettability evolution in case of a preserved sample (a) and an unpreserved sample (b)

A) Difficulty of estimating wettability *in situ*, alteration of core wettability

One major difficulty is that no method is currently available for direct estimation of wettability *in situ*. This data cannot be estimated by any type of log analysis or well test. We might expect that, in the long term, measurements deduced for example from nuclear magnetic resonance (§ 1-3.3.2, p. 258) may provide a means of approximating this notion.

The only "experimental" method to verify the assumption made on wettability lies in the coherence between the production history and the modelling calculation based on wettability data (history matching). This proof which comes *a posteriori* is debatable since numerous other causes may be put forward.

We must therefore rely on the results obtained on cores. Unfortunately, numerous factors may disturb the wettability state of a rock sample between the reservoir and the laboratory. During the coring and storage operations, the rock is subjected to three main types of process which may have a serious impact on wettability.

- *Flushing by drilling mud.* This is one of the most obvious cases of wettability alteration. The muds contain additives (e.g. soda) which may have a drastic effect. Techniques are available, however, to control the degree of invasion, especially by using
  markers. By adding small quantities of deuterium, tritium or special salt to the mud,
  the depth of penetration of the filtrate inside the core can be measured on small core
  samples. It is therefore possible to demonstrate that if coring is carried out with a suitable mud pressure, the core centre is often free from any contamination.
- *Deposition of organic molecules precipitated during depressurisation.* When the core
  barrel is brought up to the surface, the pressure drops suddenly and the dissolved gas
  returns to gaseous phase, suddenly "flushing out" the liquids remaining in the core.
  More importantly, it may also induce the precipitation of high molecular weight molecules (e.g. asphaltene) which have a major impact on wettability. This phenomenon
  can be limited by carefully controlling the speed at which the core barrel is brought to
  the surface. The ideal solution would be to have a pressurised core barrel, in order to
  bring up the core without any pressure variations, or at least without lowering the
  pressure below the bubble point. Since this aspect represents a major financial stake
  (more for preservation of saturations than wettabilities) technical solutions have been
  proposed while still, apparently, remaining outside the scope of routine applications.
- *Drying and alteration during storage and transport.* If left untreated, the core characteristics are soon modified. It is easy to remedy this situation, however. As soon as
  they are removed from the core barrel, "native" samples are placed in sealed contain-
  ers, refrigerated or sometimes even frozen. They therefore reach the laboratory under
  optimum conditions.

B) "Restoring" wettability in the laboratory

The modifications suffered by the cores are such that in order to relate the laboratory results to the conditions prevailing in the reservoir, the state of the samples considered must be clearly specified for the various petrophysical measurements. Three types of sample will be used:

- **Preserved samples.** These samples are carefully protected on leaving the core barrel.
  Their hydrocarbon content has been significantly modified (degassing) but hopefully,
  the connate water has been preserved. The qualifier "connate" designates the water
  always present in hydrocarbon reservoir rocks, at least in very small proportions
  (§ 1-1.2.2.C, p. 58, § 1-2.3.3, p. 186) (this term may be becoming obsolete). However,
  it is difficult to machine the core fragments without disturbing them too much and
  very difficult to remove the residual hydrocarbons without touching the connate
  water, making the experiments much more complicated. In practice therefore, samples in preserved state are rarely used, except for measuring water saturation in the
  reservoir (S_{wi}).

– **Cleaned samples.** There are two quite different types of cleaning:

+ Soft cleaning. The hydrocarbons and brine are removed from the samples received in the laboratory in true "preserved" state by "soft" cleaning, using mixtures such as alcohol/toluene which do not attack the very long chain organic molecules. We might expect that the samples cleaned in this way retain on the surface of the porous medium a large proportion of the polar organic molecules brought by the oil and deposited over geological times. Studies on the "wettability/geology" relation can be conducted on this type of sample, assuming that some of their "acquired" wettability has been preserved (see below).

+ Hard cleaning. This type of cleaning seems to be more rarely used. Unlike the previous case, the samples are washed very vigorously using special solvent mixtures. The aim is to remove all the organic molecules to create a perfect water-wet state (assumed initial state).

– **Restored samples.** The sample is then returned to its "initial" water saturation state with oil from the reservoir (preferably "live oil" recombined with its gas content). The saturated sample then undergoes temperature ageing for several weeks. An experimental approach based on wettability measurement (see above) at different times indicates that a stable wettability state is reached after a "certain time" (a few weeks?). The wettability assumed to exist in the reservoir is therefore restored. This method is extremely practical for laboratory experiments since it can be used to obtain samples close to the most "likely" wettability state. However, under no circumstances does it provide an indication of wettability *in situ*. This reserve must be borne in mind to avoid any misunderstanding or confusion.

C) Geological causes of wettability variations

The main difficulty when studying wettability is due to the fact that it is directly related to the surface condition of the wall of the porous space which is itself the result of interaction between the minerals forming the rock and some molecules brought by the hydrocarbons (oil fields) or produced by the microorganisms (soil, rock outcrops). When considering the geological causes of wettability variations, we must therefore consider the distinction between "innate" and "acquired".

a) Characteristics specific to the porous medium

Amongst the characteristics specific to the porous space, we will mention two which may have a significant impact on wettability:

– The mineralogical nature of the constituents may govern the wettability directly by speeding (or slowing down) the adsorption of polar molecules. Calcite, for example, has been considered as being particularly favourable to these phenomena, which would explain the oil wettability trend of limestone rocks. Similarly, some clay minerals (kaolinite) are considered to be more oil-wet. In actual fact, this point is not as clear as it would appear *a priori*, since in the absence of precise physical measurements, it may be difficult to accurately determine the effect of the mineralogy.

– The roughness may also play an important role. The effect of roughness was mentioned with respect to the capillary pressure hysteresis. It seems likely that it has a direct impact on wettability, for example through its control on the "irreducible" water layer. The roughness of the pore walls may vary considerably. Two contrasting examples of roughness, observed using epoxy pore casts (§ 2-2.1.3, p. 331) are illustrated on Figure 1-2.33. The roughness parameter is difficult to quantify absolutely (numerous microscopic analysis would have to be made on images of the type shown on Figure 1-2.33). However, since the roughness is directly related to the nature of the

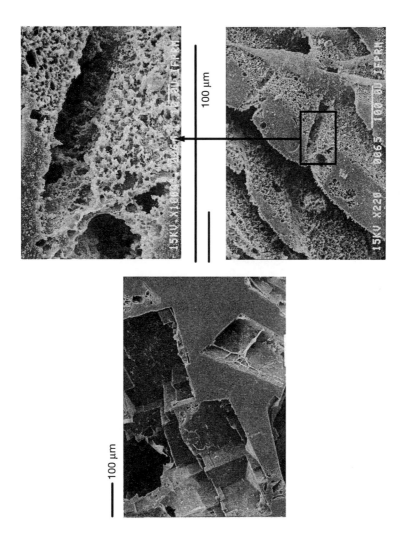

Figure 1-2.33 Examples of roughness on pore walls observed
using the epoxy pore cast method (pore cast, see § 2-2.1.3)

a) Inner wall of foraminifers, high roughness.
b) Dolomite crystal faces, zero roughness.

constituents, it is likely that qualitative conclusions based on a petrological analysis of the roughness could be obtained relatively quickly. In limestone rocks, roughness is probably directly related to the nature of the allochems and especially the bioclasts. In sandstones, it is likely that type of clay covering the pore walls plays the main role. This could possibly represent a better way of accounting for wettability variations depending on the geological context.

b) Geological history: condition and duration of the oil-rock contact

This effect of geological history has been clearly demonstrated by Hamon [2000] who shows the wettability variations of reservoir rocks with respect to their structural positions. The important parameter is the distance to the water level. Since secondary migration (§ 1-1.2.6A, p. 102) takes place mainly under the floor of the caprock, a reservoir level fills from top to bottom and the higher the zone above the water level, the longer it has remained in contact with hydrocarbons. Figure 1-2.34 drawn from data of G. Hamon shows that the oil-wettability (wettability measured according to a method derived from Amott) of the rock increases with the distance from the water level.

We will make a methodological comment which although apparently obvious has serious consequences. Making this type of observation involves the use of samples which have suffered the least perturbation possible (i.e. soft cleaning) and whose "acquired" characteristics have not been erased by extremely thorough cleaning of the porous media walls, after which there should be nothing left to see. What is the situation, however, concerning the true representativeness of core analysis measurements routinely taken on "hard cleaned" samples? This is a typical example of the contradictions which are not always easy to resolve in these wettability studies.

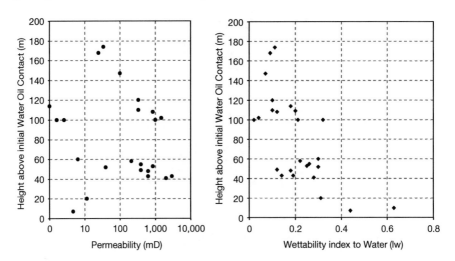

Figure 1-2.34 Evolution of oil wettability against the distance from the oil-water contact.
Sandstone reservoir. Wettability index measured using a method derived from Amott,
samples subjected to "soft" cleaning then restoring. Experimental values of G. Hamon [2000]

The relation permeability/distance from the water level indicates no special relation between these two parameters.

The oil type plays a central role in wettability modifications since the oil carries polar molecules which disturb wettability by adsorbing onto the solid surfaces. It is therefore important to emphasise that the chemical nature of the oil (often observed through physical values: relative density, viscosity, bubble point, etc.) may significantly vary within a given reservoir, depending on the geographic or structural position. There may be a true "natural distillation" column found in particular in the multilayer reservoirs of recent deposits (Tertiary) which exhibit fast dysmigration. When attempting to put the previous notions into practice, the possible variations in oil type will obviously have to be taken into consideration.

Similarly in the geological zones where tectonic processes were active at the same time as the secondary migration, or after, structural shifts frequently disturb the water levels and therefore the apparent chronology of the rock/hydrocarbon contacts. This is another factor which must be borne in mind during the analysis.

Conclusion on wettability

Wettability plays a major role in the production of oil reservoirs due to its determining effect on the oil residual saturation (the proportion of oil trapped in the reservoir before implementing special tertiary recovery methods) and on the oil mobility (determining the recovery rate). This parameter is implicitly expressed through the capillary pressure and relative permeability (KrPc) results.

The difficulty when studying wettability is that it depends both on the petrological type of the reservoir, the chemistry of the crude oil and of the brine, the *in situ* oil saturation (governed by the water-oil contact distance) and the history of the crude-rock interaction. This complexity explains the large number of publications on this subject.

1-2.3 RELATIVE PERMEABILITY AND END POINTS

To model the flows in reservoirs where two or three non-miscible fluids cohabit, the mobilities of these fluids must be known. For a fluid, this mobility, which results from the interaction between three main forces: capillary, viscous and gravitational, varies with the relative content of this fluid in the porous space (saturation). The relations between mobility and saturation form the basis of the relative permeability concept.

1-2.3.1 Overview

A) Definition of relative permeability

In the chapter on *"intrinsic permeability"* (§ 1-2.1), we emphasised the need for one particular condition, necessary to take a correct measurement: total saturation of the medium by a single fluid phase. This gives the measurement its intrinsic character by "cancelling" the effect of the fluid through normalisation by its viscosity. In nature, however, and especially in hydrocarbon reservoirs, porous media are frequently saturated by two or three non-miscible phases. Consequently, the notion of *"intrinsic permeability"* can no longer be used since each fluid disturbs the flow of the other, sometimes considerably. We then define a *"relative permeability"* which depends both on the geometry of the porous space **and** the saturation state.

As with capillary pressures, we will base our description on a case of perfect water wettability.

If we force the joint circulation of a volume flow rate of water and (non-wetting) oil in a porous medium we obtain for each flow, after a certain stabilisation time, a steady state characterised by the saturations, the flows and the pressure gradients for each fluid.

This ideal experiment is schematised on Figure 1-2.35.

If, by convention and for simplification purposes, we apply Darcy's law to each fluid as defined in § 1-2.1.1, p. 126:

$Q = (K/\eta) \, S(\Delta P/\Delta l)$ we obtain:

$Q_W = (Kw/\eta_W) \, S(\Delta P_W/\Delta l)$

and $Q_O = (KO/\eta_O) \, S(\Delta P_O/\Delta l)$

where: Q_W and Q_O: respective flow rates of water and oil,

η_W and η_O: respective viscosities of water and oil,

ΔP_W and ΔP_O: pressure difference in water and oil,

S: sample cross-section, Δl: distance between the pressure tapping points in the fluids.

At each water saturation state (the complement being oil saturation), we could measure the (apparent or effective) water and oil permeabilities $Kw_{(Sw)}$ and $KO_{(Sw)}$.

By dividing these values by a "reference" permeability of the medium considered (K), we obtain the *"relative permeabilities"* for a fluid and, once again, a given saturation state:

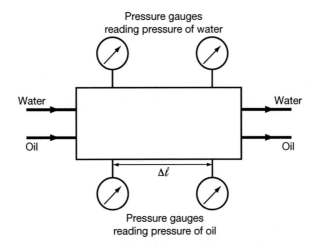

Figure 1-2.35 Diagram of an ideal experiment to define relative permeabilities, according to Marle [1981]

$KRW_{(Sw)}$ *"water relative permeability"*, $KRO_{(Sw)}$ *"oil relative permeability"*, $KRG_{(Sw)}$ *"gas relative permeability"*, at saturation state S_w.

In practice, the reference used by most petrophysicists is not the intrinsic permeability but the non-wetting fluid (hydrocarbon, air) permeability at the so-called "initial" (S_{wi}) water saturation point (see below) $KO_{(Swi)}$. Using the previous conventions, we therefore have:

$$KRO_{(Sw)} = KO_{(Sw)}/KO_{(Swi)} \text{ and } KRW_{(Sw)} = Kw_{(Sw)}/KO_{(Swi)}$$

During discussions between "production geologists", "petrophysicists" and "reservoir engineers" responsible for digital modelling of the reservoir dynamics, care must be taken regarding this convention since it could be a source of error. The error is minimised by the fact that frequently, the permeability of oil in the presence of water at "initial" saturation (at maximum capillary pressure) is equal or nearly equal to the oil single-phase permeability.

We will therefore also use $KO_{(Swi)}$ during a displacement of oil by gas and $KG_{(Swi)}$ when displacing gas with water. When displacing water with gas, K_w, the water single-phase permeability will be used.

The notion of relative permeability is a pragmatic concept quantifying multi-phase flows at macroscopic scale. This generalisation of Darcy's law has no theoretical basis but it is overall confirmed by the laboratory measurements... and to date, there are no other expressions sufficiently operational to be included in the reservoir evaluation digital models.

B) Initial definition of "End-Points"

Note firstly that, as defined above, the experiment only makes sense between two saturation states determining the saturation range for which the two fluids are mobile. These saturation limits have already been mentioned in the paragraph on capillary pressure (§ 1-1.2.2, p. 56): they are the so-called "initial" saturation at maximum capillary pressure for water and "residual" saturation for oil. We will return to these points in the next paragraph (§ 1-2.3.3).

In initial water (the wetting fluid) saturation state, the water is found in very thin films or inside menisci of very small capillary radii. It is relatively immobile. The water effective permeability (and therefore also $KRW_{(Swi)}$) tends to zero. Note that due to its position in the smallest anfractuosities of the porous space, the water does not obstruct the flow of oil, the oil effective permeability ($KO_{(Swi)}$) will therefore be close to the single-phase permeability (Fig. 1-2.36).

We have seen (§ 1-1.2.2) that in an imbibition process (increase of water saturation) in the cases of strong water wettability, there is a so-called "residual" oil saturation for which the entire oil phase consists of discontinuous globules or "ganglions" due to capillary trapping of the non-wetting phase. In this saturation state the oil effective permeability ($KO_{(Sor)}$) is virtually zero. Concerning the second fluid (water), the situation is opposite to that observed with the previous one, but it is not symmetric: the oil globules occupy the central areas of the porous network (Figure 1-2.36; see also photographs on Figures 1-1.32, p. 62 and 2-2.13, p. 340), considerably obstructing the water flow. In this situation, the water *"relative permeability"* ($KRW_{(Sor)}$) is therefore well below 1. Consequently, the *"relative*

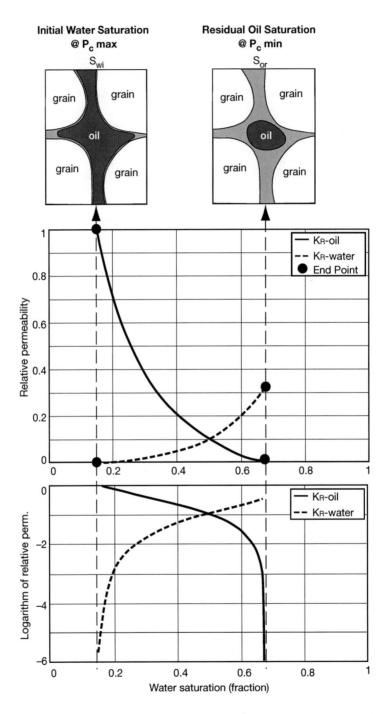

Figure 1-2.36 Diagrammatic *"relative permeability"* curves in case of strong water permeability, using arithmetic and semilogarithmic scales

permeabilities", considered in the most ordinary case, will lie between four points (corresponding to the two saturation states mentioned above). These limits are generally known as the "End-Points".

a) Remark on "relative permeabilities" outside the end-points

Note that it is always possible to devise experiments for which the saturations lie outside the "natural" limits defined above. For example, by drying, we can obtain a water saturation less than the so-called "initial" saturation, then replace the drying gas by oil. It is even easier to obtain non-wetting fluid (oil or gas) saturations much lower than the "residual" saturation. We can therefore plot "*relative permeability*" curves over the entire water saturation interval (between 0 and 1). It is therefore not surprising *a priori* to see such curves... except that, sometimes, these "*relative permeability*" curves, continuous over the space 0, 1 of S_w, are the result of illegitimate extrapolations between the End-Points and 0 or 1. Obviously, such practices should be avoided.

Lastly, note that this description concerns the case of strong water wettability. Remember that most reservoirs exhibit intermediate (or fractional) wettability, or even a stronger affinity for oil. The definition of these "end-points" (zero permeability of the second fluid) is less precise. It becomes more a practical and conventional notion. For example, when displacing oil by water, in conventional production, it is accepted to redefine the "end-point" S_{or} as being the point where the oil relative permeability is equal to a given value, e.g. 10^{-4} (see below 1-2.3.3). The representation using a semilogarithmic scale (Figure 1-2.36) therefore becomes much more meaningful.

C) Shape of relative permeability curves, Corey exponent

The simplest description of "*relative permeabilities*" using a linear scale corresponds to the so-called "X" curves connecting the "end-points" in pairs. For simplified models, straight lines between the "end-points" are sometimes used. In practice, (Fig. 1-2.36, top part), the relative permeability curves have an upward concave shape (permeabilities in arithmetic values). The concavity of the curves is a very important point in the description of "*relative permeabilities*". It is also an adjustment parameter used when modelling two-phase flows at reservoir scale. The effect is complicated by the highly difficult scaling process. This is all the more true since the effect of the size and geometry of the heterogeneities (up-scaling as such) and the effect of digital dispersion inherent to computer simulation must both be evaluated.

This highly complex domain is not discussed in this book and we can only recommend that readers should refer to the specialists in modelling multi-phase flows in porous media.

An easy way to represent the relative permeability curves is to use a power law with exponents nw and no, the so-called Corey exponents for water and oil respectively [Corey, 1954]. The variable is the water saturation (estimated between the end-point S_{wi} or S_{or}) and divided by the saturation value between these end-points:

$$KRW_{(Sw)} = KRW_{(Sor)}[(S_w - S_{wi})/(1 - S_{wi} - S_{or})]^{nw}$$

$$KRO_{(Sw)} = KRO_{(Swi)}[(1 - S_w - S_{or})/(1 - S_{wi} - S_{or})]^{no}$$

$KRW_{(Sor)}$ and $KRO_{(Swi)}$ are the end-point relative permeabilities.

Hysteresis

We have seen on several occasions that, in the two-phase displacement phenomena, the history of the saturation process (hysteresis) could play a highly important role and that various fluid location shapes could correspond to the same saturation value. The same observations should therefore be made regarding relative permeabilities.

The diagrams of Figure 1-2.36 correspond to the evolution history most frequently found in petroleum practice: imbibition (increase in water saturation). Data in the literature show that for drainage curves (reduction in S_w), KRO tends to be above and KRW slightly under. The difference would be greater for the non-wetting fluid branch than for the wetting fluid branch. However, these data are much less frequently encountered that might be expected, the main concern being that these variations often lie within the margin of experimental error, which could be large, depending on the methods implemented (see below). The detailed shape of relative permeability curves therefore remains a subject of debate for the specialists.

Initially, when the processes used involve the displacement of oil by water (or gas by water), the "imbibition" branches must be considered, whereas in case of a displacement of oil by gas, the "drainage" branches are concerned.

1-2.3.2 Tentative simplification: search for the most important points on the relative permeability curve

The study of relative permeabilities is disconcerting for the non-specialist who mainly observes the contradictory trends between highly advanced isolated experiments and a real difficulty in finding general conclusions (for example in the previous paragraph on the true situation and possible shape of a hysteresis).

For non-specialists, we suggest trying to identify the points which are most important to understand the behaviour of reservoirs and seeing whether this could provide a way of reducing certain difficulties for a very general analysis.

It is not always necessary to know the "*relative permeabilities*" over the entire saturation space and the importance of these values may vary considerably depending on the production mode and during the life of the field (see Dake for example [1978]).

We will take the very simple (and therefore probably highly debatable) example of a homogeneous layer produced by waterflooding (active water level). Figure 1-2.37 shows the saturation profiles as a function of the distance to the producing well (assumed located in D_P). Profile 1 corresponds to the start of production. Profile 2 corresponds to a late phase when a large proportion of the reservoir is assumed to be in residual oil saturation state.

The basic production calculation data is the head loss between the end D_E and the production well D_P. The well flow will be related to this head loss by the average calculated permeability on the distance $D_E - D_P$. Throughout the first life phase of the deposit (Fig. 1-2.37, profile 1) and assuming that the transition zone T_Z between the flooded part and the part with oil is small, the greatest length will correspond to the initial water saturation state (S_{wi}). The oil permeability in this saturation state ($KRO_{(S_{wi})}$) will therefore

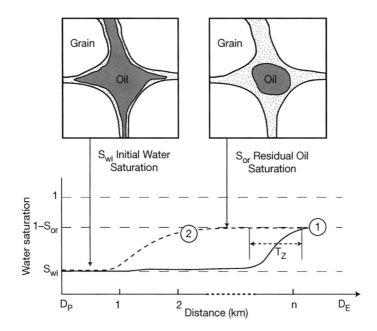

Figure 1-2.37 Diagrammatic saturation profiles for two production phases of a layer (theoretical case of strong wettability to the displacing fluid). Producing well assumed located in D_P

be the most important. We can see that in this case, the relative permeability has relatively little impact on the production model.

This early production phase was the most critical when calculating the financial amortisations. This may explain why the production engineers put up with the high uncertainties observed in the measurement of relative permeabilities. This situation is currently changing in the operating companies.

In a later production phase of the field, when a large part of the reservoir has been waterflooded (Fig.1-2.37, profile 2), the oil permeability in the flooded zone is nearly zero and the important parameter is the water *"relative permeability"* in residual oil saturation state ($KRW_{(Sor)}$). We have seen that this value, always well below 1, could in some cases be very low. This is important data when modelling mature fields. The shape of the curves when approaching the residual saturation is very useful information.

Summing up, the conclusions drawn from our simplified example show that, if the transition zone is small, the impact of the *"relative permeability"* values for intermediate saturations is quite low and that we must focus on the "End Points". We consider that focusing on the "end points" is a pragmatic way of dealing with the relative permeabilities in the very general context we set ourselves. However, we considered the simplified case of perfect wettability (water being the wetting fluid in a system where oil is displaced by water).

This simplification will also allow us to study the effect of wettability on relative permeabilities.

1-2.3.3 More information about the "End Points", wettability effects

On several occasions (§ 1-1.2.2, p. 51, this section, etc.), we have mentioned the "End Points" to designate initial water saturation or residual hydrocarbon saturation. Generally, we have used the quotation marks to emphasise the inaccuracy of these empirically defined terms. We will review the definitions of these saturation states.

Note firstly that the symmetry between initial saturation and residual saturation is only apparent since the capillary phenomena involved are different (drainage with high capillary pressure for one, imbibition for the other) and that concerning the oilfield applications, the equilibrium times are quite different: geological times for S_{wi}, exploitation times (several decades) for S_{or}. Since in experimental practice these saturations are defined at laboratory time, we can appreciate all the underlying problems.

A) Initial water saturation

a) Initial saturation vs. irreducible saturation: a difficult terminology problem

It has been common practice for many years to define the saturation state reached at end of drainage under high capillary pressure as the "irreducible water saturation". This definition assumes the existence of a saturation value for which the water effective permeability would become zero. This value would be a petrophysical characteristic of the sample. In reality, there does exist a water saturation range for which the saturation variation gradient related to the capillary pressure is very low and it is tempting to define this saturation range as a parameter of the reservoir rock. Although the corresponding water saturation values are generally low, there are major exceptions such as the shaly rocks. The equilibria of the water phase at very low mobilities are only reached over "geological" times. At laboratory scale, only saturations at "non-equilibrium" states can be observed. When considering geological times, the notion of "irreducible" saturation becomes very vague.

This incorrect use of the notion of "irreducible" saturation has led to numerous misunderstandings and even errors when it has been applied without sufficient forethought to reservoir calculations. It must be replaced by the notion of initial saturation (S_{wi}) which is much more objective since we are talking about the water saturation state for a given capillary pressure, in other words the maximum capillary pressure reached during the process considered. This is the approach we have decided to adopt in this book.

b) Initial saturation in reservoirs

The main point is the initial saturation in the reservoirs. The problem can be summarised as follows: can the initial saturation in a reservoir be different from that corresponding to the capillary equilibrium defined by the geometry of the porous network measurable using mercury porosimetry (§ 1-1.2.4D, p. 79)? Remember that in the laboratory, at high capillary pressures a significant difference is often observed between the case of water (water/oil restored states for example) and that of mercury vapour (porosimetry).

A priori, one might think that the equilibria obtained over geological times would correspond to the capillary equilibrium. Only "non-hydraulic" phenomena should lead to *in situ* saturations different from those estimated by the capillary equilibrium curve.

Before considering this problem, note firstly that accurately estimating the water saturation in a reservoir is more difficult than might be expected since the resistivity log analyses are not sufficiently accurate (§ 1-3.1.2, p. 209) and that great care must be taken when analysing the measurements on core samples (estimation of invasion § 1-2.2.2, p. 174). Consequently, very few reliable data are available. They must combine quite specific core conditions (mud quality, "low invasion" type core barrel, etc.), a suitable drilling program (mud tracing) and laboratory measurements confirming the measurement of the sample water content as such by analysing the salts or fluids extracted. This is required to correct any artefacts due to invasion by the filtrate. Providing an answer to the questions posed is therefore no easy task.

In oil/water systems, it would appear that under high capillary pressure (great height above FWL, § 1-1.2.6B, p. 104), water saturations slightly greater than those which might have been predicted by the geometric analysis alone (mercury injection) may tend to exist. Various mechanisms have been put forward to explain this phenomenon. Morrow [1971], for example, suggests involving the water diffusion related to the temperature and salinity gradient. In contrast, in gas/water systems the values interpreted from the porosimetry test seem quite relevant.

In practice, we may probably consider that the effect of wettability on the initial saturation is sufficiently low. We emphasised the fact that most reservoirs had to be considered as water wet, as regards the accumulation of oil and therefore the S_{wi}. It is reasonable to think that a wettability effect could only be observed in reservoirs which were initially not water wet, therefore corresponding to very special geological conditions (formation of oil *in situ*, dysmigration followed by a further accumulation of oil, etc.). There are very few such cases and it is likely that there is very little reliable statistical data concerning the consequences of this wettability state on the initial water saturation.

As regards most oil reservoirs, whose wettability evolves towards an often intermediate state and even sometimes towards greater oil affinity, this transformation occurs precisely in the hydrocarbon zone defined by the S_{wi} zone. Experience shows that this could lead to a redistribution of fluids at pore scale, but to date nothing indicates a modification of the saturation at macroscopic scale.

c) Importance in reservoir characterisation of determining the initial saturation

Determining the water initial saturation is extremely important in reservoir characterisation since it is a major parameter in the determination of hydrocarbon accumulations. Remember that the notion of accumulation, which corresponds to the total quantity of oil *in situ*, is quite different from that of exploitable reserves which necessarily depend on a technico-economic production model of the field.

The accumulation is equal to the integration over the entire reservoir of the product of the porous volume by the value $1-S_{wi}$. A good approximation of the porous volume is generally obtained by the geological description (area, thickness) and the porosity log analyses; we can therefore appreciate that the parameter S_{wi} plays an important role. This explains the care taken by the production geologists to determine this parameter as accurately as possible. Note, however, that the consequences of the uncertainty on Swi must be estimated numerically on 1-S_{wi}, which puts the problem into perspective somewhat, since S_{wi} is generally less than 0.5. The importance of the uncertainty on S_{wi} when determining reserves compared with the uncertainties on the other modelling parameters must therefore not be overestimated.

B) Residual oil saturation

a) definition of residual saturation

In theory, we have seen that the residual saturation is the saturation state for which the oil effective permeability would be zero ($KO_{(Sor)} = 0$). In practice, it is extremely difficult in an experiment to guarantee that the value of a variable is zero, especially when considering a logarithmic scale (Fig. 1-2.36). The long exploitation time scale (decade) must also be taken into account. To define S_{or} therefore, it is impossible to speak of zero effective permeability. We must set a lower limiting value for the relative permeability. In practice, 10^{-4} is often chosen as limiting value of $KRO_{(Sor)}$ when defining S_{or}. This limiting value must nevertheless be modulated by the reservoir absolute permeability value. The permeability of some reservoirs may exceed 10 darcy. In this case, a relative permeability of 10^{-4} leads to oil effective permeability of the order of the millidarcy, which is far from being negligible!

When studying capillary pressures in case of strong wettability, we described the phenomenon of capillary trapping (§ 1-1.2.2.D, p. 59) and the parameters of the porous space governing this trapping. We will now make an observation which we did not develop in this paragraph. Pushing the systemisation of the mechanism proposed to the extreme, in case of perfect wettability, we should tend towards total trapping of the non-wetting fluid (oil, gas), i.e. obtain a value of S_{or} equal to 1-S_{wi}. As soon as we consider a sufficiently large sample, there is a high probability that on the path of the oil towards the exterior, there exists an access radius R_{PT} smaller than the radius of curvature of the oil phase R ($R_{PT} < R$, using the notation shown on Figure 1-1.33).

In practice, the oil is never completely trapped (fortunately for the oil production industry!). This observation does not contradict the process described in § 1-1.2.2.D. It is just that non-capillary phenomena (e.g. dynamics in relation with phase volume variations) are involved, and also that wettability is never perfect and that in the event of imperfect wettability, the access radii which are small with respect to the pore radius can no longer play their role of capillary "scissors" so well. We can therefore appreciate the full effect of wettability on the residual saturation, as opposed to what was mentioned for S_{wi}.

However, this consideration of the correlation between the increased size of the sample studied and the increased probability of there being a small access acting as capillary scissors must remind us that we cannot ignore the importance of scale effect and heterogeneity on S_{or} (once again, unlike S_{wi}, which is relatively insensitive to these parameters).

b) Effect of wettability on residual saturation

We have already mentioned the importance of wettability on residual saturation. Once again, this subject has been discussed in numerous specialised publications, with sometimes conflicting results. We will restrict ourselves to a brief summary based on the description of the "U" curve of Jadhunandan and Morrow [1991]. The curve on Figure 1-2.38 corresponds to values measured on high-porosity, high-permeability Berea sandstones (about 0.22; 1 000 mD). The wettability of numerous sandstone samples from the same origin has been modified using various "ageing" protocols by varying the type of oil (crudes, synthetic oil), of brine, the initial water saturation and the temperature. The result of these various preparations was observed by using a variant of Amott's method (§1-2.2.1B, p. 171) and the residual saturation was measured after flooding with a volume equivalent to 20 times the porous volume of the sample.

This "U" curve clearly shows the increasing saturation (increasing trapping) with the trend in strong wettability, whether oil or water wettability. This result agrees with the intuitive interpretation given in the previous paragraph, at least as regards the positive wettabilities. We must beware of intuition in this domain, however! The "U" curve seems to be an interesting qualitative approach, while stressing the qualitative aspect: the values shown on the graph must not be compared directly against values measured on reservoir samples.

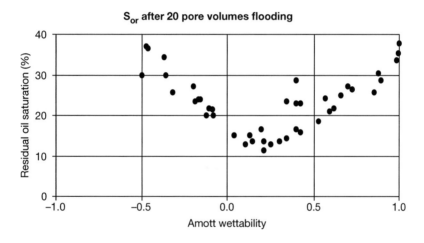

Figure 1-2.38 Relation between wettability and the residual saturation measured after flooding with a volume equivalent to 20 times the pore volume of the sample. Berea sandstone. The values are taken from Jadhunandan and Morrow [1991]

c) Importance of residual saturation in reservoir characterisation

A priori, residual saturation seems to be an extremely important parameter in reservoir characterisation since it determines the proportion of oil which can be extracted by conventional production methods. Note firstly that this parameter is involved in the reserve calculation process but not in the accumulation calculation process.

Pursuing the comparison/opposition with the parameter S_{wi}, we observe that S_{or} is much more difficult to estimate than S_{wi} since there is no simple and reliable log analysis method. The estimation of residual oil in the area near the bore hole, after invasion by the mud filtrate

("S_{xo}" of the log analysts) is not very representative for numerous reasons (dynamic effects related, for example, to the filtrate penetration rate and the wettability). In addition, a borehole is rarely available in a waterflooded area and in this privileged case, log analysis of the saturation is even more difficult than at S_{wi} saturation state. An observation well (non-producing) is required in order to take repetitive measurements using Thermal Decay log analysis.

Although accurate measurement of S_{or} is possible in the laboratory, we cannot be sure, due to problems of porous space heterogeneity and restoration of wettability, that these measurements are representative.

The residual saturation strongly depends on the wettability state and this parameter cannot easily be correlated with petrographic characteristics that can be obtained via a direct geological approach. In view of this observation, together with the fact that it is not required during the initial reservoir definition stage (accumulation calculation) and that it is extremely difficult to obtain reliable values, the petrophysical parameter S_{or} is not currently taken into account in reservoir geology.

C) Effects of wettability on relative permeabilities

Although a matter of concern for the special core analysis (SCAL) professionals, we will only mention this point briefly since it is still very difficult to establish generalities. We will limit ourselves to a qualitative logic. Starting from the "strong water wettability" situation schematised on Figure 1-2.36, the oil wettability increase will result in a decreasing tendency of the oil fraction to accumulate in the centre of the pores. For a given saturation, the water flow will be less obstructed by the "more wetting" oil accumulations: the water relative permeability will increase with the oil wettability.

This qualitative conclusion corroborates the traditional experimental results [Owens and Archer, 1971] and the generally accepted observation whereby the water and oil relative permeability curves intersect ($KRO_{(Sw1)} = KRW_{(Sw1)}$) at water saturation values S_{w1} generally greater than 50% in case of water wettability and less than 50% in case of oil wettability. These points are schematised on Figure 1-2.39.

In these cases of intermediate wettability, the usual situation encountered in oilfields, it is obvious that summarising the two-phase dynamics at the "end-points" is no longer sufficient, even to provide a simplified description.

Oil production becomes far from negligible after the "water breakthrough" at the production point. Referring to Figure 1-2.37, unlike the theoretical state of perfect wettability, the oil saturation continues to evolve behind the waterflooding front. There is a difference between the oil saturation at breakthrough So_{BT} and the final oil saturation So_{min} (equal to S_{or}?). We observe an additional problem here: this effect is similar if the oil is very viscous (§ 1-2.3.4B, below), irrespective of the wettability.

In numerous reservoir exploitation configurations, the shape of the relative permeability curves near the "end points" at S_{or} is as important when evaluating recovery as the values of these points themselves.

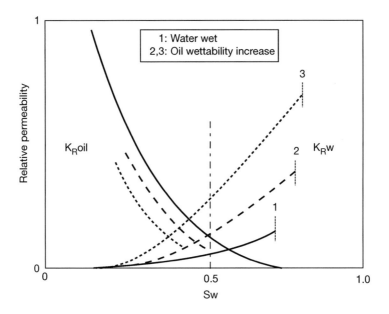

Figure 1-2.39 Highly diagrammatic illustration of the effect of wettability on relative permeabilities. There is no full consensus regarding this representation

1-2.3.4 Relative permeability measurement principle

A) Measurement principles

Measurement of relative permeability is an important experimental field, it is the hub of special core analysis (SCAL). Returning to the definition of relative permeability, the measurement principle is quite simple: allow two fluids to flow simultaneously in the sample at a known flow rate and measure the pressure gradient in each fluid in order to calculate the effective permeabilities. The saturations can be calculated from a matter balance. In practice, it is more difficult!

There are two types of difficulties:

- Experimental difficulties related to setting up the flows, taking the pressure measurements whilst respecting the capillary phenomena, and measuring the saturations.
- Theoretical difficulties related to inversion of flow/pressure/saturation data in terms of relative permeability. To stay in line with the spirit of this book, we will not address this particular aspect, numerous publications being available in this field [e.g. Christiansen, 2001].

We will restrict ourselves to a qualitative description. There are two main experimental directions:

- limit the experimental difficulties (equipment, duration) by only considering unsteady states, in which case the calculations based on flow modelling will be more complicated;

– set up permanent flows in the sample: the experiments will be longer but the interpretation easier (and the quality of the results better?).

Modern techniques for direct measurement of saturations during the experiment by gammagraphy or X-ray absorption considerably simplify the experimental procedure and allow heterogeneity to be introduced in the experiment calculation simulation.

a) Steady state or Penn-State method

This method (Fig. 1-2.40) is conceptually the simplest. Its schematic diagram is similar to the theoretical figure described in the introduction to this chapter (Fig 1-2.35). It is extremely difficult to carry out in practice, however, if the side effects are to be eliminated.

We must ensure in particular:

– that we have eliminated the capillary effects developing at the ends of the sample and of the end pieces (when present) and at the contacts between these pieces and the sample to be measured;
– that the fluids circulating in the sample are homogeneously distributed, and
– that measurement of the volumes injected and produced is perfectly controlled.

Consequently, despite the apparent simplicity of the experiment, it is rarely carried out by the service laboratories or the operating companies, especially since acquisition is very long even if all the above conditions are satisfied. We must wait until the pressures reach equilibrium at each flow stage. This may take an extremely long time in the event of intermediate wettability. The pressure measurements in each fluid are taken using sets of ceramics made perfectly wet to water and oil respectively.

Modern techniques provide a means of accessing saturation profiles by electromagnetic radiation absorption (X-ray, gamma-ray). However, these measurements must be carefully calibrated. *They must always be calibrated against a matter balance at the sample output.*

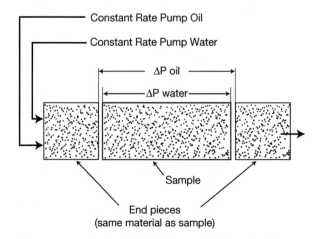

Figure 1-2.40 Schematic diagram of the steady state method

These experiments, whose representativity turns out to be difficult to guarantee, must be reserved for a few special cases. These include in particular the case of samples that are heterogeneous in the direction of flow, at laboratory scale (e.g. vuggy carbonates), especially if oil wettability is suspected and a process involving oil recovery by waterflooding is to be studied.

b) Unsteady state methods

In these methods, the relative permeabilities are deduced from a test where one fluid is displaced by another. If we take the typical example of oil displacement by waterflooding, the parameters recorded are:

- The oil and water productions in the procedure at constant pressure gradient. Depending on the gradient chosen, the fluid flow rate will be greater or smaller and the relative influence of the capillary and viscous effects will vary at the same time.
- The oil production and the pressure gradient in case of waterflooding at constant rate.

In the first case, the experiment can be conducted at a relatively fast rate "crushing" the capillary effects at the ends of the sample and providing a first result of relative permeabilities, using an analytical interpretation (the so-called Welge or "JBN" [Johnson-Bossler-Naumann, 1959] method). The values obtained may contain significant errors, however, since they do not take into account capillary effects, especially as we move away from water wettability.

To study the wettability effects, a complementary experiment is conducted at a rate which is of the order of magnitude of that found *in situ* (a few centimetres or decimetres per day). The capillary forces spread the saturation front and develop significant effects at the ends of the sample. The results of the experiment (fractional pressure and rates) are simulated by digital modelling taking into account capillarity and gravity.

Consequently, the resulting relative permeabilities are not derived from a direct interpretation of the experiment but from the numerical simulation, which simultaneously produces the two relative permeabilities and the capillary pressure involved. In the example described here, of oil displacement by waterflooding, the imbibition branch will therefore be concerned. The shape of this "Dynamic Pc" capillary pressure curve is assumed to be similar to that of the "Static Pc" curve (obtained using the Restored State method: § 1-1.2.4A, p. 73). Although reserved for experienced laboratories - very few, in fact - this experiment gives much more reliable results (over 2/3 of the oil reservoirs exhibit only poor water wettability).

The various phases of an experiment involving oil accumulation and oil displacement by waterflooding are illustrated on Figure 1-2.41.

Experience shows that the data acquired before the water breakthrough are relatively insensitive to numerical adjustment while the two-phase production phase provides a considerable amount of information. This corroborates the considerations expressed earlier on the importance of the shape of the curves between saturation at breakthrough and S_{or}.

Non-destructive implementation of the saturation profile acquisition in virtually real-time improves the interpretation and may possibly remove the need for the first step described above. It remains essential, however, to calibrate these profiles on the matter balance. It is still necessary to physically record the oil and water productions of the sample.

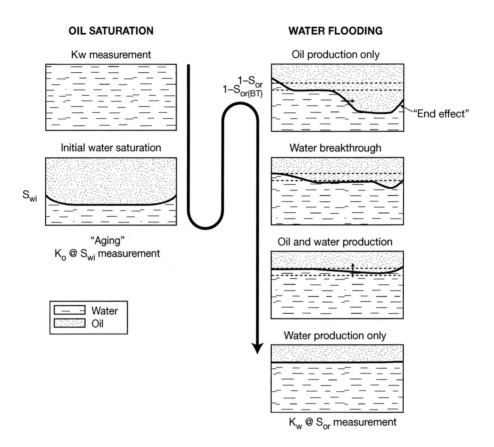

Figure 1-2.41 The various phases of oil displacement by waterflooding.
Saturation profile inside the sample. The two graphs on the left concern the oil accumulation

c) Some additional definitions

The definitions of a few notions frequently used when interpreting relative permeabilities and evaluating reserves are given below.

– the water-cut rate f_w: $f_w = Q_w/(Q_w + Q_o)$
where: Q_w and Q_o: are the respective flow rates of water and oil.

– the mobility and the mobility ratio M:

$$M = (K_{RD}/\eta_D)/(K_{R\_d}/\eta_{\_d})$$

where: K_{RD}: is the relative permeability of the displacing fluid behind the waterflooding front and defined at the average saturation of the displaced zone

η_D: is the viscosity of the displacing fluid;

K_{R_d}: is the relative permeability of the displaced fluid in the zone in front of the waterflooding front

$\eta_{\_d}$: is the viscosity of the displaced fluid.

For example, for oil displacement by water in case of strong water wettability:

$KRD \approx KRW_{(Sor)}$ water relative permeability at S_{or};

$KR\_d \approx KRO_{(Swi)}$ oil relative permeability at S_{wi}.

d) Overview on the various production processes

Emphasis has been placed on oil displacement by water since this is the most common process and it illustrates the difficulties related to wettability. It may be necessary, however, to study the symmetric process: displacement of water by oil, especially in case of oil accumulation in the reservoir (secondary migration), which in this case is a "drainage" situation raising questions related to hysteresis.

Concerning the production as such, we must consider:

- the gas/oil processes in the presence of S_{wi}:
    1. In "drainage": displacement of oil by gas in oil reservoir: expansion of the cap gas at reservoir scale, expansion of the dissolved gas at microscopic scale.
    2. In "imbibition": appearance of condensates below the dew point in gas reservoir.
- The gas/water processes:
    1. In production of gas displaced by an active aquifer ("imbibition" branch).
    2. In underground gas storage management where the two phenomena ("drainage" and "imbibition") are involved in turn.

Some of these examples require a specific experiment (expansion of dissolved gas, depletion of condensate gases) and special interpretation. It remains within the configuration described above, however. Others cannot avoid taking into account the compositional aspect (flow of condensate gases). In this case, the experiments are extremely complex and require large samples (association of full size samples involving the experimental expertise in capillary contacts). Both the laboratory and the interpreter must be highly skilled in thermodynamics (PVT).

Conversely, some aspects of the processes such as gas/water flooding can be addressed using simple experiments (S_{gr}), although care should systematically be taken with processes implementing gas, evaporation and vapour pressure equilibrium phenomena.

B) Experimental conditions and difficulties; problem of reproducibility

The choice of samples, the choice of fluids and the methods are critical, and are a subject of heated debate amongst the special core analysis (SCAL) professionals.

a) Choice of samples

Many Service Companies use 4 cm, 1.5 or 2 inch diameter plugs sampled perpendicular to the core axis, claiming that they were working on so-called "horizontal" samples and avoiding the serious problems of vertical permeability anisotropy (§ 2-1.2.2, p. 283). There are many disadvantages with this choice, however, since the sample length is limited, which implies:

- a small porous volume, therefore low accuracy on the measurement of the fluids produced;

- a small pressure difference, practically impossible to measure at field rate *in situ*;
- large end capillary effects compared with the sample length;
- heterogeneity in the direction of flow.

To partially overcome these difficulties, some laboratories place several samples end to end, to obtain samples of sufficient length (known as composite samples). The capillary contact at the sample junctions is a difficult point to control however. Secondary capillary effects develop at these junctions.

The best compromise is therefore the use of full size core pieces, trimming them down if necessary to remove the possibly invaded outer ring. The advantage with this type of sample is that it offers a suitable porous volume and sufficient length, avoiding the need to join pieces end to end. Despite an apparent contradiction (flow along the core axis), the experiment is representative. Any heterogeneities are perpendicular to the imposed direction of flow. They are therefore easy to model with the digital simulation used, often limited to 1-D.

In all cases, a heterogeneity study (e.g. CT Scan, § 2-2.3, p. 342) must always be carried out before the experiment, to select the most suitable method or reject a sample with too many defects.

Lastly, remember that from a practical point of view, for relative permeability measurements it is important to choose samples in various structural positions (and therefore of variable initial water saturation) and especially to ban the use of samples taken in the water level, whose wettability state would be meaningless.

b) State of samples

Since wettability is an essential parameter, a sample state which is as representative as possible of the *in situ* state must be found. Readers can refer to § 1-2.2.2, p. 173.

- Some laboratories are tempted to use samples in their fresh state or native state. This is not recommended since:
  • the geometric and quantitative distribution of the fluids is modified (possible invasion, decompression and degassing as the core is brought up to the surface),
  • the wettability may have been affected: mud additives, deposit of asphaltenes sometimes difficult to dissolve again in oil;
- The samples can be used immediately after cleaning the core. The aim is to reach a reference "water wet" state. In this case, a refined oil is used (to test the effect of viscosity). The problem concerns the type of solvent to be used. It is generally recommended to use sequences of complementary solvents (polar then apolar) which are not too aggressive to avoid extracting the "insoluble" organic phase included in the rock matrix (iso-propanol/toluene mixtures of increasing toluene content are often used). Other common solvents also include non-polar hexane or heptane, chloroform and methanol, but the latter interacts with the clay phases. Highly aggressive basic solvents are rarely used. Aromatic solvents can be used to extend the action of toluene, but their use is limited due to their toxicity.
- The big laboratories use the samples after cleaning (above) and ageing in crude oil in restored state for 2 to 4 weeks. This is considered to be the best method to represent the *in situ* wettability state. At this stage, it is extremely important that the sample

water saturation should be equal to the *in situ* Swi. Remember that this initial saturation depends on the position of the sample in the reservoir.

Care must be taken regarding the terminology of the sample "states", since the meanings of the various adjectives used by some Service Companies are different from those described above.

c) Fluid type

In experiments where oil is displaced by water it is important to choose the right type of fluid. There are several possibilities concerning oil and brine:

- The following types of oil can be used:
  - A refined oil: these oils are easy to implement, their characteristics (density, viscosity) are known and stable.
  - A dead crude oil: absence of dissolved gas makes the experimental setup much simpler It is assumed that the wettability is not affected.
  - A live crude oil: recombining the oil with its gas makes the experiment and its preparation much more complicated.
  - It is important to preserve the *in situ* viscosity ratio.
- Concerning the brine in experiments where oil is displaced by water, restoring the brine salinity by simply adding salt (NaCl) often proves insufficient. At the very least, the divalent cation/monovalent cation ratio must be restored. The effect is certain on the clay phase, if any, but it is also possible on the wettability. Ageing the brine in the porous medium is also recommended before restoring the initial saturation. But above all, it is essential to deoxygenate the brine. Not only to avoid degassing during the experiment, but also since wettability is sensitive to the redox conditions. Reservoir conditions are generally anaerobic.

The duality between the simple and easy-to-implement "experiments under laboratory conditions", which can be used to obtain numerous data in order to precisely identify the reservoir behaviour, and the more representative "experiments under reservoir conditions" is not as clear-cut as that we have seen for other petrophysical parameters. We will return to this point in § 2-1.1.2, p. 266.

One thing is certain for field application, the simpler experiments (strong water wettability conditions, simulated by rough cleaning of the sample, executed on plugs and interpreted by analytical method: Welge or JBN) are absolutely ruled out. Obviously this does not apply to the methodological research aspects where these experiments are extremely useful. The argument whereby these measurements are taken to acquire numerous data to identify behaviour groups (sort of dynamic Rock-Type) is difficult to justify in this case, since the effect of wettability is often more important than the rock intrinsic characteristics.

However, we still need to decide whether to:

- conduct the experiments under "total" reservoir conditions, including with the combination of fluids, which leads to large experimental setups and tests spanning several months, or even a year;

– or conduct experiments under "light" reservoir conditions focusing on the parameters considered *a priori* to be the most important. In the example of oil displaced by water, emphasis is placed on washing and wettability restoration by ageing. The parameters which appear to be least active (use of storage oil, absence of gas which considerably simplifies the experimental environment) are ignored by simulating the major second order repercussions (oil viscosity being adjusted by adding a light cut). As a result, the experiment durations and costs are "reasonable": a few weeks or months, including preparation of the sample and the fluids, then interpretation by calibration on digital model.

In conclusion, these are difficult experiments, and only very few laboratories have the know-how required to control the operating conditions minimising the artefacts. Their interpretations are equally complex and are the result of calibrations on digital model. There are no fast and inexpensive tests in this field.

Conversely, there is no point in systematically implementing experiments attempting to exhaustively reproduce the full range of reservoir conditions. This leads to prohibitively expensive services, whose execution delays are often incompatible with the constraints of the reservoir engineer, confirming his tendency to consider relative permeabilities as the main adjustment variable of his digital model, free from experimental physical constraints!

Finally, we will give a few "rules of thumb":

At the scale of a sample (or a model mesh):

– the greater the oil viscosity, the greater the water cut rate f_w and the lower the water saturation at breakthrough;
– the greater the reservoir layer dip, the lower the water cut rate f_w and the greater the saturation at breakthrough;
– the greater the total flow rate, the greater the water cut rate f_w and the lower the water saturation at breakthrough.

At reservoir macroscopic scale:

– the lower the mobility ratio, the more stable the displacement (fingers are avoided) and the greater the flooding efficiency;
– the heterogeneities of a reservoir (organised heterogeneities: stratified layers, etc., or random heterogeneities: lenticular structures, diagenesis, etc.) increase the effects of both wettability and the forms of the relative permeabilities on the flooding efficiency.

CHAPTER 1-3

Log and Geophysical Analysis

Knowledge of the **Electrical Properties** (formation factor and saturation exponent) and **Seismic Properties** (elastic wave velocity) of the rocks is essential for log analysis. **Nuclear Magnetic Resonance** applications have developed considerably since the 90's. The broadest principles concerning these topics are described in this third chapter, without going into log analysis as such.

1-3.1 ELECTRICAL PROPERTIES: ESTIMATION OF HYDROCARBON SATURATION

The electrical resistivity of rocks, used in well logging (electric logs) is a petrophysical parameter allowing to estimate the hydrocarbon saturation. We will first consider the case of rocks totally saturated with an electrolyte, in order to define the "Formation Factor", an intrinsic parameter used to characterise resistivity. We will then describe the case of two-phase saturation in electrolyte and insulating fluid. Lastly, we will discuss the special case of clay formations.

In the description, we will often switch from conductivity to resistivity. It would be better to keep to conductivity, for example, which is often done in non-petroleum publications, but use of the term resistivity has become common practice in log analysis!

1-3.1.1 Resistivity of rocks saturated with electrolyte: Formation factor

A) Conductivity of the matrix and conductivity of the solid surface.

a) Conductivity of the matrix

Two cases must be identified, depending on the conductivity of the solid matrix:

- the solid matrix is conductive:
  - When, independently of the presence of electrolytes, the matrix contains minerals which are themselves conductors (oxides, sulphides, etc.). This situation is quite

rare in common rocks and this aspect will not be discussed except to mention that the presence of conductive minerals, detected for example when analysing matrix density data (§ 1-1.1.5, p. 26) must be seriously taken into account during log analysis.

- If clay minerals are present, with specific conductivity but dependent on the equilibrium with the saturating electrolyte. This additional conductivity can in fact be attributed to the surface properties of the clay minerals. This case, which is highly important from a practical point of view, must be dealt with separately (§ 1-3.1.3).

  – the solid matrix is insulating; this is the most general case for non-clay rocks. Resistivity will therefore only depend on the resistivity of the electrolyte saturating the porous space.

b) Surface conductivity

Even for perfectly insulating minerals the solid surface exhibits a certain degree of conductivity, due in particular to the individual properties of the various elements adsorbed on this surface. This conductivity, proportional amongst other things to the specific area (and therefore related to the granulometry) plays a central role in some phenomena such as the streaming potential [Journiaux *et al.* 2000]. Concerning the study of electric logs, this surface conductivity only becomes detectable for brines of resistivity greater than 1 ohm/m, which corresponds to NaCl contents of a few grams per litre. This phenomenon will not be taken into account.

B) The formation factor: definition and measurement

a) Definition of the formation factor

The formation factor (F) is defined as the ratio of the resistivity of the brine saturated rock (R_o) to the resistivity of the saturating brine (R_w).

$$F = R_o/R_w$$

If the matrix is non-conducting, and ignoring the surface conductivity, the formation factor does not depend on the brine resistivity, it is an intrinsic property of the porous medium.

b) Measurement of the formation factor

The principle consists in measuring the resistivity of a sample saturated with a brine of known resistivity. Correctly choosing the measurement current frequency avoids the risk of electrode polarisation due to the presence of electrolyte. A frequency of 1000 Hz is considered as ideal (Fig. 1-3.1a). Some practical precautions concerning the geometry of the measurement device are required. Devices with four electrodes are often used (two for current injection, two for the measurement, Fig. 1-3.1b).

The formation factor is one of the least expensive petrophysical parameters to measure.

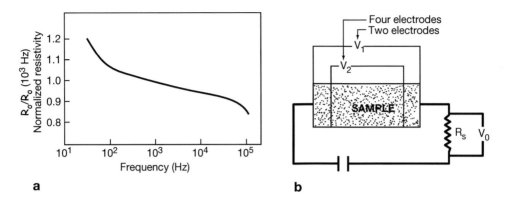

Figure 1-3.1 Measurement of the formation factor
a) Effect of the current frequency on measurement of sample resistivity.
b) Diagram with two and four electrodes. According to Giouse [1987].

C) Geometric characteristics of the porous space affecting the formation factor and simple models

a) The straight capillary or parallel fracture model

Several important conclusions can be drawn from the highly simple case of the straight capillary model, already discussed for permeability (§1-2.1.3A, p. 138). In a porous space formed from straight conducting capillaries in an insulating matrix, the resistivity (R_0) measured parallel to these capillaries is equal to the resistivity of the medium (in this case R_w) divided by the cross-sectional area of the conductors (S). And this conducting surface divided by the total cross-section of the medium is equal to the porosity

$$R_0 = R_w/S = R_w/\phi \text{ and } F = 1/\phi$$

Although obvious, the conclusions must be repeated: the formation factor is independent of the absolute value of the dimension or shape of the cross-section of the conductors. The result for parallel plane fractures is therefore the same as for straight capillaries. Two homothetic models have the same formation factor. The contrast is striking when comparing these results with those obtained for permeability: the formation factor will depend mainly on the porosity and little on the geometry of the porous space.

b) Equations calculated for media formed from spheres

Several physicists (and since a long time ago!) have calculated the conductivity (and therefore the formation factor) of media formed from spheres, in contact or not and immersed in an electrolyte. For example:

Maxwell's equation: $F = (3 - \phi)/(2\phi)$.

Fricke's equation (generalisation of Maxwell's equation to "spheroids":

$F = [(c + 1) - \phi]/c\phi$ where $c = 2$ for the spheres and $c < 2$ for the spheroids

Starwinsky's equation (spheres in contact): $F = (1.3219 - 0.3219\phi)^2/\phi$

These values are shown on Figure 1-3.2. They form a "base line" applicable to unconsolidated sands. These results (m = 1.2 to 1.5) have been corroborated experimentally [Jackson *et al.,* 1978]. When the grain shapes deviates from "spheroids", the exponent m (see below) increases up to 1.9 for plates, for example.

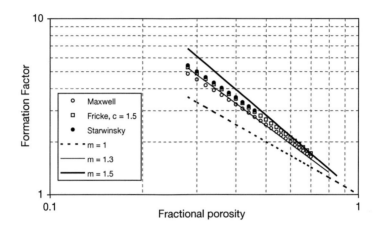

Figure 1-3.2 Relation between Formation Factor and Porosity,
equations calculated for packings of spheres or spheroids in contact or not

c) Notion of tortuosity

The notion of tortuosity (τ) was mentioned earlier in the description of the Carman-Kozeny equation (§ 1-2.1.3, p. 140). The equivalence between the various tortuosities is not obvious and we must therefore speak of electrical tortuosity (τ_e). This notion has led to numerous, more or less successful, developments. It is an imprecise notion since its objective definition can be obtained in terms of F and ϕ using at least two different equations:

$$\tau_e = F \cdot \phi \text{ [Wyllie, 1957] or } \tau_e^2 = F \cdot \phi \text{ [Carman, 1937]}$$

The interpretative definition of tortuosity must be related to the mean developed length of a current line (l_d) joining the two ends of the model, with respect to the true length of the latter (l_m):

$$\tau_e = l_d/l_m$$

The tortuous capillary model (Fig. 1-3.3) of total length l_d is used to connect the two types of definition. To calculate the straight section of the capillary corresponding to the unit section of the model, it would appear normal to use the value $\phi(l_m/l_d)$ which respects the total volume of the capillary, leading to the result $\tau_e^2 = F \cdot \phi$. If we choose the value ϕ for this straight section, the volume expression is no longer valid but we obtain $\tau_e = F \cdot \phi$.

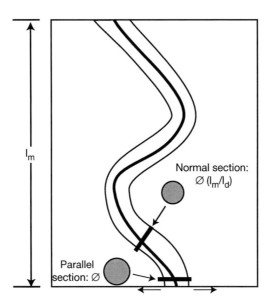

Figure 1-3.3 Diagram of the tortuous capillary model

The resistance of the model of length l_m is in fact equal to that of the capillary, i.e. $l_dR_w/[\phi(l_m/l_d)]$ or $l_dR_w/[\phi]$], depending on the assumption, and the formation factor is calculated by simply dividing by l_mR_w, resistance of equivalent volume of electrolyte.

A dimensionless number with a certain intuitive meaning (tube model), tortuosity may appear *a priori* as a useful parameter to characterise a porous medium. However, if we refer strictly to the objective definition $\tau_e = f(F \cdot \phi)$, it brings nothing more than the exponent "m" described below. Note that tortuosity is used to model numerous transport phenomena: hydraulic (§1-2.1.3, p. 140), electrical (this paragraph) and even acoustic concerning Biot's theory and the "slow" compression wave P2 (not discussed in this book, refer to Bourbié *et al.* [1987] for example). In the true world of natural porous media, however, these various tortuosities do not correspond to the same notion.

D) Experimental relations between Formation Factor and Porosity

Two formulae resulting from experimental observations are used to express the F *vs.* ϕ relations in rocks:

a) Archie's formula: $F = \phi^{-m}$

On a graph in logarithmic scale (Fig. 1-3.4), Archie's formula corresponds to a set of straight lines of gradient $-m$ passing through points F and fractional ϕ equal to 1.

Parameter m, commonly called the cementation exponent, is a characteristic of the sample since a value $m = -\log F/\log \phi$ can always be calculated. For unconsolidated subspherical packings, Archie [1942] measured values of m close to 1.3 showing good

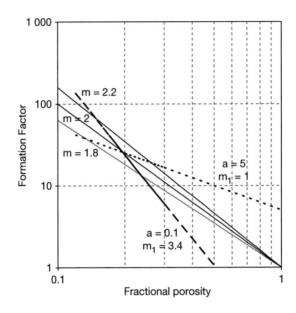

Figure 1-3.4 Examples of F, ϕ relations in logarithmic scale corresponding to Archie's formula (thin full lines) and to the Humble formula (extreme values of a and m_1); bold dashes

correspondence with the theoretical values given above. The values for consolidated rocks are centred around 2. Note that the term "cementation" exponent (suggested by Wyllie?) may lead to confusion since its relation to cementation in the petrographic meaning seems rather fuzzy.

b) Humble formula: $F = a\phi^{-m_1}$

Defined by Winsauer *et al.* [1952] this formula involves two parameters which, in practice, are not independent (see below). The values of the coefficients most frequently observed are $a = 0.62$ and $m_1 = 2.15$. The Humble (Humble Oil Company) formula corresponds to an infinite number of straight line families characterised in logarithmic scale by their gradient - m_1 and the formation factor "a" for the fractional porosity 1 (Fig. 1-3.4). This limit, not equal to 1 when the porosity tends to 1, has no physical meaning.

Dependence between a, m_1 and m

In practice, parameters a and m_1 are not independent. The F, ϕ regression lines (in logarithmic space) are calculated from a limited number of points corresponding to a porosity interval which is often relatively narrow. It is therefore interesting to calculate what the relation between a and m_1 would be for the infinity of Humble type lines passing through a point (F, ϕ) of cementation exponent (Archie) m.

By definition $m = -\log F/\log \phi$. To simplify, let $\gamma = 1/\log \phi$; ($m = -\gamma\log F$)

We obtain $m_1 = -(\log F - \log a)/\log \phi = -\gamma\log F + \gamma\log a$ and therefore

$$m_1 = m + \gamma\log a$$

If there is any doubt as to the choice of Humble or Archie, we can propose, for the same set of data, several formulae with two parameters related as above.

Some authors have compiled the parameters a and m_1 of numerous Humble type relations published in the literature and have calculated the regressions between these two parameters. For example, Gomez -Rivero [1976] gives the following results:

Sandstone $m_1 = 1.8 - 1.29 \log a$

Carbonates $m_1 = 2.03 - 0.9 \log a$

Note the parallelism between these two regressions and the previous formula with equivalents of m equal to 1.8 and 2.03, very close to the mean values proposed for Archie's formula (the fractional porosities corresponding to the coefficients are respectively 0.17 and 0.08).

This agreement is not an evidence but it leads to a conclusion corroborated by the observations of experimental results in the next paragraph: Archie's formula corresponds more to the general trend of the F *vs.* ϕ relation, whereas it is more likely that the Humble formula corresponds to the change in a given rock type over a small porosity interval. Caution is therefore required when using the Humble formula. Before assigning to m_1 and a values which are sometimes quite far from the mean (but dependent), the representativeness of the line chosen must be carefully criticised with respect to the location of experimental points.

E) F *vs.* ϕ relations in rocks: the example of carbonates

a) General relation

Limestones form an excellent medium on which to study the effect of the porous space microstructure on the formation factor. Since they are generally clay-free, the analysis is not likely to be disturbed by problems of mineral conductivity.

The formation factors of about 500 limestone samples (measured under a differential pressure of 10 MPa) are shown on Figure 1-3.5.

We mainly observe (by comparison with the relation studied for permeability § 1-2.1.4B, p. 149) the relatively low dispersion of values and the apparent independence of the results with respect to the "microstructure". Concerning the formation factor, mudstones are not especially different from the other types of limestone since, as already mentioned, the resistivity is independent of the pore throat dimensions: two homothetic porous media have the same formation factor.

For a given porosity, it is the geometric distribution of the conducting fraction which governs the formation factor. This point can be illustrated by an example taken in the oolitic limestones.

b) Oolitic limestones

Figure 1-3.6 shows the ϕ *vs.* F relation in two types of oolitic limestone. The structure of the porous space of these limestones is illustrated by stereo photographs of epoxy pore casts (§ 2-2.1.3, p. 332). In Chaumont limestone (outcrop, France) the porosity is concentrated

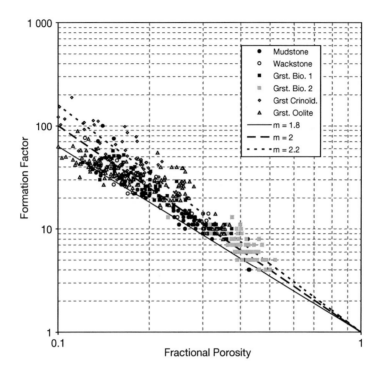

Figure 1-3.5 F *vs.* φ relation in limestone rocks of varied petrography

inside oolites (intergranular microporosity). The oolites are surrounded by a calcite cement (void on pore cast). The contacts between oolites are extremely small (low contiguity index) and we can appreciate that the conductivity of this electrolyte-saturated rock is relatively low. Considered according to Archie's formula, the cementation exponent is well above 2.2.

The porosity of Vilhonneur limestone (outcrop, France) is also primarily intergranular microporosity, but before the (partial) calcite cementation, initial diagenetic compaction caused interpenetration of the oolites, resulting in a high contiguity index and therefore good contact between the various intraoolitic microporous lumps. At equal porosity, conductivity is therefore better than for Chaumont limestone. For most samples, the cementation factor is well below 2.

This is a clear example of the impact of porosity distribution at microscopic scale on the value of the formation factor. We must emphasise the fact that this "heterogeneity" of the porosity distribution is observed at a scale much smaller than that of the Representative Elementary Volume generally taken into account for the porosity (§ 2-1.2.1, p. 273) which corresponds to a few grain diameters. These oolitic limestones are perfectly homogeneous, at sample scale. These microstructure effects play an even greater role in case of two-phase saturation (see below).

Figure 1-3.6 shows the automatically calculated regression curves. They are characteristic of the Humble type. We can see qualitatively on this example the relatively subjective nature of this relation when the "cloud of points" is dispersed.

CHAUMONT OOLITE

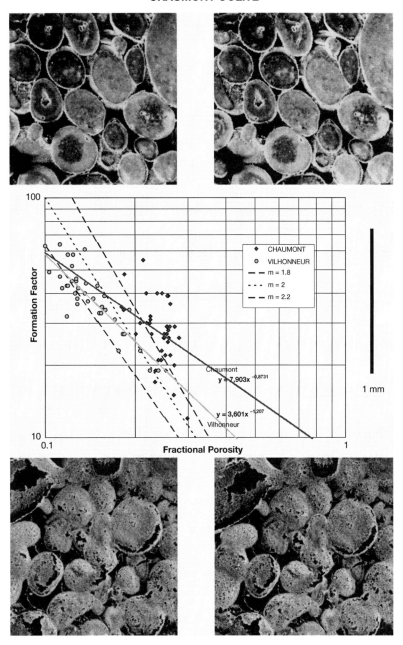

VILHONNEUR OOLITE

Figure 1-3.6 F *vs.* φ relation in two type of limestone.
Corresponding epoxy pore cast (stereographic SEM photograph)

1-3.1.2 Resistivity of porous media saturated with a two-phase mixture

When the porous medium is saturated with a mixture of non-miscible fluids, one of them being a non-conductor (a hydrocarbon for example), the situation described above is significantly modified. The total resistivity of the medium (R_t) varies considerably with the insulating phase content (saturation).

A) Archie's formula: resistivity index, saturation exponent

Archie [1942] presented a first analysis of the effect of two-phase saturation on resistivity, thereby initiating the quantitative interpretation of resistivity logs in terms of hydrocarbon saturation.

The rock resistivity in two-phase saturation state R_t (with one non-conducting phase: oil or gas) is compared to the resistivities R_w (of brine) and R_o (of the brine-saturated rock).

From an empirical analysis, Archie defined the resistivity index:

$$I = R_t/R_0$$

and this index is related to the brine saturation by the formula:

$$I = S_w^{-n}$$

where n is known as the saturation exponent or Archie's exponent.

We can see why this formula is so useful: by knowing the porosity (basic log analysis parameter), the F *vs.* ϕ relation (see above), the resistivity of the formation water (estimated for example in the zones where $S_w = 1$) and the exponent n, an approximation for the water saturation – an extremely important parameter – can be calculated.

Determination of n is a major petrophysical problem which we will discuss again later. In a first analysis, we may consider that n is close to 2 in brine-wet porous rocks.

B) Qualitative approach of Archie's formula

Being so useful, Archie's formula generated numerous developments around the formula itself or its derivatives (see next paragraph). A highly qualitative approach of the phenomenon, using the simple example of a porous medium whose matrix is non-conducting, provides a better understanding of certain aspects.

Initially, we will consider that the total water saturation state is not fundamentally different from the two-phase saturation state, since there is a conducting phase: the brine, closely mixed with an insulating phase obtained by adding the rock matrix and the insulating fluid fraction (oil for example). Ignoring any topological considerations, to interpret the resistivity, we may therefore consider that we are in the presence of a porous medium totally saturated with brine of resistivity R_w and of conducting porosity $\phi_{co} = \phi S_w$. With this new parameter, we can express I in a form similar to the formation factor, using for the latter the Archie type expression:

$$F = \phi^{-m} = R_0/R_w; \quad R_0 = \phi^{-m} R_w \text{ and } Rt = R_0 S_w^{-n} = R_w \phi^{-m} S_w^{-n}$$

assuming m = n = 2, we obtain $I = \phi_{co}^{-2}$.

Note that m and n are "only" the exponents of a power law relating the conductivity of a medium to the conducting fraction content of this medium. Using this formulation, it is easier to justify the fact that m et n depend only on the "topology" of this conducting phase. We are using the term "topology", probably incorrectly, to express the fact that only the relative positions of the conductive volumes and their connections determine the values of these exponents, the dimensions as such not being involved. We may therefore return to the analysis of simple models started for m: in a straight capillary model and parallel to the capillaries, m and n are equal to 1. In contrast, in a model of conducting spheres buried in an insulating medium, as long as the percolation threshold is not reached (§ 1-2.1.3D, p. 142), the exponent is not defined (infinite R_t) then takes very high values near this threshold. For this type of model, the percolation threshold is close to a 30% content (ϕ_{co}).

The large dispersion of values possible for m and n therefore corresponds to the infinite variety of geometries which may be encountered in the porous media and the conducting phases. A given value of m cannot characterise a petrographic structure in case of large porosity variation. More important, it is difficult for the exponent n to remain constant, for a given porous medium, when the saturation tends towards limiting values (S_{wi}) with the resulting major topological modifications.

We may try to illustrate the previous observations using a very simple model consisting of a two-dimensional electrical network formed from a grid (square mesh of dimension 1) of rectangular conductors (surface resistivity rs) of small size e (Fig.1-3.7). The conductor content (similar to ϕ_{co} or S_w) is equal to $2e$ and its resistance R, parallel to the main directions of the mesh and per unit length, is equal to rs/e. We are in the two-dimensional equivalent of the straight capillary model and I (or F) $= (rs/e)/rs = e^{-1} = (\phi/2)^{-1}$. The exponent associated with ϕ (similar to n or m) is indeed equal to 1.

If square conductors of side C are positioned at each node of the mesh, it is easy to calculate the conducting porosity $\phi_{co} = 2e(1 - C) + C^2$ and the corresponding resistance $R = rs[(1 - C)/e + C/C] = (1 + e - C) \, rs/e$.

To vary the conducting porosity, we can either vary C, the dimension of the squares (Fig. 1-3.7 left hand graph), or the thickness e of the connections (right hand graph). These simplistic models are obviously remote from the reality of porous media but they confirm that m and n may vary significantly. Amongst other things, we observe that for small variation in the accesses (e) to the conductive lumps (C), the global resistivity depends very little on the conductor content.

C) Value of the saturation exponent in reservoirs

a) Remarks concerning measurement of the saturation exponent

As suggested by the previous remarks, we observe significant variability in the values of n measured in the laboratory or inferred from log analysis results.

Before giving a brief summary of some data in the literature, we must emphasise one simple but sometimes neglected point. The formation factor depends only on the geometry of the porous space, it is an intrinsic characteristic of the rock. We may speak of the "rock formation factor". However, the resistivity index depends on the geometry of the porous

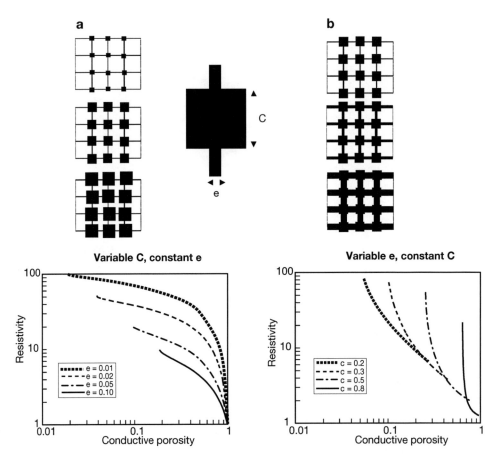

Figure 1-3.7 Example of highly simplified conducting network and Resistivity *vs.*
Conducting porosity relation in the case where the width of the connections is kept constant
and the dimension of the squares variable (a) and vice versa (b)

space but also (and perhaps especially) on the distribution of the conducting and insulating phases in the porous space. We cannot speak of the "rock resistivity index", we must consider the resistivity index of a rock/fluid assembly in a fixed capillary equilibrium. The wettability conditions have a major influence on the resistivity index. It is not uncommon to see two samples from the same reservoir, with the same porosity and facies, and saturated with the same fluids, exhibit dissimilar resistivity index curves. Variations in wettability should be analysed first to explain this phenomenon.

For a given saturation, the values of *n* vary depending on whether drainage or imbibition is occurring (§ 1-1.2.2B, p. 56). But most experimental data are acquired during drainage.

Although the formation factor (rock totally saturated with brine) is one of the simplest petrophysical parameters to measure, the situation is quite different when measuring the resistivity of partially saturated rocks, since the result may sometimes be disturbed by side effects which the experimenter is unaware of. For example, the measurement could be seriously corrupted by selective desaturation of the sample near the electrodes. To measure the saturation exponent correctly, the sample resistivity must be measured for as many saturation states as possible, in order to plot the best curve of R_t *vs.* S_w. This saturation must be varied extremely carefully to avoid any possibility of local artificial heterogeneities inside the sample, while respecting pressure/temperature stability conditions to prevent wettability changes.

b) Porous media in strong water- wet state

Clean sandstones

This is clearly the simplest case since we are dealing with an intergranular medium within which the wetting/non-wetting phases are distributed according to a geometric structure which is relatively stable during the saturation variation. It is under these conditions that Archie's formula gives the best fit. The exponent *n* is close to 2. Values measured in the laboratory fluctuate more or less around this mean but, in the light of experience, log analysts consider that the constant value 2 is sufficient in practice.

We have mentioned on several occasions the effects of hysteresis on fluid movement. It is therefore logical to expect to see these effects as regards Archie's exponent. Numerous results have been published on this subject (for a list of references, refer to Al-Kahabi *et al.* [1997]). As we have seen for relative permeabilities (§ 1-2.3.1C, p. 184), however, it is difficult to propose general rules.

Carbonates

Numerous measurements of the carbonate reservoir saturation exponent have been published, concerning in particular the Middle East oilfields where carbonates prevail. The results are often quite surprising, they show that the value of *n* may vary significantly for the same sample depending on the saturation interval considered and that these values may be noticeably lower or higher than 2, even when the sample cleaning conditions dismiss the assumption of oil wettability [Bouvier and Maquignon, 1991], [Dixon and Marek, 1990].

This situation is mainly due to the coexistence in carbonate rocks of at least two types of porous network: a macroconnected network, saturated with oil even at low capillary pressure and a microconnected network likely to retain interstitial water up to high capillary pressure (see Fig. 1-1.55, p. 113). In two-phase saturation state, the macroaccessible porosity corresponds to the insulating phase and the microporosity containing brine to the conducting phase. We can see intuitively that the relative position of these two phases has more influence on the overall resistivity than the absolute value of the brine saturation. If the macroporous phase "encircles" microporous zones (Fig. 1-3.8), a relatively low drop in S_w, corresponding to invasion of some of the macroporosity by oil, could induce a relatively high variation in the resistivity index. The opposite is true if the macroporosity likely to be oil-saturated consists of "isolates" inside a mainly brine-saturated microporous phase. We have adopted the same argument to explain the formation factor differences in oolitic limestones of Fig. 1-3.6.

In an outstanding article, Sen [1997] applies the effective media theory to quite realistic petrographic examples of limestone rocks and provides a quantitative justification of the previous qualitative deductions.

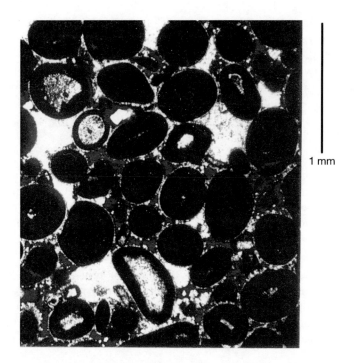

Figure 1-3.8 Characteristic example of double-porosity limestone rock with structural segregation
of the two phases. The oolites (in black) exhibit abundant microporosity (see Fig. 1-3.6)
and are surrounded by equally abundant intergranular macroporosity

c) Porous media in intermediate wettability or oil-wettability state

Very high saturation exponents ($n > 4$) would generally be associated with the case of oil-
wet reservoirs [e.g. Donaldson and Siddiqui, 1989]. This is intuitively true if we consider
that, in the event of low water wettability, the conducting electrolyte tends to reduce its area
of contact with the solid by forming globules which are poorly connected, or even
disconnected in case of very high non-wettability, when the residual saturation in non-
wetting fluid is reached (equivalent to S_{or} for water wettability). As soon as we move away
from the case of very high electrolyte saturations, the resistivity may increase sharply,
leading to very high n exponents. We may legitimately question whether the use of Archie's
formula is appropriate in such a case.

A paradox is traditionally put forward when comparing the approaches taken by
laboratory petrophysics and log analysis with respect to wettability. We will consider the
wettability estimated on core from the measurements given in § 1-2.2.1, p. 169. Reservoirs
of intermediate wettability are quite frequently encountered and strongly oil-wet reservoirs
are not uncommon. In log analysis, however, the numerous discussions on the value of
Archie's exponent very rarely concern values greater than 3 whereas the distribution of
fluids in a poorly water-wet reservoir is such that n should be greater than 3, as explained
above. The most likely explanation is that log analysis generally concerns a pristine

reservoir which has retained the geometric distribution of the fluid fractions induced by the original water wettability. In practice, through lack of information, the log analyst tends to neglect this wettability effect.

1-3.1.3 The special case of clay media

A) Specificity of clay media

The presence of a clay phase induces greater conductivity than that estimated by Archie's law, whether in case of total or partial water saturation (Fig.1-3.9)

The precise nature of the physical processes generating this electrical conductivity specific to the clay media is still open to controversy. There is a clear indication, however, that this effect is due to cations fixing to the surface of the clay crystallite. They compensate for the electrical imbalance of the crystalline lattice (see Fig.1-1.9, p. 19). These cations (Ca, Mg, K, Na, etc.) are exchangeable and their relative proportions depend on their concentration in the water over the mineral (although not equal to it). The additional conductivity is therefore clearly a surface property of clay minerals. Depending on the authors, it is attributed either to the mobility of the compensating cations themselves, or to the hydration protons of these cations.

This surface property extends over a certain distance into the water surrounding the mineral. This is the double-layer theory, frequently mentioned in books on clay minerals. Firstly, there is a condensed layer along the mineral interface whose thickness depends on the volume taken up by the compensating cations possibly accompanied by molecules of their hydration water. Secondly, there is a "diffuse" layer whose cation concentration decreases exponentially and whose thickness is inversely proportional to the square root of the concentration of the brine saturating the porous medium.

These points are mentioned here just because they justify certain interpretation models. In practice, it turns out that the thickness of this "double layer" would hardly exceed some 10 Å under normal conditions. Since this "double layer" is similar to the notion of "bound water", we immediately see the relativity of this notion in terms of volume compared with the total porous volume of the clay materials usually encountered. The "bound water" would only represent an infinitely small fraction of this total volume.

This explains the practical applications of the various formulae proposed in terms of saturation interpretation. Many of them are in fact based on a porosity specifically linked to the clay phase. The difference between total porosity and this special porosity would define the notion of effective porosity (see § 1-1.1.3E, p. 20). The identification which could have been made initially between the water associated with clays and the volume occupied by the bound water is therefore faulty. The difference in the orders of magnitude prevents this identification.

Relevant clay content parameters

– Based on experience and according to virtually all authors, the Cation Exchange Capacity (CEC) is the best clay content parameter to discuss the subject. In addition, it is easy to acquire, for example using colorimetry back-titration with cobaltihexamine.

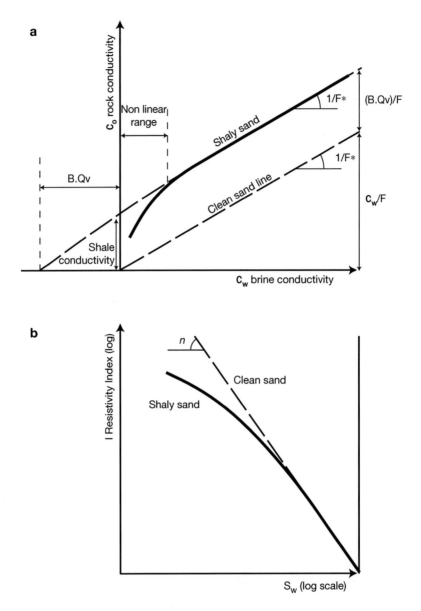

Figure 1-3.9 Specificity of clay media. (According to Clavier *et al.* [1984] and Giouse [1987])
a) Sample conductivity (C_0) *vs.* brine conductivity (C_w) relation.
b) Resistivity index (I) *vs.* Water saturation (S_w) relation.

– Although Specific Area, highly developed in clay media, is also a fundamental parameter, it seems less efficient in explaining the electrical conductivity deviations with
respect to Archie's law.

- The clay content quantifications expressed in contents by volume or weight of the clay species are less significant. The same is true of indirect measurements such as the granulometric cuts (at 2μ for example) on the weight or volume granulometric curves. *A fortiori*, clay content estimations made using thin sections are also inefficient in this context.
- In concrete terms, there is little advantage in knowing the type of mineralogical assemblage, compared with the global measurement of the CEC. Once again, this is a recurrent remark already made earlier for other physico-mechanical properties of clay media. Many authors emphasise the importance of smectite quantification. In practice, knowledge of a global parameter such as the CEC is sufficient to account for the effect of this "mineralogical variable". Moreover, the CEC parameter is a great deal easier to quantify.
- If ever another parameter deserved to be evaluated to complete the CEC measurement, it would definitely be the type of clay phase organisation. This textural effect will be discussed below.

B) Some examples of interpretation models

a) Case of laminar shales

The main types of shale distribution, as designed by log analysts, are illustrated on Figure 1-1.4, p. 14. Laminar shales correspond to a relatively simple model of conductivity additivity (where R_{sh} and R_{sd} are the resistivities of shales and sandstones respectively and V_{sh} is the shale volume fraction). This conductivity additivity is directly related to the assumption of parallel conductors for the sand/shale layers, the current lines implemented in the main resistivity tools being sub-horizontal.

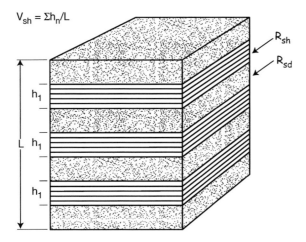

Figure 1-3.10 Diagrammatic representation of the laminar shale model

$$1/Rt = V_{sh}/R_{sh} + (1 - V_{sh})/R_{sd}$$

with the sand resistivity: $R_{sd} = (F_{sd} \cdot R_w)/ S_w^n$

hence $1/Rt = V_{sh}/R_{sh} + ((1 - V_{sh}) S_w^n)/ (F_{sd} \cdot R_w)$

and $S_w^n = (1/Rt - V_{sh}/R_{sh}) \cdot (F_{sd} \cdot R_w)/(1 - V_{sh})$

This is Poupon's first equation [1954].

The so-called structural clays (in the sense used by log analysts) are interpreted in the same way.

Note the relative importance of clay and sandstone conductivities in this equation. If the clay is highly conductive (water with a high ion concentration) and the sandstone highly resistant (oil saturation), then the clay conductivity prevails, possibly leading to a serious underestimation of the oil saturation in the clastic layers.

b) Case of dispersed clays

Including the few laminar shale equations, Worthington [1985] draws up a list of some thirty different models classified according to the parameters involved in expressing the conductivity of the clay material C_t: the water conductivity C_w; a clay content term; and/or a crossed term C_w/clay content. Some models are based on the notion of double conductors (in particular and amongst the most widely used: the "Waxman and Smits" model, taken up by Juhahsz [1981], the "Dual Water" model of Clavier *et al.* [1984], and other models are based on the notion of "double porosity". Giouse [1987, in French] adapts, comments and adds relevant information to this review. In practice, very few authors rely on actual experiments and many use the data of Waxman and Smits [1968] which remain the reference.

We will give an overview of these models below. The following notations are used:

C_o, C_t, C_w: the conductivities, respectively of brine-saturated rock, partially brine-saturated rock and brine

F*   formation factor of the W&S model; F_a apparent formation factor
ϕ_t total porosity; ϕ_e effective porosity   n* saturation exponent of the W&S model
Sw_t saturation related to total porosity; Sw_e "effective" saturation (related to ϕ_e)
CEC cation exchange capacity (meq/100 g)
Qv   cation exchange capacity per unit pore volume (meq/cm^3)
$Qv = (CEC \cdot \rho_s (1 - \phi_t)/(100 \cdot \phi_t)$
B equivalent conductance of clay exchangeable cations (ohm^{-1} m^{-1} meq^{-1} cm^3 in the W&S formula)

Example of the 'Waxman and Smits' (W&S) model

Fig. 1-3.9a shows the physical meaning of some parameters used in this model. The conductivity of the totally brine-saturated medium is:

$$C_o = (1/F^*) \cdot (C_w + B \cdot Qv)$$

When the pore volume is partially brine-desaturated:

$$C_t = (Sw_t^{n^*}/F^*) \cdot (C_w + B \cdot Qv/Sw_t)$$

This is a model with two parallel conductors: the first is the porous medium of the rock "conventionally" quantified by the notion of formation factor F*, the second results from the contribution of exchangeable cations.

We observe:

- that the formation factor (therefore the tortuosity) is the same for both conductors;
- the second term is proportional to the saturation is we assume n ≈ 2.

These remarks apply as soon as the brine conductivity is high enough to leave the non-linear range. The non-linear range is modelled by an exponential form not explained here.

Parameters F*, **Qv** (i.e. the CEC) and **B.Qv** are easy to access experimentally by relatively simple and inexpensive routine core analysis measurements. The log analysis calibration can be backed up by the "core petrophysical log" approach (see § 2-1.3.1F, p. 305).

The model slightly overestimates the effect of clay content at low CEC but underestimates it at high clay contents.

Overall, the higher the clay content (estimated by **Qv**), the higher the "cementation factor" of argillaceous sandstones. It extends from 1.5 for clean media to more than 2.5.

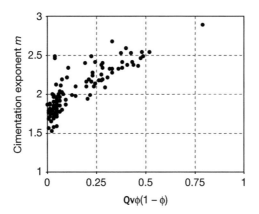

Figure 1-3.11 Cementation exponent *m* as a function of **Qv**$\phi(1 - \phi)$ calculated for the Waxman and Smits data [1968] and for the Waxman and Thomas data [1974] (according to Clavier *et al.*, [1984])

The "Dual Water" model [Clavier et al., 1984]

This model can be used to correct some faults of the W&S model, distinguishing between two "waters" in each term of the W&S equation.

The second term is developed by considering that it represents the conductivity added by the "double layer" mentioned at the start of this paragraph.

The conductivity of the saturated medium becomes:

$$C_o = (1/F_o) \cdot ((1 - \alpha \cdot V_q \cdot Qv) \cdot C_w + \alpha \cdot V_q \cdot Qv \cdot C_{cw})$$

where $\alpha = 1$ in "non negligible" salinity or $\alpha < 1$ if the salinity is very low

V_q: coefficient related to the thickness of the diffuse layer (bound water) $(meq^{-1}\,cm^3)$

C_{cw}: apparent conductivity of "clay bound" water $(S\,m^{-1})$

F_o: formation factor of the "dual water" model

And when the pore volume is desaturated:

$$C_t = (Sw_t^{n^*}/F_o) . ((Sw_t - \alpha . V_q . Qv) . C_w + \alpha . V_q . Qv . C_{cw})/Sw_t$$

This is the model implemented in the most widely used analysis programs. It fits the experimental data better but still remains unsatisfactory in the non-linear range of low-concentration brines.

The "normalised Qv" model [Juhasz, 1981]

It is based on the W&S model, adapting it to an absence of core data. Taking the shale zones as reference, its purpose is to normalise **Qv** to a maximum **Qv** evaluated in a 100% shale zone.

normalised **Qv**:   **Qvn** = Qv/Qv$_{sh}$

where **Qv$_{sh}$**:   **Qv** of shale

we then demonstrate that: $Qvn = V_{sh} . \phi_{sh}/\phi_t$

where V_{sh}: shale volume content (porosity included)

and ϕ_{sh}:   shale porosity

and ϕ_t: total porosity

thus:

$$C_t = (Sw_t^{n^*}/F^*) . (C_w + (V_{sh} . \phi_{sh})/(Sw_t . \phi_t) . (C_{wsh} - C_w)$$

Where: $C_{sh}/C_{wsh} = \phi_t^m$

Observing the equivalence between the normalised-**Qv** and the term $V_q . Qv$, Raiga-Clémenceau *et al.* [1984] point out the similarity between the Juhasz equation and the "dual water" model equation, in spite of the different approach. From a practical point of view, application of this model seems to give more satisfactory results.

The few soil physicists (in the agronomic sense) studying electrical conductivity also model the behaviour of these soils by two parallel conductors. They propose a slightly different formalism. It relates to the void index E ($E = \phi/(1 - \phi)$) and not to the porosity ϕ: see § 1-1.1.2-B, p. 11). Strangely enough, we observe that the models proposed are empirical and that the CEC concept is not used.

Double-porosity models

The three most frequently employed models, those above, are two-conductor models: The "double-porosity models", designed to be more "physical" are used to specifically parameterise the clay phase.

Poupon's Indonesian formula, mentioned at the start of this chapter, is considered as the prototype. Advanced models are proposed, for example:

– The Raiga-Clémenceau *et al.* model [1984]

– The Giouse model [1987] characterised by the fact that is uses a morphology parameter of the porous medium obtained from the porosimetry curve, comparing the microporosity with the phase associated with the clay matrix (see bimodal sandstones on Fig.1-1.46, p. 90).

The double porosity model can be used to explain the impact of pore wall roughness (e.g. fig. 1-1.47, p. 93) on the saturation exponent n (Fig 1-3.12). Diedrix [1982] using the textural parameters of the clay phase, interpreted the variation of n with the saturation in the Rotliegend sandstones from the North Sea.

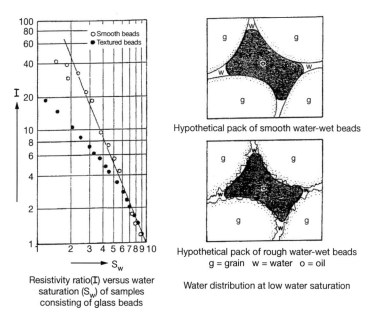

Resistivity ratio(I) versus water saturation (S_w) of samples consisting of glass beads

Hypothetical pack of smooth water-wet beads

Hypothetical pack of rough water-wet beads
g = grain   w = water   o = oil

Water distribution at low water saturation

Figure 1-3.12 Effect of texture on the resistivity index [Diedrix, 1982]

These double-porosity formulae, however, have not yet become standard practice in log analysis.

In conclusion, the increasing number of models clearly illustrates the difficulty in using electrical measurements to estimate the saturation in clay medium. Some parameters are still poorly understood, such as the influence of salinity, especially at low concentrations and that of the cation ratio, especially the monovalent/divalent cation ratio.

Whatever the case, it is now generally accepted that it is primordial to think in terms of total porosity and not in terms of effective porosity (log analysis), which is considered too ambiguous.

Calibration on a representative cored layer, including shale levels, remains essential both to estimate the interpretation parameters (ϕ_t, CEC, V_{cl}, m and n, etc.) and to adjust the saturations using direct measurements, whenever possible.

1-3.1.4 Effect of pressure and temperature on electrical properties

A) Effect of stresses on the formation factor

Concerning the effect of stresses (which only affect the geometry of the porous space), we will discuss the distinction between "pore porosity" and "crack porosity" later (§2-1.1.2, p. 267). The case of the formation factor is much simpler than that of permeability, however, since a possible variation in pore dimension has no consequences.

- With "pore porosity", the main effect of stress is to reduce the porosity (pore compressibility). Since the change expected on the topology of the porous network is very small, we may reasonably consider that the factor m does not vary with differential pressure. To make a pressure correction on formation factor measurements taken in the laboratory without confinement, the porosity variation (if known) induced by the stress would simply have to be applied in Archie's formula or the Humble formula. Note that applying Archie's formula at small porosity variations leads to $dF/F = -2d\phi/\phi$. Since the usual porosity variations under the effect of differential pressure are at most a few percent in normally porous rocks (§ 1-1.1.6, p. 43), the effect on the formation factor generally remains minor. Note that for practical applications, the absolute values of the formation factor do not really matter, the only parameter of interest being the exponent m. The need to correct this factor seems of secondary importance, compared with other causes of uncertainty.
- With "crack porosity", the effect of differential pressure on the topology of the porous space is much more sensitive (closure of the pore throat), but we are dealing with low-porosity media where the formation factor values are high and measurement difficult (e.g. true state of the sample/measurement device contacts). It seems reasonable to think that, even with these types of pore geometry, the effect of differential pressure variations on the measurements is not the main factor to be taken into consideration.
- If a stress correction on exponent m is required, experimental data tend to demonstrate that m increases as the stress increases.

The previous remarks could probably be applied to the resistivity index, given the fact that the stress is only expected to have a slight effect on the capillary properties.

B) Effect of temperature on electrical properties (characteristics of saturating fluids)

As for the other petrophysical properties (§ 2-1.1.2, p. 268) the effect of temperature variations (remaining within the range of variations generally encountered in sedimentary basins) should only be studied through the variations induced in the characteristics of the saturating fluids. This effect on the saturating fluids, in relation with the electrical properties, is extremely important due to:

- mainly, the brine resistivity variation with temperature;
- the effect of temperature on wettability and the capillary properties in general (relative viscosity, surface tension).

The effect of temperature on brine resistivity has no impact on the formation factor (by definition). In contrast, the wettability/capillarity effect on the geometric distribution of fluids may be very large. Temperature may therefore have a predominant effect on the resistivity index and the saturation exponent.

We find once again the marked difference, mentioned earlier, between the formation factor, an intrinsic property of the porous space and directly related to its geometry, and the resistivity index, in which the nature of the capillary equilibrium between the phases plays a major role.

1-3.2 SEISMIC PROPERTIES: EFFECT OF POROSITY, LITHOLOGY AND NATURE OF THE SATURATING FLUIDS ON THE ELASTIC WAVE VELOCITY

The seismic properties of rocks (mainly the elastic wave velocity) are important for two related fields of application: geophysical prospecting and log analysis (sonic logs). Nowadays, sonic logs are practically no longer used as porosity logs and they are mainly recorded as an aid to seismic analysis.

The study of seismic properties of rocks is a special area in the science of petrophysics since the analysis methods, based mainly on applications of the wave equation and elasticity, have little in common with fluid mechanics, the "pillar" of petrophysics in the ordinary sense of the term. The main reason for this relative separation is due to the fact that the potential users of the results are geophysicists who, in the traditional petroleum structures, are rarely in contact with the reservoir engineers, the main "customers" of petrophysics.

This situation is changing rapidly, however, due to the increasingly important role played in the monitoring of reservoir production by 4D seismics (3D + time), the new name for repetitive seismics, whose objective is to detect fluid movements in a reservoir through very slight observable differences between two seismic records performed in the same way and separated by a production phase.

The preparation ("feasibility") and interpretation of these often costly operations requires good knowledge of the impact of fluids on seismic properties as well as close cooperation between geophysicists and reservoir engineers.

This new branch of petrophysics is sometimes called "petroacoustics".

To remain within the scope of this book, we will describe qualitatively the main notions pertaining to the elastic wave propagation in porous media. For a more rigorous presentation, interested readers may refer to the numerous books published on this subject: [e.g. Bourbié *et al,* 1987; Mavko *et al* 1998]. We will mainly focus on the notions involved in 4D seismics feasibility studies: the effect of differential pressure and the effect of saturating fluids.

Lastly, although petroleum 4D seismics currently represents the major financial application of "petroacoustics", in the future we may expect to see the applications of these methods extended to environmental problems such as monitoring the water table level or pollutant concentrations in the ground.

1-3.2.1 Simplified definitions: moduli, P and S waves, velocities, attenuations

The seismic waves we will discuss in this chapter correspond to propagation of a mechanical disturbance in the material, resulting in local deformation of the medium. These mechanical waves are qualified as elastic. As the waves pass, the medium is considered to react according to the theory of elasticity. Note that there may be a certain amount of terminological confusion between the adjective "acoustic" and "elastic". Strictly speaking, acoustic waves are pressure variations propagating in a fluid medium (i.e. not bearing the shear and therefore unable to propagate S waves (see below). Sound waves are acoustic waves of frequency within the human audible range ($10^2 - 10^3$ Hz). Elastic waves are stress variations propagating through a solid. The term acoustic is sometimes used for P waves and elastic for all P and S waves.

A) Hooke's law, Moduli of elasticity.

The elastic (reversible) behaviour of a body may be described by Hooke's law which states that the applied stress (σ) is directly proportional to strain produced (ε), the proportionality coefficient being the modulus of elasticity (M):

$$\sigma = M\varepsilon$$

Since strain is dimensionless, the modulus has the same unit as stress (Pa). The usual unit is the gigapascal (GPa). For isotropic materials (i.e. all directions are equivalent, § 2-1.2.2, p. 281), two constants are sufficient to characterise elastic behaviour. Physicists use Lamé's parameters (λ and μ), but the disadvantage with λ is that its definition does not correspond to a simple experiment. Concerning wave propagation, the parameters mainly used are the bulk modulus K, which corresponds to an isotropic compression experiment, and the shear modulus μ, which corresponds to a pure shear (Fig. 1-3.13). Note that in fluids, where (by definition) $\mu = 0$, a single parameter is sufficient: generally K, the reciprocal of compressibility which is frequently used in fluid mechanics. It is important to note that the compressibility to be taken into account in acoustics is the isentropic compressibility, since the wave passes quickly enough to prevent the temperatures from reaching equilibrium, whereas the compressibility most frequently used (in reservoir modelling for example) is the isothermal compressibility. In some cases, this confusion may lead to errors when calculating K (see § 1-1.1.6C, p. 45).

Another elastic parameter often used, the Young modulus, corresponds to a simple compression experiment. Obviously, one-to-one relations exist between these parameters considered in pairs. They can be found in tables in the general books mentioned above.

B) P waves, S waves, propagation velocities

By combining Hooke's and Newton's laws, we can demonstrate that two wave types may propagate in solids:

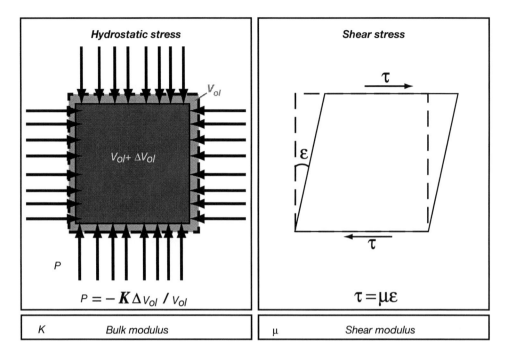

Figure 1-3.13 Definition of bulk modulus (K) and shear modulus (μ)

- Waves corresponding to the propagation of volume expansion, the P *(Primary)* waves observed first since they travel more quickly at a velocity $Vp = \sqrt{\dfrac{K+4/3\mu}{\rho}}$, ρ being the density of the propagation medium.
- Waves corresponding to movements with no change of volume: shear waves or S *(Secondary)* waves, since they travel more slowly and arrive after the compression waves at velocity $Vs = \sqrt{\dfrac{\mu}{\rho}}$.

With P waves, the direction of polarisation, i.e. the direction of displacement of material particles as the wave passes is parallel to the direction of propagation of the wave, whereas with S waves, the direction of polarisation is perpendicular to the direction of propagation. To illustrate the P and S waves macroscopically, for P waves we use the analogy of a spring with a disturbance moving along its length, and for S waves the analogy of a vibrating string (Fig. 1-3.14).

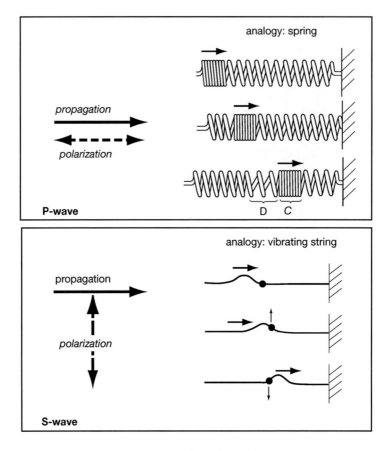

Figure 1-3.14 Analogy of P and S waves

C) Attenuation, viscoelasticity, velocity dispersion

a) Attenuation

As the waves propagate their amplitude decreases. They are numerous reasons for this attenuation, but they can be divided into two broad types:

– **Intrinsic attenuation, which is strictly related to the type of material in which the wave is propagating.** This attenuation corresponds to physical phenomena, resulting in a change of mechanical energy into heat. The main reason for this attenuation in geological materials can be attributed to water adsorbed on the mechanical microdefects. Few details of the phenomena are available, however. Another reason for intrinsic attenuation is the energy dissipation related to the differential movement of the saturating fluid and the solid, as the wave passes. This mechanism, actually observed under some conditions, gave rise to numerous studies aimed at measuring permeability through the attenuation of elastic waves. Unfortunately, the lack of rigour of some of these studies led to incorrect practical conclusions and the overall result is rather discouraging.

- **Extrinsic attenuation, which is related to the propagation geometry and the heterogeneities of the medium**. Geometric divergence is one simple example: when a wave generated by a point source is propagated in an infinite medium, the area of the wave front increased indefinitely while the quantity of mechanical energy to be shared remains unchanged. The wave amplitude will therefore decrease in proportion. Consequently, there is no loss of mechanical energy in extrinsic attenuation, but rather a defocusing of this energy.

Obviously, since one of the prime objectives in studying petrophysical properties is to characterise the materials, the only factor of interest is intrinsic attenuation. Generally however, intrinsic attenuation is lower than extrinsic attenuation. For example, in reflection seismics, the wave reverberation on the numerous strata limits ("stratigraphic filter") plays a much greater role in attenuating the propagated signals than intrinsic rock attenuation itself. This secondary aspect of intrinsic attenuation explains the limited number of applications developed using attenuation measurement, despite the research effort. For the same reason, we will not discuss attenuation, except when mentioning viscoelasticity.

b) Viscoelasticity

In the previous paragraph, we considered perfectly elastic behaviour characterised by the instantaneous nature of the reaction to the stress/strain modifications. All the energy stored as the wave passes is immediately restored. Materials with this type of behaviour are perfectly elastic, exhibiting no intrinsic attenuation. Consequently, the absence of time factor also indicates that the frequency of the wave considered has no effect on the properties of these materials. In actual fact very few materials exhibit this behaviour, aluminium being one. Most materials, especially rocks, exhibit a more or less time-dependent stress/strain "delayed behaviour". A typical example of delayed behaviour is illustrated by the creep experiment (Fig. 1-3.15). In this experiment, the sample previously at rest is subjected to a constant stress over time, resulting in strain variation (increase). When the stress is no longer applied, the strain does not disappear immediately but decreases progressively.

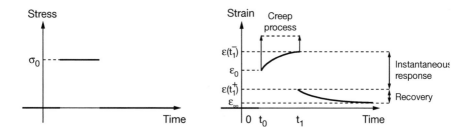

Figure 1-3.15 Creep experiment, [Bourbié *et al,* 1987]

This stress/strain time dependence induces numerous consequences. We may intuitively deduce that this dependence on the "time factor" characterised by a phase shift between stress and strain will result in dissipation of energy which is not immediately restored as the wave passes and in a frequency dependence of the mechanical properties. This viscoelastic modelling is used to account for intrinsic attenuation and velocity dispersion.

c) Velocity dispersion

In viscoelastic media, velocity depends on the wave frequency. This dependence is known as velocity dispersion (dispersion is also caused by factors other than viscoelasticity). In rocks, the velocity increases with frequency. There is a direct relation between viscoelastic attenuation and dispersion. This dispersion can be calculated using viscoelastic models. To provide a quantitative idea of this dispersion, we can use the formula of a model well adapted to rocks: the Constant-Q.

$$V/V_0 = (\omega/\omega_0)^{(Arctg(1/Q)/\pi)}$$

where V is the velocity corresponding to the frequency ω considered, V_0 the velocity at the reference frequency ω_0 and Q the quality factory which is inversely proportional to the attenuation. We can plot the graph of Figure1-3.16, where the value $\omega/\omega_0 = 10^2$ corresponds roughly to the case of ultrasonic experiments/sonic log or sonic log/seismic prospecting. We observe that there is very little dispersion for all rocks under confinement, i.e. buried at a certain depth. Dispersion only becomes important with some highly attenuating samples measured in condition at atmospheric pressure.

In view of the extremely low influence of the effect of dispersion on rocks subject to minimum confinement, the laboratory velocity measurements can be upscaled.

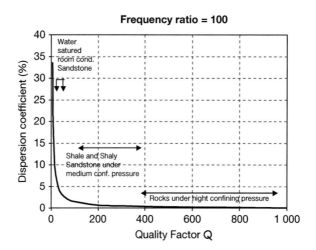

Figure 1-3.16 Relation between the dispersion coefficient (V/V0) in% and the Quality factor Q for a frequency ratio (ω/ω_0) of 100 corresponding approximately to the order of magnitude of the frequency ratio between laboratory ultrasounds and the sonic log or between the sonic log and reflection seismics. The order of magnitude of factor Q for various types of rock is shown on the figure

1-3.2.2 Scale effect: Ultrasonic, Sonic, Seismic frequency bands. Static properties, dynamic properties

The simplest way to describe a wave is to express the displacement (A) of the material point as the wave passes as a function of time through the use of a periodic function of type $A = \cos(\omega t + \varphi)$ where ω is the pulse directly related to the frequency ($\omega = 2\pi f$) and φ the phase.

Note that, with mechanical waves and unlike other physical phenomena using the wave formulation, the above formula corresponds to a situation directly perceptible by humans since, for example, the shape of a vibrating string observed using a stroboscope is sinusoidal.

Note that the Fourier transform can be used to decompose any signal into a sum of sine functions whose amplitudes and phases depending on the frequency form the signal spectrum (Fig.1-3.17). A propagating vibration of any shape can therefore be reduced to a sum of fixed frequencies periodic waves, provided that we remain in the linear domain, i.e. that the stress as the wave passes remains small enough so that the elastic moduli of the propagation medium do not vary. This linearity condition has been demonstrated on numerous occasions in the field of geophysics and log analyses.

Returning to the periodic expression, we see that the "length unit" used to define the homogenisation scale of the vibratory phenomenon is obviously the wavelength Λ, ($\Lambda = V/f$), the distance separating two identical vibratory states.

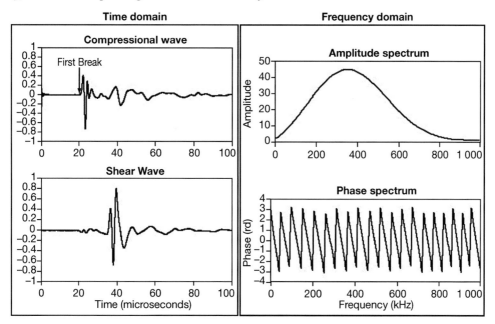

Figure 1-3.17 Example of waveform (P and S) in the time domain (on the left) and in the frequency domain for the P wave only, after Fourier transform (on the right, amplitude spectrum (top right) and phase spectrum (bottom right)

A) Wavelength and homogeneity of samples

A difficulty arises with the characteristics of elastic waves since measurement representativeness must be taken into account, considering not only the representative elementary volume (REV, § 2-1.2.1, p. 272), but also the scale of the wave itself, quantified by its wavelength.

To obtain a meaningful petroacoustic measurement, the sample must respect the conditions related to representative minimum volume (l_c characteristic length), but it must also be homogeneous with respect to the wavelength used, i.e. the dimension of the largest heterogeneity in mechanical property must be small compared with the wavelength. Comparing the characteristic length (l_c) with the wavelength (Λ), three broad domains are usually considered:

- $l_c \ll \Lambda$. In this case, the wave propagates in a medium which is continuous and homogeneous at "its scale". We are in the validity domain of the effective-medium theories. The petroacoustic measurements are meaningful in this scale domain.
- $l_c \approx \Lambda$. This domain corresponds to that of scattering. Concerning the laboratory petroacoustic measurements using ultrasound, scattering phenomena are the most dangerous since they disturb the velocity or attenuation measurement results and sometimes remain unnoticed. The path dispersion mentioned in § 1-3.2.3, is an example of low level scattering.
- $l_c > \Lambda$. In this case, we may consider that wave propagation occurs in various parts which are different propagation media which have to be considered separately. The wave propagation modes can be calculated using the ray theory. This scale domain is clearly outside the scope of petroacoustics.

However, there is no measurement representativeness condition between wavelength and sample length if the sample is representative in the general sense (§ 2-1.2.1B, p. 274): excellent measurements can be taken at low frequency (e.g. $\Lambda = 1\ 000$ m) on plugs. The wavelength corresponding to static tests tends to infinity.

We cannot emphasise enough the need to analyse the sample homogeneity with respect to the wavelength of the signals used. The objective of petroacoustics is to help improve the analysis of field records. In a large majority of cases, the frequency bands used for these records are different from those used for laboratory measurements. So that they can be upscaled, the laboratory measurements must be taken under valid homogeneity conditions, both in terms of sample and wavelength. Most petroacoustic failures can be attributed to problems in this area.

This point is taken up and developed in paragraph § 1-3.2.3B.

B) Frequency band

The range of wavelengths used in petrophysics applications is very wide and the problems may be quite different depending on the frequency band considered. Table 1-3.1 summarises the characteristics of the three broad bands used in geophysics applications.

Table 1-3.1   Order of magnitude of frequencies
and wavelengths corresponding to the three frequency bands used in geophysics

Wavelength	100 m	10 m	1 m	0.1 m	1 cm	1 mm
Frequency	10 Hz	100 Hz	1 kHz	10 kHz	100 kHz	1 MHz
	Decahertz band		Kilohertz band		Ultrasonic band	
Applications	Seismic prospecting		Sonic logging		Laboratory petroacoustics, Non-destructive testing	

Note that, unfortunately, the interest and importance of geophysical applications increase from right to left on the table, whereas the ease of implementation increases in the opposite direction since ultrasonic frequencies are the easiest to implement.

C) Static properties, dynamic properties

In our description we defined the elastic moduli (implicitly with respect to a mechanical experiment: compression, shear, etc.) and we saw that the wave velocities were directly related to these moduli.

Considering the theory of elasticity, a pure shear experiment and an ultrasonic S wave propagation experiment are strictly identical. In the first, the stress is applied progressively over several minutes (or hours, etc.), the frequency corresponding to this disturbance tends to zero. We speak of a static experiment (or measurement, or modulus). With ultrasonic wave propagation, the disturbing cycle lasts only about one microsecond (megahertz frequency). In this case, we speak of a dynamic experiment (or measurement, or modulus).

We have to point out a terminological ambiguity. The signification of the terms static and dynamic used here has absolutely no relation with the one of the same words used above to characterize fluid accumulation and flow in reservoir rock.

In practice, a dynamic elasticity measurement is several orders of magnitude faster (and less expensive) than a static measurement. We can appreciate the value of comparing the results of the two types of experiment and therefore of comparing the static moduli and the dynamic moduli. A highly simplistic summary is proposed below:

- In perfectly elastic media, of which aluminium is the most common example, there is very good agreement between the moduli measured in a mechanical experiment and those deduced from a wave velocity measurement. The situation is quite similar in crystals which generally exhibit excellent elasticity. The numerous values given in crystallography books for the elastic modulus have in fact been determined from wave velocity measurements. These are values of "dynamic" moduli, but in this case the distinction is unnecessary.
- In rocks, the situation is quite different since noticeable differences may be observed between static moduli and dynamic moduli. Rocks are not perfectly elastic, a point we mentioned earlier very briefly. The main cause accounting for the differences between static and dynamic properties should probably not be attributed to a frequency effect

(modelled by a viscoelastic theory, for example) but much more to a strain value effect.

– In wave propagation the strain, a dimensionless value equal to the displacement of the material point divided by the wavelength (A/Λ), is very low. It is typically less than 10^{-8} and experimenters in the field of elastic non-linearity [Johnson and Rasolo-fosaon, 1996] find it extremely difficult to obtain strain greater than 10^{-6}. We find the opposite situation in static tests since, due to the technological constraints in measuring displacements, strain of less than 10^{-4} are difficult to measure. The sometimes large differences observed between static and dynamic moduli, especially in highly porous or microcracked rocks, are therefore most likely the result of this large difference in the strain value.

Solving these problems of static moduli/dynamic moduli relation would probably require the development of experimental devices which would fill this technological "gap" in the field of intermediate strain.

1-3.2.3   Laboratory velocity measurement principle and "path dispersion" problem

A) Velocity measurement principle

In the laboratory, measurement of ultrasonic mechanical wave velocities is based on the use of transducers whose active part is a piece of piezoelectric ceramic. A variation in the electrical potential across the ceramic produces a mechanical deformation, and vice versa. It is therefore extremely easy to generate and analyse ultrasonic frequency mechanical vibrations.

The simplest method to measure velocity is to record the time taken by the first vibratory energy to cross the sample (first break) as shown on the example of Figure1-3.17). If a standard is available (made from aluminium for example) in which the propagation time (transit time) of the signal is accurately known, it is easy to calculate the transit time in the sample studied, by subtraction, and therefore the propagation velocity. This method is not always suitable, however, if good accuracy is required. If the material is not perfectly elastic (as is the case with rocks) the acoustic signal used for the measurements, which is not a single frequency, deforms during propagation by selective filtering of the high frequencies and, in theory, cannot be compared with an elastic reference in the time domain.

In practice, the first break is always easier to pick up for the P waves which, by definition, are the first to appear over the very low background noise. In contrast, it is sometimes very difficult to distinguish the precise arrival of an S waveform (S first break) over the noise related to the reflected P waves and the various P-S converted waves.

Above all, however, if the material is no longer homogeneous at the scale of the wavelength (Λ), which is frequently the case when using ultrasounds (centimetric Λ), scattering effects qualified as "path dispersion" [e.g. Cadoret *et al.* 1995] may seriously disturb the measurement.

B) Path dispersion

A simple way of explaining this "path dispersion" is to say that some of the acoustic energy (generally small) takes faster or slower paths, depending on the rock heterogeneities. This phenomenon is schematised on Figure 1-3.18 illustrating an ultrasonic propagation model [Cadoret, 1993, in French] in a heterogeneous sample. As opposed to the case of the homogeneous medium, the acoustic signal propagating in a heterogeneous medium undergoes time defocusing: some of the energy arrives before (precursors) and some after (secondary arrivals, codas, etc.) the "main energy packet". The precursor arrival time is measured by the first break.

Emitter   Receiver

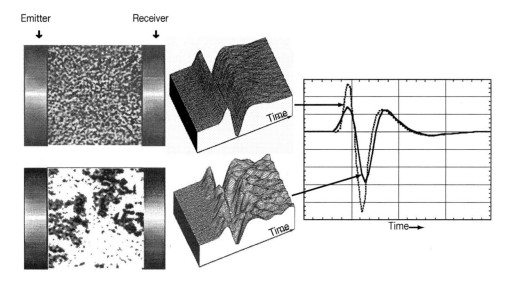

Figure 1-3.18 Path dispersion modelling (according to Th. Cadoret [in French, 1993])

This path dispersion may even be encountered in rocks whose "petrographic granulometry" is small compared with the wavelength and where there seems to be no danger of diffraction phenomena. This is due to the fact that "physical granulometry" (impedance), the important factor for wave propagation, may be radically different from petrographic granulometry. An example of this situation is shown on Figure 1-3.19 which represents Brauvilliers oomoldic limestone both in thin section and in X-ray tomography.

The latter method (§ 2-2.3, p. 344) can be used to map the density and porosity contrasts in this monomineral rock. We clearly see that the physical granulometry (tomography) is noticeably coarser than the petrographic granulometry (thin section, § 2-2.1.2, p. 327). This phenomenon is frequently observed in limestone rocks [Lucet and Zinszner, 1992].

The best way of minimising this highly disturbing effect is to measure the velocity "frequency by frequency". This can be done after applying a Fourier transform to the signal recorded (see Figure 1-3.17) to compare it, in the frequency domain, with a reference signal

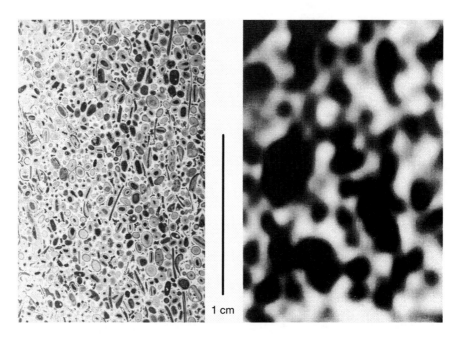

Figure 1-3.19 Example of scale difference between a petrological quantity (grain granulometry)
and a physical quantity (density granulometry)

On the left, photograph of a thin section of rock injected with epoxy resin.
On the right, X-ray tomography.
Brauvilliers oomoldic limestone.

recorded in a homogeneous medium of perfectly known velocity (e.g. aluminium). This is
the phase spectral ratio method (phase velocity). If we consider the frequency spectrum of
the phase of the signal studied and of a reference sample, we can easily calculate the phase
difference between these two signals. The phase difference ($\Delta\varphi$) *vs.* frequency (f) relation
will be approximately linear over the frequency intervals for which the quality factor Q
(viscostatic attenuation) remains constant. We see that this assumption of constant Q is
often true on the frequency passbands of the usual ultrasonic transducers. Since the phase
velocity in the reference sample is known, it is easy to calculate the phase velocity in the
sample studied from the gradient of the straight line $\Delta\varphi$ *vs.* f. This method provides excellent
results provided that the quality of the acoustic signal is good. Dampened transducers are
therefore required, which may sometimes make it more difficult to construct devices for
measurement under confining pressure. Obviously, the result obtained depends on the
quality of the spectra calculated, in other words how the signal portion studied was chosen.
This phase velocity method is therefore not quite as suitable for automated measurements as
the first break method or the signal correlation method (not discussed here).

1-3.2.4 Effect of differential pressure on velocities: Hertz coefficient

A) Definition of the Hertz coefficient

In rocks, the elastic wave velocities depend on the differential pressure (P_{diff}) applied to the sample. The differential pressure is equal to the difference between the confining pressure (P_{conf}) and the pore pressure (P_{pore}). In some cases, this pressure effect is very strong. The reason for this dependence is well known [Birch 1960, Walsh and Brace 1966]: it is mainly due to the ubiquitous mechanical microdefects (microcracks, grain contacts, etc.) which close under the effect of differential stress, thereby increasing the rigidity of the material.

Experimentally, we observe that the velocity *vs.* differential pressure relation obeys a power type law, at least over a limited pressure interval. Figure 1-3.20 shows that the

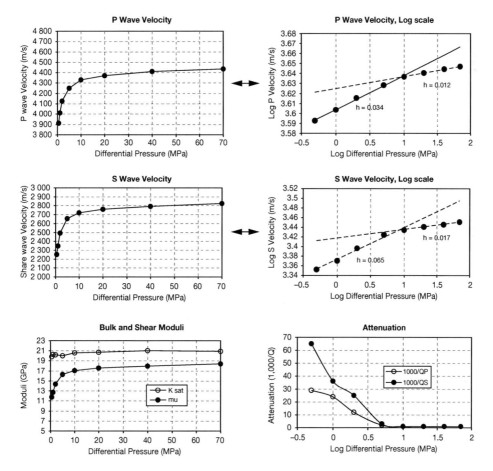

Figure 1-3.20 Effect of differential pressure on the velocities,
moduli and ultrasonic attenuations of Fontainebleau sandstone ($\phi = 0.206$, Kair = 2900 mD)
Linear scale (on the left), Logarithmic scale (on the right).

formula $V = kP^h$ is perfect for the P waves and very satisfactory for the S waves, as long as pressures less than and greater than 10 MPa are considered separately. Note that the change in the rock behaviour around 10 MPa is also highly visible when we consider the attenuation (expressed in $1\ 000/Q$). Above 10 MPa, attenuation is too low to be estimated using the spectral ratio method. In this clean sandstone, attenuation and velocity variation both have the same cause: the grain contacts (an extreme example is shown Fig. 1-2.16, p. 148) which are virtually all closed under a differential pressure of 10 MPa. This limit varies considerably depending on the rock type. In Vosges sandstone with low clay content, this limit is above 40 MPa.

Hertz demonstrated by calculation that in a packing of isodiametral spheres, in elastic contact, the velocity varied with pressure according to a power function and that for the P waves, the exponent was 1/6. The exponent h is therefore known as the "Hertz coefficient".

To evaluate the effect on wave velocities of a differential pressure variation in a reservoir rock, we simply need to know the Hertz coefficient. It is then easy to calculate the proportionality factor of the power function and apply it to the values supplied by sonic logging.

B) Hertz coefficient measurement and core measurement representativeness

It is extremely easy to measure the Hertz coefficient on sample: simply measure the velocity variation under the effect of the differential pressure and calculate the gradient of the linear relation V *vs.* P_{diff} using bilogarithmic axes. The most important point is to accurately measure a relative velocity variation, which corresponds to the simplest case.

To our knowledge, no method is available to measure the Hertz coefficient *in situ*. We are therefore forced to rely completely on laboratory measurements for which there is a problem of core mechanical representativeness.

During coring operations, the rock undergoes sudden stress variations (relaxation) (see § 1-1.1.6, p. 41). The mechanical damage suffered by the rock may be increased by alteration of some minerals (especially clays) due to drying. We may therefore legitimately question the mechanical representativeness of the cores. It is considered that returning the cores to the confining pressure (sometimes to values greater than those found in the reservoir) may compensate this effect for velocities, for example. What is the situation, however, as regards the Hertz coefficient which precisely characterises the sensitivity to pressure?

Some experimental studies [Ness *et al.*, 2000] demonstrate this damage to the samples. We tend towards a similar conclusion by statistically comparing the values of the Hertz coefficient measured on cores and on outcrop samples. The outcrop samples have undergone very slow decompression (over geological time scale) and have therefore been protected from this cause of damage. A long weathering process started, however, on approaching the surface.

The histograms of Figure 1-3.21 show that, for sandstone samples, the Hertz coefficients are statistically much larger for core samples than for outcrops, both for P waves and S waves.

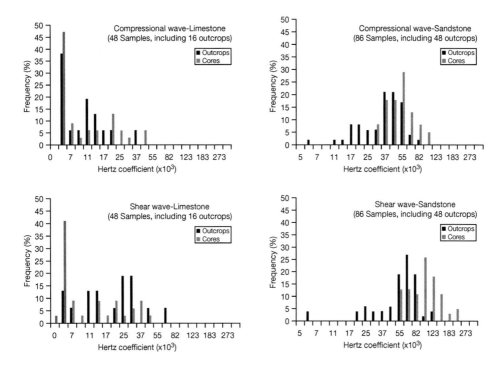

Figure 1-3.21 Distribution of Hertz coefficients in sandstones and limestones.
Samples from outcrops and cores

C) Values of Hertz coefficients in rocks, the limestone-sandstone contrast

There is a striking contrast between the Hertz coefficients of limestone and sandstone. Figure 1-3.21 shows that the Hertz coefficient for more than $^3/_4$ of the limestones reported is less than that of the sandstones, for both P and S waves, to such an extent that there is practically no overlap between the histograms. This is a quantitative illustration of the well-known general observation: velocities in outcrop limestones are much less sensitive to differential pressure than in sandstones.

The reason is quite simple: in limestones, mechanical microdefects are much less frequent than in sandstones, since calcite cement forms much more rapidly. Not enough data are available on dolomites to obtain a statistical value, but the Hertz coefficients seem much higher in sucrosic dolomites (intercrystalline contacts) than in the other carbonate rocks.

Note on Figure 1-3.20 that differential pressure has virtually no effect on the bulk modulus K_{sat} of Fontainebleau sandstone. The slight fluctuations at low pressures lie within the margin of experimental uncertainty. The effect of differential pressure corresponds entirely to the variation in shear modulus. This is frequently observed in clean and very low clay content sandstones. In shaly sandstones as such, however, the bulk modulus may vary considerably with differential pressure. This – at least partial – insensitivity of K_{sat} to the differential pressure explains the distribution of HertzS/HertzP ratios in sandstones, where the modal value is about 1.5.

D) Notion of terminal velocity

This dependence of velocity on pressure (sometimes very high) poses a practical problem: what differential pressure state should be chosen to define the characteristic velocity of a rock in order to determine velocity/porosity relations, for example? In practice, we use the notion of terminal velocity, based on a number of converging observations. Firstly, the relation between velocity and pressure is such that, for high pressures, the relative velocity increase drops sharply. For geophysical applications, the maximum differential pressures exerted on porous rocks are in the region of a few tens of MPa. If a porous rock is subjected to excessive differential pressure, the structure may be destroyed by the pores being crushed (§ 1-1.1.6A, p. 42). This phenomenon is quite clear in some limestones. For each rock therefore, there is a "limiting" upper velocity known as the "terminal" velocity. This velocity will be used to characterise the material.

Although the definition may appear shocking due to its lack of precision, it nevertheless remains a useful example of pragmatism to characterise the properties of consolidated rocks. This notion is of no use, however, if the material undergoes significant variations under the effect of even moderate stress. This is the case with clays or surface sediments undergoing mechanical "consolidation".

1-3.2.5 Effect of the saturating fluid: Gassmann equation and its linear approximation as a function of K_{fl}

A) Biot-Gassmann equation

a) Complete expression

Numerous publications have been dedicated to the Biot-Gassmann equation, used to quantify the effect of a variation in saturating fluid on seismic velocities. For a detailed study, readers may refer to the general books mentioned above [Bourbié *et al.*, 1987, Mavko *et al.*, 1998]. We will restrict ourselves to a brief summary.

Gassmann [1951] proposed an explicit expression of Biot's macroscopic parameters in terms of petrophysical parameters of direct practical interest to calculate the effect of fluid on velocities. We consider two different states for a fluid-saturated porous medium:

- the drained or dry state in which the pressure of the saturating fluid remains unchanged during the mechanical stress (local variations in fluid volume induced by the vibratory deformation being "compensated" by the outside);
- the non-drained or saturated state in which the local variations in fluid content are zero during the mechanical stress. Some of the vibratory stress will be taken up as a pressure variation in the saturating fluid.

Gassmann calculated, for a virtually static stress, the difference in elastic modulus between a drained porous medium and the same medium in non-drained state. This theory assumes a continuous and homogeneous porous medium, but implies no conditions on the pore geometry.

In order to apply the Biot-Gassmann equation to wave propagation (dynamic case), the frequency must be low enough so that the variation in saturating fluid pressure induced by passage of the wave is homogeneous in the porous medium (no gradient).

The Biot-Gassmann equation establishes a relation between the dry (K_{dry}) and saturated (K_{sat}), bulk moduli, using parameters:

- specific to the porous medium: porosity ϕ and incompressibility of the mineral forming the grains (K_{grain}). Values of K_{grain} are given in Table 1-3.2;
- specific to the saturating fluid: K_{fl}, fluid bulk modulus.

The dry bulk modulus (K_{dry}) is associated with that of the rock saturated with gas under low pressure (dry rock) since the fluid bulk modulus (K_{fl}) is equal to zero. It is important to note that, experimentally, the modulus measured on a dry rock may be different from Biot-Gassmann's modulus K_{dry} since, in terrestrial atmosphere, a "dry" rock always contains adsorbed water which may have a considerable influence on the capillary forces at the grain contacts (*cf.* Kelvins's formula, p. 77). These capillary forces modify the rock rigidity in a way which is totally impossible to include in a Biot-Gassmann approach and taking into account moduli in this saturation state may seriously corrupt the results. To minimise these special capillary effects, the water content of the rock must be sufficiently high (a few percent saturation?).

A practical expression of the Biot-Gassmann equation involves the Biot coefficient β [1941]:

$$K_{sat} = K_{dry} + \beta^2 M, \text{ where}$$

$$\beta = 1 - (K_{dry}/K_{grain})$$

$$\text{and } M^{-1} = [(\beta - \phi)/K_{grain}] + \phi/K_{fl}$$

Remember that since non viscous fluids do not bear the shears, the dry (μ_{dry}) or saturated (μ_{sat}) shear moduli are equal.

b) Linear simplification

In the previous expression, we can see that the first term of the expression in M^{-1} is often much smaller that the second since K_{grain} is always much bigger than K_{fl}. Neglecting this first term, we obtain a very simple expression:

$$K_{sat} \approx K_{dry} + (\beta^2/\phi) K_{fl}$$

The error generated can be estimated by calculating the difference between the two expressions for values corresponding to K_{dry} and K_{grain} parameters typical of limestones and sandstones [Rasolofosaon and Zinszner, in French, 2002]. For limestones (high K_{grain} of calcite) of porosity greater than 0.2, when considering usual fluids, the difference between the result of the simplified formula and that of the complete expression remains well below 0.05, in terms of elastic modulus (and therefore 0.025 in terms of velocity). For porous sandstones, it is about 0.05 and therefore negligible compared with experimental uncertainties. The difference between the two expressions may only become significant in the case of a low porosity reservoir (0.1), especially if it is sandstone.

B) Principle used for experimental verification of the Biot-Gassmann equation

The above simplified equation forms a basis for a simple experimental verification of the Biot-Gassmann equation (although more complicated, the procedure remains valid if the complete formula is used). By replacing fluids of varied incompressibility in a porous medium and measuring the compression (Vp) and shear (Vs) wave velocities, it is easy to calculate $K_{sat} = \rho[Vp^2 - (4/3)Vs^2]$ where ρ is the rock density (see § 1-1.1.5, p. 26). The simplest and most useful expression of our measurements will therefore be the graph of K_{sat} vs. K_{fl}, which, in the general case where the Biot-Gassmann equation applies, will be a straight line of gradient β^2/ϕ intersecting the y-axis at K_{dry}.

If the porosity is accurately known, as it is often the case, it will be easy to calculate the Biot coefficient and derive the experimental value of K_{grain}. One can demonstrate that the value obtained for K_{grain} is highly sensitive to experimental uncertainties, the slightest fluctuation on the gradient of the K_{sat} vs. K_{fl} regression line generating high variations in the value of K_{grain}. We also know the rock type and therefore the order of magnitude of K_{grain}. K_{grain} is sometimes known *a priori* accurately in the case of monomineral rocks (e.g. limestones). However, even if there is a certain degree of uncertainty (as with shaly sandstones), the calculated value of K_{grain} is an excellent way of checking the quality of the experimental results.

Figure 1-3.22 shows the results obtained for an Estaillades bioclastic limestone (Lubéron, France, $\phi = 0.285$, $K_{air} = 235$ mD) which has undergone numerous fluid substitutions.

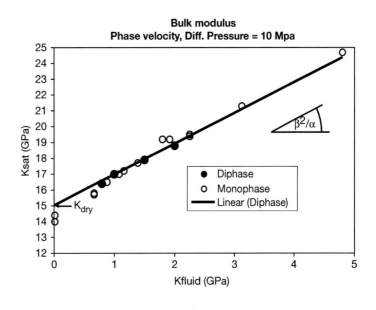

Saturation	K_{fl} (GPa)
Wet	0
Vacum.	0
Water	2.25
Ethanol	1.08
EthylGlyc.	3.12
Glycerine	4.8
Pentane	0.67
Heptane	0.88
Soltrol	1.16
Kerosene	1.39
Albelf	1.8
Polyal	1.92
Swi - Pentane	0.79
Swi- Heptane	1
Swi- Kerosene	1.51
Swi-Polyal	1.98

Figure 1-3.22 K_{sat} *vs.* K_{fl} relation in Estaillades bioclastic limestone for various monophase and diphase saturations with $S_{wi} = 0.26$

The quantitative results of the linear interpretation are summarised below:

Saturation type	Number of experimental points	β	K_{dry} (GPa)	K_{grain} (GPa)
All saturations	21	0.79	14.2	68
Monophase	17	0.79	13.9	67
Diphase	4	0.75	14.9	58

This sample consists exclusively of calcite whose K_{grain} is about 70 MPa. The results on K_{grain}, considered as a control parameter, are therefore extremely good (we must not forget that the parameter is highly sensitive).

Using the phase velocity method we obtained, in limestone rocks, numerous results of this quality which in our opinion represent an extremely convincing experimental verification of the Biot-Gassmann equation, in limestones. In spite of what geophysicists often think, the Biot-Gassmann equation is largely verified experimentally in carbonate rocks. Failures of this verification are due to the disturbing effect of the path dispersion (§ 1-3.2.3B, p. 231) being incorrectly taken into account.

1-3.2.6 Porosity/lithology/velocity empirical relations

In this paragraph, we will give a very brief description of the porosity/lithology/velocity relations in rocks. The first objective is to propose a few values which can be used to build simple models, e.g. to understand repetitive seismic operations. For more detailed information, readers can refer to Mavko *et al.* [1998].

A) Choice of elastic parameters to be studied: Vp, Vp/Vs, K_{dry}, K_{grain}

The elastic parameters whose relations with porosity and lithology are to be studied will depend on the particular applications concerned. We will therefore have to consider several expressions of the same phenomenon, bearing in mind that choosing the most suitable expression may considerably simplify the practical solution.

The initial studies conducted on elastic wave velocities in rocks focused mainly on sonic log analysis in terms of porosity. This explains why the Vp *vs.* φ relations are the most numerous. The greater difficulty in measuring S waves and the lack of S velocity operational log analysis until around the 70's, limited the number of publications on S velocity values. We explained in the previous paragraph that the "dry" bulk modulus K_{dry} of the Biot-Gassmann theory was an elastic constant highly adapted to rock characterisation. We will therefore recalculate the Vp relations as K_{dry} relations. We obviously require the shear modulus to do this and we are faced with the problem of relative shortage of data on S velocities. The solution to this problem is simplified by the use of the Vp/Vs ratio, which depends much more on lithology than porosity.

Concerning the Vp/Vs ratio (see Table 1-3.2), note that in the theory of elasticity, there is a one-to-one relation between this ratio and the Poisson coefficient. As Thomsen [1990] said, however, "Poisson was not a geophysicist!" and the Vp/Vs expression is more suited to geophysical analysis. We will therefore restrict ourselves to this expression only.

We provide highly summarised results on the K_{grain} values which obviously depend on lithology only, then on Vp and Vp/Vs, enabling us to calculate the equivalent results on K_{dry}.

B) Values of K_{grain}, μ_{grain} and "matrix" P and S velocities

Even in case of highly precise mineralogy, the choice of K_{grain} values is not as easy as it seems. Firstly, the basic values available correspond to the elastic tensor of the mineral. Since the usual minerals forming rocks are highly anisotropic, we must calculate an average value corresponding to a set of crystals whose axes would be randomly distributed in space. Several calculation methods can be used, but the result is always the same. Consequently, the method used is not responsible for the dispersion of values observed on the compilations [Mavko *et al.* 1998]. There are probably numerous causes. We cannot even be certain that all the measurements were taken according to professional crystallographic standards. The situation can be summarised as follows: apart from quartz, for which numerous and redundant measurements are available, probably due to the fact that quartz is used in applied physics, there is a real uncertainty regarding the crystal modulus values. As regards calcite and dolomite, the practical consequences resulting from this inaccuracy are not too serious. For clay minerals, however, no reliable values are available (difficulty in obtaining measurable crystals, variation of physical state, water content?). Once again, we appreciate the difficulties of conducting petrophysical studies on clays. Note nevertheless that the moduli of clay minerals in a normal hydration state are remarkably low and that the effect of these minerals on velocities in sandstones is very strong.

Table 1-3.2 below shows average values of K_{grain} and μ_{grain}. We consider that the accuracy for carbonates is sufficient for Biot-Gassmann type applications. For clays the values are not reliable.

Table 1-3.2   Average values of bulk and shear moduli in the main minerals forming sedimentary rocks and corresponding P and S velocities

	kg/m³	GPa	GPa	Values calculated from K and μ		
	ρ	K	μ	Vp	Vs	Vp/Vs
Quartz	2 650	37	45	6 050	4 120	1,47
Calcite	2 710	70	30	6 370	3 330	1,91
Dolomite	2 870	80	50	7 150	4 170	1,71
Siderite	3 960	120	50	6 870	3 550	1,94
Clay Estimations	2 750	25	9	3 670	1 810	2,03

C) Vp-Lithology-Porosity relations

a) Theoretical modelling and formula

Numerous authors have proposed formulae to calculate the elastic moduli of materials formed from mixtures of bodies with known characteristics (a porous material can always be considered as a mixture of minerals and fluid). For identical proportions and elastic characteristics, the result may greatly depend on the geometric distribution of the various phases (see for example the effect of microcracks). Readers interested in these models can refer to Mavko *et al.* 1998, who provide a clear, precise and detailed description of these problems. We will discuss two points briefly:

- The Voigt, Reuss and Hashin-Shtrikman limits, sometimes mentioned in the context of Velocity-Porosity studies. They are the extreme values (maxima/minima) of the mixture moduli, calculated without making any assumptions regarding the geometric distribution of the phases. These formulae are interesting and useful for theoretical mechanical developments, but we consider that in most cases, the difference between the extreme values calculated in this way for a given porosity value is too large to allow practical application of these limits.
- The critical porosity of Nur and co-workers [Nur, 1992]. This is a simple and attractive interpretation of the general form of the Velocity-Porosity relations considered over a large porosity range (therefore including the unconsolidated zones). In our opinion, the interest of this interpretation is more due to the fact that it generates discussion on the rock structure than to its actual usefulness in estimating velocities.

  This notion of critical porosity is based on the observation of two contrasting types of acoustic behaviour. For low porosities, the V *vs.* ϕ relation is roughly a straight line of negative gradient. For the very high porosities, in other words in actual fact for muds or suspensions in which only the liquid phase offers mechanical continuity, the velocity depends very little on the porosity. It is zero if we consider the S waves and low (liquid) for the P waves. The limit between these two domains corresponds to the critical porosity.

  It is obviously difficult in practice to measure the change of V *vs.* ϕ on domains concerning both the consolidated sediments and the suspensions – experiments could be conducted on clean sandstones and sand suspensions. The experimental curves of the consolidated domain can always be extrapolated, however, to estimate this critical porosity.

  Figure1-3.23, which is the diagrammatic transposition of data supplied in Mavko *et al.* [1998], shows that the value of this critical porosity may vary significantly. It is less than 0.4 for sandstones and more than 0.6 for foam lavas (pumice). For cracked crystalline rocks, the value of the critical porosity appears to be very low (0.05). In actual fact, this observation is only valid for very low differential pressures.

When studying mechanical characteristics, this notion of critical porosity is useful to illustrate the need to separate the case of consolidated rocks from that of poorly or averagely consolidated sediments.

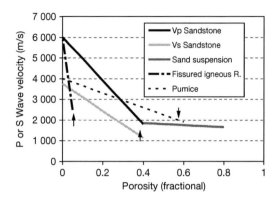

Figure 1-3.23 Notion of critical porosity [Nur, 1992]. Vp or Vs *vs.* φ relation for various types of rock, schematic diagram using data by Mavko *et al.* [1998]. The definition of critical porosity (change of acoustic behaviour domain) is clearly shown during the change from sandstone to sand suspension

b) Empirical or semiempirical formulae for (consolidated) reservoir rocks

We can never overemphasise the sometimes highly important effect of differential pressure (P_{diff}) on seismic wave velocities. All the velocities mentioned in this paragraph are terminal velocities as defined in paragraph § 1-3.2.4.

Wyllie's equation

Wyllie's equation [e.g. Wyllie *et al.*1958] is the oldest and most well-known of the Vp *vs.* φ relations. It is sometimes called the time-average equation since it is based on the weighted average of transit times (Δt) in the solid (Δt_{ma}) and the fluid (Δt_{fl}):

$$\Delta t = (1 - \phi)\, \Delta t_{ma} + \phi \Delta_{fl}$$

The transit time values to be used are summarised in Table 1-3.3.

Table 1-3.3 Matrix and fluid velocity values to be used with Wyllie's equation [Schlumberger, 1989, modified]. The metric equivalents have been rounded quite roughly, with no consequences on the results (see below)

	Matrix (V_{ma}) or fluid (V_{fl}) velocity		Transit time in the solid (Δt_{ma}) or in the fluid (Δt_{fl})	
	Metric (m/s)	**US (ft/s)**	**Metric (µs/m)**	**US (µs/ft)**
Sandstone	5 500-6 000	18 000-19 500	182 – 168	55.5 – 51.3
Limestone	6 400-7 000	21 000-23 000	156 – 143	47.6 – 43.5
Dolomite	7 000-7 900	23 000-26 000	143 – 126	43.5 – 38.5
Freshwater	1 520	5 000	660	200
Water mud	1 615	5 300	620	190

Note the substantial differences between the "matrix" velocities in Table 1-3.3 and the corresponding velocities in Table 1-3.2 which reflect the true situation with compact rocks more closely. We must not try to attach any special significance to this observation since an extraordinary aspect of Wyllie's equation is that it is physically "totally" false (it would only be valid if the wavelength was small compared with the pore dimension, which is in contradiction with the assumption of homogeneous medium) and that it is quite representative of the experimental values. It must therefore be considered as an empirical formula whose coefficient Δt_{fl}, coincidentally, corresponds quite well to the actual transit time in water (some authors even suggest applying the formula to S waves, assigning a special value to the transit time of S waves in water… which is physically absurd). Note that these remarks in no way detract from the empirical value of the formula which was proposed following the study of a large number of laboratory measurements. Wyllie's equation, which is extremely easy to remember, remains an efficient way of determining the order of magnitude of the P wave velocity in a rock.

Raymer (or RHG) equation

This equation [Raymer et. al, 1980] expresses the parameters in Wyllie's equation in a different form. It is proposed for non-clay rocks of porosity less than 0.37:

$$V = (1 - \phi)^2 V_{ma} + \phi V_{fl}$$

The "field observation transform"

This relation is calculated on a large number of field observations (log analysis) [Schlumberger, 1989] using a unique constant $C = 0.7$ (or $C = 0.67$) and a unique matrix velocity value V_{ma} for a fixed lithology. These values are shown in Table 1-3.4.

$$V = V_{ma}(C - \phi)/C$$

Table 1-3.4 Matrix velocity values to be used for the "field observation transform"

	Matrix velocity (V_{ma})		Transit time in the solid (Δt_{ma})	
	Metric (m/s)	US (ft/s)	Metric (μs/m)	US (μs/ft)
Sandstone	5 450	17 850	184	56
Limestone	6 250	20 500	160	49
Dolomite	6 950	22 750	144	44

Note that the results (Fig.1-3.24a) are very close to those given by the Raymer equation used with the same values of V_{ma}. Due to its simplicity (linearity) and its unambiguous empirical nature, we suggest that this formula should be used rather than the others.

c) P velocities in poorly consolidated sediments

For the poorly consolidated sediments, the notion of terminal velocity used for consolidated rocks is no longer meaningful since the state of the poorly consolidated sediment depends directly on its compaction level, which in turn strongly depends on the differential pressure

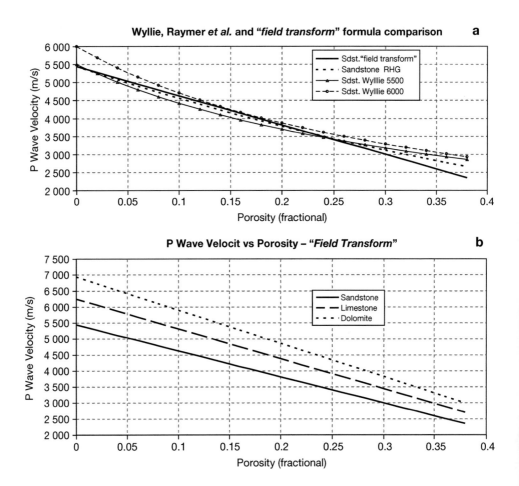

Figure 1-3.24 P Velocity-Porosity relation in non-shaly consolidated rocks, empirical formulae
a) in sandstones, comparison of the Wyllie (for two extreme values of V_{ma}), Raymer *et al.* and "field transform" equations.
b) "field observation transform" for the various lithologies.

exerted on the sediment. In a first analysis, it is therefore better to relate the P velocity value to the compaction level. Although the compaction level is itself directly characterised by the porosity, it will be less appropriate to link Vp and ϕ than was the case with consolidated rocks, if only due to the often central role played by the clay minerals in the poorly consolidated sediments.

We have seen (§ 1-1.1.3, p. 12) the difficulties related to the notion of porosity in these minerals. To respect the practical objectives we set, we provide a very brief description of the P velocities in these sediments according to two parameters: the density (equivalent to the porosity) and the burial depth (equivalent of the degree of compaction).

Figure 1-3.25 shows trend curves plotted using results from the literature. The data of Faust [Faust, 1951], although old, should still be taken into consideration since they provide a summary of a large number of geophysical results.

Concerning the Velocity *vs.* density relation, the formula $Vp = (\rho/310)^4$ [Gardner *et al.*, 1974] is useful for densities greater than 2000 kg/m3 but it does not take into account the notion of critical porosity which corresponds to densities of about 2000 kg/m³.

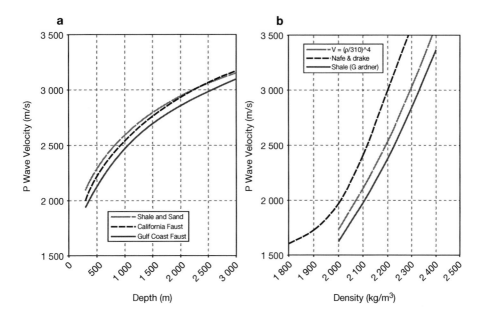

Figure 1-3.25 P wave velocity in poorly consolidated sediments
a) Vp as a function of burial (depth), shale and sand trend curves [Domenico, 1977]; Post-Eocene tertiary sediments [Faust, 1951].
b) Vp as a function of density (ρ); regression $V = (\rho/310)^4$; trend curves [Gardner *et al.*, 1974], [Nafe and Drake, 1963].

D) Vp/Vs ratio: relation with lithology and porosity

The generalisation of S velocity log analysis is more recent than that of P velocity, which explains why fewer *in situ* measurement results are available in public literature. Numerous data of Vp/Vs ratios are available, however, often corresponding to ultrasonic measurements. Note (see above § 1-3.2.5) that Vp and Vs depend on the fluid content. The presence of gas in the pores tends to reduce Vp (dominant compressibility effect) and always increases Vs (density effect alone). The presence of gas therefore always reduces the Vp/Vs ratio. Using Gassmann's equation it is easy to calculate this effect in each special case. We will therefore restrict ourselves to the case of water-saturated rocks.

To simplify matters, we will separate the cases of consolidated rocks (porosity less than 0.3), for which the Vp/Vs ratio is relatively independent of the porosity, and the case of poorly consolidated sediments (sands and shales) where Vp/Vs is highly dependent on the compaction. Note that at the extreme case (transition towards muds, Nur critical porosity), the S velocity tends to zero and Vp/Vs to infinity.

a) Vp/Vs in water-saturated consolidated rocks

A fairly large amount of data is available in public literature [Castagna *et al.*, 1993]. Users requiring the best possible accuracy will find a detailed summary in [Mavko *et al.* 1998], to determine values of Vp/Vs for their applications.

The highly simple observations already formulated by Pickett [Pickett, 1963] may often be sufficient, however, considering that the Vp/Vs values depend exclusively on lithology with the values of Table 1-3.5.

Table 1-3.5   Vp/Vs ratio according to lithology. Approximate average values

Rock	Very clean sandstone	Clean sandst.	Shaly sandstone	Compact shales	Limestone	Dolomite	Salt	Anhydrite
Vp/Vs	1.55	1.6	1.65 – 1.7	1.6 – 1.8	1.9	1.8	1.75	1.85

The results published on the Vp/Vs ratio must be kept in perspective since they are generally obtained by ultrasonic measurements in the laboratory; we have seen to what extent these measurement may sometimes prove to be unrepresentative, mainly due to heterogeneity effects. Figure 1-3.26 shows the results of several hundred ultrasonic measurements from public literature and the IFP database. This gives us an idea of the result dispersion.

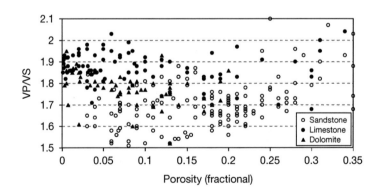

Figure 1-3.26 Vp/Vs ratio in water-saturated consolidated rocks.
Ultrasonic measurements, compilation from public literature and IFP data

b) Vp/Vs in water-saturated unconsolidated sediments

As mentioned earlier, since the S velocity tends to zero in unconsolidated sediments, the Vp/Vs ratio will increase sharply in these formations. To estimate this ratio, we suggest using the trend curves extracted from data of Brie *et al.*, 1995 (Fig. 1-3.27). The results are drawn from log analysis data (dipole sonic tool). This ratio is given as a function of the P velocity (or its reciprocal, Δt, the transit time, according to log analysis practice).

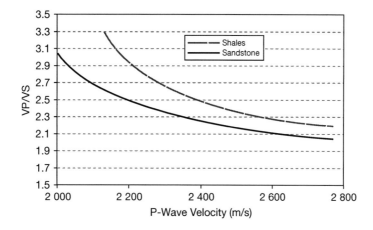

Figure 1-3.27 Values of the Vp/Vs ratio (as a function of Vp) in poorly consolidated sediments. Trend curves plotted using data of Brie *et al.* [1995]

E) Summary of empirical equations: K_{dry}-Lithology-Porosity relations in reservoir rocks

We have seen that K_{dry} is an elastic parameter especially suited to characterisation of a porous medium for applications related to fluid content variation. In practice, the parameter K_{dry} and the ratio Vp/Vs may therefore represent the two elastic parameters required to characterise a rock assumed to be isotropic.

To summarise this paragraph on empirical relations, we can calculate the K_{dry} *vs.* ϕ relation for characteristic lithologies.

For the P velocities of the water-saturated rock, we take the values given by the "field observation transform" (§Cb) and for the matrix density (ρ_{ma}), the values given in Table 1-3.2. The densities then the moduli μ and K_{sat} (§ 1-3.2.1.B) are therefore easy to calculate. Finally, we use Gassmann's equation (§ 1-3.2.5.A) to obtain the values of K_{dry}.

Obviously there is an analytical expression $K_{dry} = f(\phi)$ for each lithology (ρ_{ma}, $Vp = g(\phi)$, Kgrain Vp/Vs, etc.), but it may be relatively complicated due to the form of the Gassmann equation. In practice therefore, it is probably easier to decompose the various calculation steps, using a spreadsheet for example. The values on Figure1-3.28 were calculated using this method. The three lithologies chosen: clean sandstone, limestone and dolomite correspond respectively to Vp/Vs ratios = 1.6, 1.9, 1.8 and K_{grain} moduli = 37, 70, 80 GPa. On the figure, the values corresponding to the extreme porosities (<0.04 et >0.36) are not shown since the Vp *vs.* ϕ relation is not very reliable. For users wanting to reproduce these curves as simply as possible, we give the following polynomial approximations which, obviously, have no special physical meaning. K_{dry} is expressed in GPa and ϕ in fractional value.

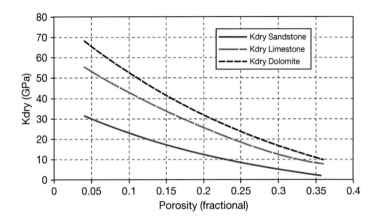

Figure 1-3.28 K_{dry} *vs.* porosity relation for three typical lithologies.
This relation was calculated using the simplest empirical formulae (see text)

Sandstone $K_{dry} = 163.6\phi^2 - 158\phi + 37.6$
Limestone $K_{dry} = 238\phi^2 - 245.5\phi + 65$
Dolomite $K_{dry} = 291\phi^2 - 300\phi + 80.5$

1-3.3 PETROPHYSICAL APPLICATIONS OF NUCLEAR MAGNETIC RESONANCE

1-3.3.1 Nuclear magnetic resonance (NMR), general principles

Nuclear Magnetic Resonance (NMR), discovered in 1946 by Bloch and independently by
E. M. Purcell (who incidentally was the brother of the petrophysicist W.R. Purcell [1949]),
has led to major developments. The most important applications can be divided into two
fields: chemical analysis in the broadest sense (NMR spectroscopy) and magnetic resonance
imaging (MRI) used mainly in medicine.

Applications to petrophysics were proposed as far back as 1956 [Brown and Fatt]. The
main developments in this speciality took place in the 80's, however, when the logging
companies undertook major research studies to develop NMR logging tools. Logging is
currently the main petrophysical application of NMR, imaging still being relatively rare. As
with the electrical and acoustic properties, we will not discuss logging as such. We will limit
ourselves to a brief description of the general principles of NMR application and focus on
laboratory experiments, which give an idea of the possibilities and limitations of log
measurement.

A) Overview of NMR principles

A full explanation of NMR requires a knowledge of nuclear physics and quantum
mechanics. Analogies are often used to introduce NMR to non-physicists and, although

highly useful, their actual bases are generally "incorrect". We will not be an exception to this rule. For a much more detailed and rigorous approach, we recommend interested readers to refer to Dunn *et al.* [2002], dedicated to petrophysics applications, or to Slichter [1990], which is more general.

a) Protons in a permanent magnetic field

Numerous atomic nuclei, especially hydrogen (proton), can be assimilated to magnetic dipoles spinning (Fig.1-3.29a) around an N-S pole axis. In a chemical body in the simplest state, the dipole axis is randomly oriented (Fig.1-3.29b), such that the overall magnetic effect of this set of "atomic" magnets subject to thermal motion is zero. If a static magnetic field B_0 is applied to the body considered, however, the dipole axes line up parallel to this field and the magnetisation of the system will reach the equilibrium value M_0 (Fig.1-3.29c)

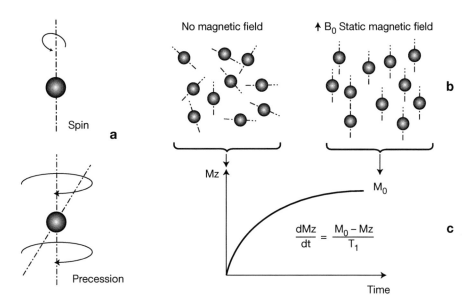

Figure 1-3.29 Diagrammatic representation of:
a) spin and precession.
b) orientation of protons in the static magnetic field B_0.
c) change of magnetisation over time.

b) Precession of spins, gyroscope analogy, Larmor frequency, resonance

Due to the existence of angular momentum, the application of a second magnetic field B_1, perpendicular to the first, results in precession of the dipole axis. The traditional analogy with the gyroscope provides a better illustration of this phenomenon.

The gyroscope (or more simply the spinning top) consists of an inertial mass evenly distributed around an axis. When set into rotation around this axis, if the axis is parallel to the lines of force of the gravitational field, the only movement is the spinning around the axis. The spinning top appears at first glance to be perfectly stationary. However, if the axis of the spinning top is displaced from its equilibrium position by a disturbing force, it generates in space a cone whose axis is parallel to the initial direction.

This rotation movement, much slower than the spin, and in the opposite direction, is known as precession (Fig.1-3.29a). Simplifying the problem, the precession frequency depends on the inertial characteristics of the spinning top, the spin frequency and the gravitational field.

Similarly, the atom with an angular momentum could undergo precession at a frequency which depends mainly on the magnetic field B_0 and a parameter characteristic of the atom: the gyromagnetic ratio. This frequency is known as Larmor frequency (f_L). For the proton, it is about 42.5 MHz per Tesla. Since it corresponds to the radiofrequency band, the magnetic fields exciting NMR are often called "*rf*". The Larmor frequency of the proton therefore depends mainly on the field B_0. We nevertheless observe that this frequency varies slightly depending on the position of the proton in the molecule. This very slight frequency variation (counted in parts per million – ppm) is referred to as the "chemical shift". Extremely important in NMR spectroscopy, it is not used in petrophysical applications.

When excited at the Larmor frequency by an electromagnetic signal (*rf*), the protons can exchange energy and "start to resonate". This interaction, between some chemical elements subjected to a magnetic field and a *rf* signal of precise frequency, is the basic observation underlying NMR.

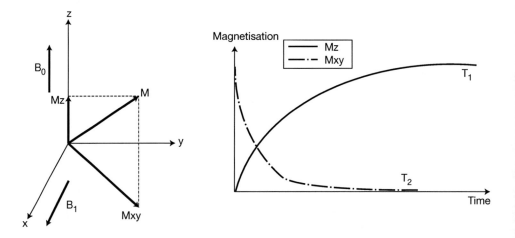

Figure 1-3.30 Representation of magnetisation (in this representation, the *xyz* coordinate system is rotating around the *z*-axis at the Larmor frequency), and definition of relaxation times T_1 and T_2

c) Relaxation times T_1 and T_2

The magnetic field/proton interactions are not instantaneous but subject to time constants (relaxation time) which play a major role in petrophysical applications. Two main parameters are defined:

Relaxation time T_1 (longitudinal relaxation, spin-lattice relaxation)

This constant is related to the variation in time of the magnetisation along the z-axis, under the effect of the static magnetic field B_0 (Fig. 1-3.29c and 1-3.30).

$$dMz/dt = (M_0 - Mz)/T_1$$

T_1 depends on the energy exchanges between the molecular lattice and the protons. For a given spin system (protons in our case), this is a property related to the environment of this system.

Relaxation time T_2 (transverse relaxation, spin-spin relaxation)

This is the time constant related to the variation of the magnetisation component Mxy (Fig 1-3.30). This relaxation corresponds to spin-spin energy exchanges during excitation variations at the resonance frequency, by the field B_1. It is expressed in the form:

$$dMxy/dt = -Mxy/T_2$$

This simple expression is only true in an xyz coordinate system rotating around the z-axis at the frequency f_L.

The main cause of this relaxation is the inhomogeneity of the static field B_0. A spatial variation in B_0, even very low, induces a correlative variation in f_L and therefore a phase shift in proton precession. This special component of T_2 is often called T_2^*. Various experimental procedures are available to estimate the loss of magnetisation caused by macroscopic inhomogeneity of B_0. Generally therefore, T_2^* is not taken into account in the expression of T_2.

The other relaxation causes are related to variations in the local magnetic field induced by the presence of neighbouring nuclei (spin-spin relaxation). In solids, where the nuclei cannot move, local variations related to the magnetic fields of neighbouring nuclei have maximum effect and generate very fast relaxations (very small T_2) even if the static field is perfectly homogeneous. In liquids, however, the nuclei move so quickly that local magnetic fields are on average cancelled out and the only relaxation effect is due to the return of magnetisation on the z-axis. In liquids therefore, T_2 tends to T_1, but never exceeds it.

B) Schematic principle of the laboratory experiment

The schematic diagram of the laboratory experiment is shown on Figure 1-3.31. An electromagnet is used to create the "static" field B_0. The sample to be studied is placed between the poles of the electromagnet, surrounded by a detector coil to record the *rf* induced signals caused by proton resonance excitation. Another coil (the same coil could be used in fact) transmits radio-frequencies inducing the alternating field B_1 perpendicular to B_0 and exciting the resonance.

The *rf* transmitting coil is excited by alternating current pulses at the Larmor frequency. The pulse duration determines the angle through which the proton axis moves away from the direction of the static magnetic field. When the *rf* pulse has stopped, precession of the protons around the static field induces an *rf* transmission whose record, called the Free Induction Decay (FID), forms the basis of the analysis.

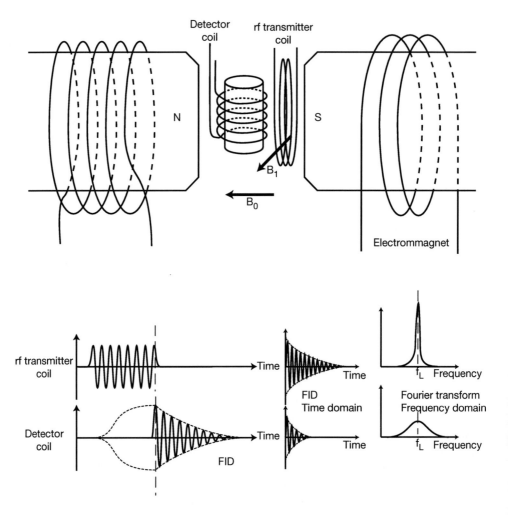

Figure 1-3.31 Schematic diagram of the laboratory experiment

Calculation of the resonance characteristics (T_1, T_2) is based on procedures involving special excitation sequences used to produce axis inclinations through precise angles (90°, 180°) and to process the corresponding FID signals in the frequency domain (Fourier transform). On the highly simplified example of Figure 1-3.31c, we observe that a long FID corresponds to a narrow peak at the Larmor frequency whereas a short FID corresponds in the frequency domain to a very wide spectrum indicating a heterogeneous distribution of the magnetic fields, as is the case in rocks.

C) NMR characteristics of bulk fluids

a) Water

At room temperature and under low or medium magnetic fields, relaxation times T_1 and T_2 for pure water are greater than one second. T_2 is extremely sensitive to the presence of ferrous or paramagnetic atoms in solution. Dissolved oxygen may have a significant effect on the relaxation times and the presence of manganese, even in very small quantities, drastically reduces T_2. This property has sometimes been used to improve NMR logs by "erasing" the NMR signal from drilling mud by addition of manganese salt.

b) Oil

Pure oils (e.g. alkanes) have quite characteristic relaxation times which depend on their formula. The orders of magnitude are quite close to those of water for carbon chains of moderate length.

It is important to note that the number of protons per unit volume, which governs the strength of the NMR response, is fairly similar in light oils and water. The ratio of the number of protons per unit volume divided by the number of protons per unit volume of pure water, called the hydrogen index is close to 1. This is an important remark for the validation of NMR porosity measurements.

Since crude oils are mixtures of a large number of different molecules (and dissolved gases), their relaxation times no longer correspond to precise values but to more or less wide spectra.

The most important point as regards petrophysical applications is the variation of these relaxation times with viscosity. Figure 1-3.32 (data by Morriss *et al.* [1997]) shows a clear relation using logarithmic scale between T_2 and the viscosity. To give an idea of the orders of magnitude of viscosity values, note that at room temperature the viscosity of water is 1 cP and that the container of a body of viscosity 10 000 cP can safely be turned upside down for a short amount of time.

c) Gas

Unlike the more general case of solids and liquids, where the relaxation is due to the dipole/dipole interactions, in gases the relaxation is mainly caused by the interaction between the magnetic moment of the protons and the magnetic fields created by the rotation of molecules, relaxation times decreasing as the molecules rotate more easily under the effect of temperature increase or density decrease. For instance, the relaxation time of methane increases as the pressure is raised and the temperature lowered.

Note also that the hydrogen index of gases is less than 1, sometimes much less than 1, at low pressure. This point has a major impact on porosity measurement. These consequences are also observed on porosity measured by neutron log.

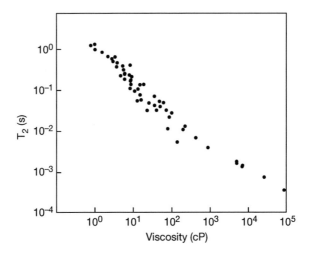

Figure 1-3.32 Relation between transverse relaxation time T_2 and oil viscosity.
According to Morriss *et al.* [1997]

1-3.3.2 Potential applications to petrophysics

A) NMR specificities in porous medium

Although the pioneers who applied NMR to reservoir logging hoped to be able to measure oil and water saturations directly from the contrast between the NMR properties of these bulk fluids measured in the laboratory, the characteristics of proton relaxation in fluids proved to be significantly modified in porous media by the predominant importance of surface effects. One advantage resulting from this complication, however, is that highly useful information concerning the structure of the porous medium can be derived from NMR.

A very brief outline concerning relaxation in porous media is provided below. For further details, interested readers can refer to the books mentioned above or to the article of Godefroy *et al.* [2001]

a) Surface relaxation

Relaxation in porous media is mainly due to very small quantities of ferrous/paramagnetic ions on pore surfaces. In sandstones, the pore surface is often coated with iron or manganese-rich impurities. This does not seem to be so much the case with limestones. Any spin reaching the surface layer is therefore relaxed extremely rapidly. The relaxation time of a fluid mass in porous medium therefore depends mainly on how quickly the protons come into contact with this layer, of estimated thickness a few molecule diameters.

The petrophysical interpretations of NMR in terms of the geometry of the porous medium are based on the assumption of a regular distribution of ferrous/paramagnetic ions on the pore surface. In natural porous media, this regularity is not always observed, however. Confirmation of this assumption is based on the agreement between NMR results and petrophysical measurements such as mercury porosimetry (§ 1-1.2.4D, p. 79). We must remember, however, that this assumption is not always true.

b) Diffusion regime

The rapidity with which the protons come into contact with the surface layer depends on the diffusion kinetics of the protons in the fluid. Simplifying to the extreme, however, we can consider that in natural porous media where the order of magnitude of pore diameters (100 μm) does not generally exceed the diffusion path of a proton during its specific relaxation time (e.g. about 1 s in water), a fast distribution regime applies. In this case, the value of the diffusion coefficient has no influence on the relaxation times. The major parameter is a geometric parameter corresponding to the volume divided by the area of the porous space. We can appreciate the importance of this parameter V/A in the geometric characterisation of the porous space (we mentioned earlier a similar parameter in a 2D space, the hydraulic radius equal to the area/perimeter ratio, for permeability calculations (§ 1-2.1.3C, p. 140)).

An experimental proof of the validity of this diffusion regime is provided by the fact that the relaxation time in a water-saturated rock is virtually independent of the temperature, whereas the opposite is true for the diffusion coefficient.

B) Determination of various petrophysical parameters

a) *Structure of the porous network inferred from the ratio V/A, consequences on capillary properties and permeability*

The basic analytical result for application of NMR to petrophysics is the amplitude distribution curve associated with a relaxation time (Fig.1-3.33). Although this NMR spectrum looks very similar to the mercury porosimetry spectrum its meaning is quite different. In a first analysis (not strictly true), the area under the curve between two relaxation time values is equal to the proportion of porous space whose relaxation time lies between these limits. Assuming that the average surface relaxivity per pore does not depend on the pore type, i.e. that the average density of ferrous/paramagnetic impurities on the pore surface does not depend on the pore type, then this NMR spectrum corresponds to a distribution curve of the parameter V/A (volume to area ratio) in the pores.

We must emphasise the meaning of this curve, since it is fundamental for applications describing the microstructure: note once again that, although its dimension is length (we mentioned earlier its similarity with the pore radius), the parameter considered V/A is not a pore radius in the common sense and certainly not a pore throat radius. The analogy and comparison often made with the mercury porosimetry spectrum may lead to confusion. The term "NMR pore radius" may prove useful in the discussion.

When describing NMR applications to petrophysics, the interpretation of NMR in terms of permeability, capillary pressure curve and "irreducible saturation", etc. are often considered under different topics. This method does not seem appropriate since it incorrectly suggests to readers that independent interpretations can be deduced from NMR measurements. This is obviously not the case. These interpretations are derived from the same parameter V/A, the "NMR pore radius".

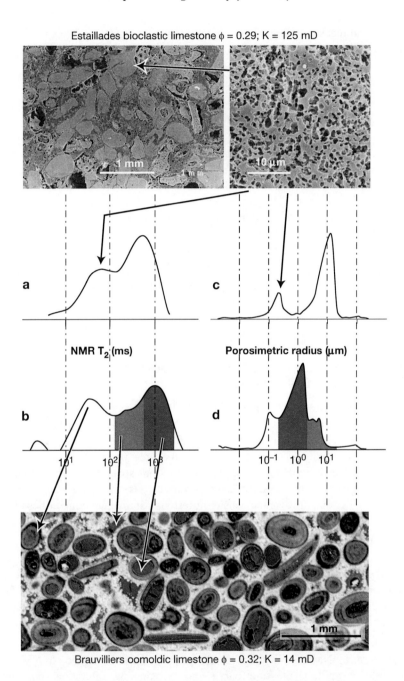

Estaillades bioclastic limestone φ = 0.29; K = 125 mD

1 mm

10 µm

a

c

NMR T$_2$ (ms)

Porosimetric radius (µm)

b

10^1 10^2 10^3

d

10^{-1} 10^0 10^1

Brauvilliers oomoldic limestone φ = 0.32; K = 14 mD

Figure 1-3.33 Comparison of amplitude spectra T$_2$ (parameter V/A) and porosimetry access radius spectra in two limestones with double (Estaillades) or triple (Brauvilliers) porous networks. For Brauvilliers limestone, note the inversion of the NMR and porosimetry peaks as regards the moldic (blue) and intergranular (red) porosities. For Estaillades, epoxy pore cast (§ 2-2.1.3B, p. 331), for Brauvilliers, thin section with two resins, shown on Figure 2-1.11., p. 338. NMR data extracted from [Fleury, 2002]

In the chapters corresponding to each of these petrophysical properties, we described the geometric parameters affecting them. They are summarised as follows:

- In case of perfect wettability, the capillary pressure curve during drainage depends only on the distribution of the pore access radii (capillary thresholds) (§ 1-1.2.2C, p. 56). The "irreducible saturation", a very vague notion which we mentioned on several occasions (especially in § 1-2.3.3A, p. 186), corresponds to very small equivalent capillary pore radius.
- The imbibition curve (§ 1-1.2.2D, p. 59) is affected by a geometric parameter involving the pore radius/access radius ratio.
- For permeability, the determining geometric parameter (§ 1-2.1.3, p. 138) is once again the pore access radius.

In this list, the "pore radius", the parameter closest to the NMR data, is only (partially) involved in the case of the imbibition capillary pressure curve alone. It is therefore not surprising that there may sometimes be a certain degree of uncertainty in NMR interpretation of porous microstructures. This interpretation is based on the assumption that there is a strong relation between the pore radius and the access radius (threshold). This assumption is generally only satisfied in the case of intergranular porous spaces, i.e. the porous spaces present in clean sandstones. Since these sandstones, frequently found in oil reservoirs, are often used as "laboratory rock", it is not surprising to see that the correlations between petrophysical parameters and NMR are statistically quite good. Nor is it surprising to find that the situation becomes more complicated in carbonates: the complex porous spaces in these rocks were described earlier, taking the example of the porosity/permeability relation (§ 1-2.1.4B, p. 149).

Figure 1-3.33 shows an example of ambiguity in the NMR interpretation of porous structures in limestone rocks. Brauvilliers oolitic limestone shows a relaxation spectrum with numerous high values of T_2, which would suggest high permeability. This is not the case since these large T_2 values correspond to the pores inside the oolites. In contrast, the permeability of Estaillades bioclastic limestone is nearly 10 times greater, for a similar spectrum. This is due to the fact that the largest pore radii correspond to well-connected intergranular porosity.

We must therefore be extremely cautious when applying NMR to "less conventional" media.

There is another reason for being extremely careful in geological cases more complicated than the clean sandstones. Interpretation is based on the assumption that the average distribution of ferrous/paramagnetic ions is similar for all pores in the medium considered. This assumption may be questioned in media containing a number of quite different porous networks whose surface states are likely to have undergone highly contrasting modifications during different evolutions occurring in the course of diagenesis (selective weathering, dolomitisation or crystallisation). This case if frequently observed in carbonates or in arkosic and clay sandstones.

b) Porosity

The porosity of a sample can easily be measured by NMR provided that the saturation state of the sample and the hydrogen index of the saturating fluid are known. The magnetisation strength is directly related to the number of protons and therefore to the quantity of fluid present in the sample, weighted by the value of the hydrogen index.

Porosity is therefore a petrophysical parameter which can be easily and accurately measured using NMR. Obviously, we must take into account certain technical constraints and, for example, avoid the use of strong magnetic fields inducing local magnetic gradients related to contrasting magnetic susceptibility in the porous medium, since these gradients could lead to very fast relaxations.

We must also be able to measure the proton magnetisation: the relaxation time must be long enough for this magnetisation to be detected. Laboratory apparatus can measure relaxation times as low as 0.1 ms. For logging tools, the limit is in the region of 1 ms. Consequently, protons placed under conditions such that the relaxation times are very short may not be taken into account in the porosity measurement. These cases of very fast relaxation include:

– Protons in solids (hydrated minerals: gypsum, silica-opal, etc.).
– Protons in solid hydrocarbons such as kerogens and bitumens up to a certain point.
– Protons located in pores with a high content of ferrous/paramagnetic bodies.

In contrast, protons in the water held in clays whose relaxation times lie between 0.1 and 10 ms are generally taken into account in the measurement. This remark makes no assumptions about the true nature of this porosity. This point was discussed in § 1-1.1.3, p. 12.

It is interesting to note the difference which may exist between the neutron porosity (see Fig. 1-1.10), which with the density log is the standard porosity log measurement, and NMR porosity. Neutron porosity corresponds to the total number of protons in the rock, including both solid and fluid fractions. NMR porosity corresponds exclusively to the protons contained in the fluid fractions. This difference may be useful in the detection of tar mats, for example.

c) Wettability

Wettability appears to have been the first petrophysical parameter studied using NMR [Brown and Fatt, 1956]. The idea behind the principle is as follows: relaxation occurs almost exclusively at the solid surface and wettability characterises the affinity of a fluid for this same surface. In two-phase state therefore, the non-wetting fluid must relax as if outside the porous medium (long T_2), whereas the wetting fluid would relax according to the kinetics specific to the porous medium (short T_2). If the proportion of the surface in contact with each fluid can be quantified, we can obtain the notion of fractional (or intermediate, or mixed) wettability.

In the laboratory method, the T_2 amplitude distribution results are compared for different saturation states: sample totally saturated in turn by the two fluids (brine, oil) and intermediate saturations (e.g. S_{wi}: water saturation at maximum capillary pressure and S_{or}:

residual saturation at minimum Pc, see § 1-2.3.3, p. 188). The spectra of bulk fluids can also be measured.

By inverting these results, it should be possible to quantify wettability according to the proportion of the surfaces in contact with each fluid. Recent studies [Looyestijn and Hofman, 2006] indicate good correlation between the results obtained in this way and those obtained by conventional methods (e.g. USBM, § 1-2.2.1B, p. 169). The big advantage with the NMR method would be its speed and its ability to produce results for intermediate states during special core analysis experiments.

C) Magnetic resonance imaging (MRI) applied to the description of fluid distribution in porous media

Significant development work is being conducted on MRI for medical applications, which represent a considerable market in the developed economies. We will give a highly schematic description of the principle underlying this method, using the elements of Lauterbur's famous experiment [1973].

If a magnetic field gradient G_x of direction parallel to the z-axis and of variable strength along the x-axis is added to the static field B_0 (Fig. 1-3.34), the total field B_0+G_x varies along the x-axis. Consequently the Larmor frequency, which is proportional to this field, also varies along the x-axis. The position of a proton in this one-dimensional space can therefore be characterised by its Larmor frequency. Imagine two capillary tubes filled with water, placed at the ends of a line segment AB and placed in the magnetic field B_0+G_x, firstly such that AB is parallel to the x-axis and secondly such that is perpendicular to the x-axis. In the first case, the Larmor frequencies at A (f_{LA}) and B (f_{LB}) are different and if the system is subjected to a wide band *rf* excitation, two peaks at f_{LA} and f_{LB} can be observed on the resonance signal received and transformed in the frequency domain. If, however, AB is parallel to the z-axis, the Larmor frequency will have the same value at A and B and there will only be a single peak in the received signal (at a frequency f_{LAB} between f_{LA} and f_{LB}).

By performing a very large number of acquisitions, changing the relative position of the magnetic gradient, we can reconstruct a tomographic image of the proton distribution. In practical applications, the proton relaxation time is also mapped.

It is interesting to note (this was in fact the primary objective of Lauterbur's experiment) that, unlike the conventional case (optical, acoustic) where the resolution power of a method is limited to the wavelength used, with MRI, radiofrequencies of metric wavelength can be used to identify infra-millimetric objects. In this case, however, we are not dealing with wave propagation disturbed by a change in a physical property, but with a change in a characteristic of the material (Larmor frequency of protons) under the effect of a magnetic field gradient. A practical consequence of this observation is that MRI resolution power does not fundamentally depend on the radiofrequencies involved and therefore on the strength of the magnetic fields.

Certain characteristics of MRI are of great interest for petrophysics since only fluids are taken into consideration (unlike the X-ray CT scan (§ 2-2.3, p. 344) where the X-rays are mostly attenuated by the solid phase) and since, using data on the relaxation times, we should eventually be able to map parameters more subtle than simple fluid contents (e.g. chemical

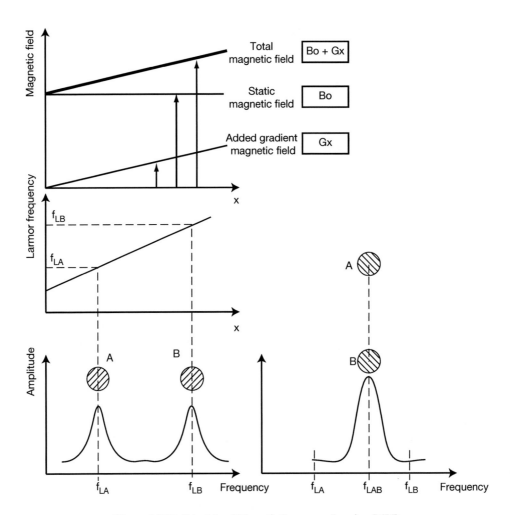

Figure 1-3.34 Principle of Magnetic Resonance Imaging (MRI)

state, wettability). Although quite a few attempts have been made to apply MRI to petrophysics (see for example Chardaire and Roussel [1992] for a description of the first attempts), none has led to routine laboratory applications. The first explanation for this relative lack of success is probably financial, since medical MRI equipment costs several times as much as an X-ray CT scanner and the operating costs are high (with, for example, consumption of liquid helium for the superconductors). The spread of X-ray tomography in the large petrophysics laboratories therefore has probably limited the interest in MRI at the current time.

This situation could change in the future with the generalisation of medical MRI (some forecasts predict the "phasing out" of X-ray CT scanners in the relatively short term) and the corresponding improvements in terms of cost and performance.

D) Perspectives and difficulties of the logging application

Considering the numerous petrophysical parameters which can in theory be derived from an NMR measurement, we could expect to be able to access most of the interesting characteristics of a reservoir using this technique and hope to see one day the emergence of a "single" log analysis method. At least we can understand why the logging companies have invested considerable research effort in this field. In practice, the method is faced with certain difficulties.

The most obvious is the transition from laboratory conditions to borehole conditions. This problem concerns all logging tools (temperature, pressure, aggressive drilling muds, etc.). With NMR, however, it is made worse by a problem due to the shape of the experimental apparatus. In the laboratory, the sample is positioned in the bore of a superconductive magnet. It is therefore easy to ensure that the static field is homogeneous and control the gradients. In the logging tool, the static field is produced by permanent magnets in the tool. The magnetic fields induced in the formation are therefore necessary variable depending on the location of the rock element considered, making analysis more complicated.

A second difficulty (partly related to the previous one) concerns the depth of investigation (DOI). It is the maximum rock thickness (around the borehole wall) for which a representative measurement is taken. With NMR tools, this DOI is between 1" (2.5 cm) and 3" (7.6 cm). These values are comparable to those of conventional porosity log analyses and raise no particular problems provided that the bulk properties of the rock solid phase are considered, hoping that they have not been damaged too much by the drilling. That is the case for porosity and structure of the porous space. The information concerning this structure (distribution of the parameter V/A), which probably represents the main contribution of NMR logging, is not especially affected by the relatively low DOI.

The situation is far less favourable, however, for the information expected concerning the fluid phase (saturation, possibly wettability). In the area around the well, there is a substantial exchange of fluids between the borehole and the geological formation. Even if the nature of the drilling mud and the pressure are highly suited to the conditions present in the reservoir, some of the liquid phase of the mud filtrate penetrates the rock, always disturbing the saturation state and probably the wettability state as well. The saturation/wettability information which may be obtained will therefore not concern the reservoir as such. Solutions have been put forward to partially remedy this situation, taking account of the invasion of the formation by a controlled fluid (doped mud, mud with perfectly known oil content, etc.)

NMR provides the most useful information when used in comparison with or in addition to other methods. It is not a "miracle logging tool" and it is regrettable that falsely optimistic statements concerning the possibilities of this method have deceived potential users, thereby delaying the generalisation of this technique.

PART 2

SCALE CHANGES AND CHARACTERISATION OF POROUS MEDIA: METHODS AND TECHNIQUES

———————————

Measurement Representativeness and Reservoir Characterisation

To determine whether a sample measured in the laboratory is representative compared with the petrophysical characteristics of the geological formation from which it was taken, we must consider various types of problem which can be grouped around two main themes:

- The effects of mechanical or chemical damage on the integrity of the sample and the effects of stresses and temperature on the petrophysical characteristics. We will discuss them briefly in § 2-1.1.
- The scale effects (of the sample, of the measurement characteristics, etc.) discussed in § 2-1.2 to define notions of representative volume, isotropy and homogeneity, and § 2-1.3 concerning scale changes as such.

To close this chapter, § 2-1.4 is dedicated to Rock Typing, in other words the procedures used to sort reservoir rocks into groups designed to simplify flow modelling.

2-1.1 OVERVIEW ON THE EFFECT OF DAMAGE, STRESS AND TEMPERATURE VARIATIONS ON PETROPHYSICAL CHARACTERISTICS

2-1.1.1 Damage to laboratory samples

Rocks studied in the laboratory come from numerous sources. Depending on these sources, various types of damage may have impaired the integrity and therefore the representativeness of the samples. There are two broad types of damage.

- Weathering due to surface elements (air, meteoric water, vegetation, etc.). This type of weathering in the conventional sense of the term mainly involves chemical changes to the minerals. Its consequences may not always be visible (with clays or highly porous limestones, for example). This weathering has an unforeseeable effect on the integrity of natural outcrops or shallow artificial outcrops (e.g. roads). It represents a serious obstacle to petrophysical studies associated with field geology.

- Mechanical change induced by the sudden relaxation of the *in situ* stresses when core samples are brought to surface (or during deep mining). This relaxation results in reactivation of the mechanical defects (grain joints, intercrystalline joints, twin lamellae) and artificially increases the number of "microcrack" type elements in the porous space of the sample. The consequences of relaxation may have a major impact on the mechanical properties (compressibility § 1-1.1.6, p. 39, seismic wave velocity § 1-3.2.4, p. 233) and are discussed in these paragraphs.

2-1.1.2 Influence of stresses and temperature on petrophysical characteristics

Laboratory measurements are easiest to perform when the experimental conditions are close to room conditions. There is no real need to perform measurements under reservoir conditions (P/T measurements) unless these parameters have a significant impact on the result. Since the importance of these effects varies considerably depending on the petrophysical characteristic concerned, they are discussed in the chapter dedicated to each property. Our objective here is to provide some general points in order to simplify the corresponding discussion in the respective paragraphs.

To obtain a better insight into the possible P/T effects, the petrophysical properties we are studying must be divided into various groups (bearing in mind that these groups are totally unrelated to the "static properties" and "dynamic properties" chosen for the plan of this book):

- Properties depending exclusively on the geometry of the porous space (intrinsic properties): the intrinsic permeability, the capillary pressure curve in case of perfect wettability, and the formation factor. In this simple case, the effect on the petrophysical property can easily be deduced by studying the effect of pressure and temperature on the geometry of the porous network.
- Properties depending on the porous geometry **and** the saturating fluids: relative permeability, capillary pressure in case of intermediate wettability, and the resistivity index. This is the "multiphase" development of the previous intrinsic properties. The situation is more complex since the variations in the capillary parameters (surface tension and wettability) may have major consequences. The main parameter likely to have significant effects is temperature. We are always faced, however, with a permanent ambiguity regarding the effects of temperature acting mainly on the properties of the fluids: it is preferable to consider the effect of surface tension and wettability on a particular property rather than the effect of temperature on this property.
- Properties depending on the porous geometry and the nature of the solid. This is the case for the mechanical properties: compressibility and elastic moduli of the dry medium (e.g. K_{dry}, μ). The P/T consequences on the porous geometry (as above) and on the mechanical properties of the grains must be studied. The stress has very little impact on the grain moduli (elastic non-linearity [e.g. Johnson and Rasolofosaon, 1996]), so the only factor which could affect them is temperature. As long as we remain within the first few kilometres of the Earth's crust, however, the effect of tem-

perature is negligible. Obviously, this is no longer true in the deep environment or in special cases (e.g. geothermy).

A) Influence of stresses on petrophysical characteristics

a) Effects of stresses on the pore geometry, "pore" and "microcrack"

To estimate the effect of stresses on the geometry of the porous spaces and the consequences on the petrophysical properties, it is important to distinguish between two types of porous object which exhibit contrasting behaviour:

Pores in the common sense (pore porosity)

The simplest example of this type of porous space is intergranular porosity in a packing of spheres (§ 1-1.1.4, p. 21). These are truly three-dimensional objects and the dimensions in the three directions, although highly varied, have the same order of magnitude. These structures display excellent resistance to the application of stresses (arching in the solid phase). They undergo relatively little deformation and this slight variation retains, in a first analysis, the structure of the porous space (homothetic deformation). Analysis of the consequences may therefore be restricted to those of small porosity variation.

This type of porosity represents virtually all the porosity of common sedimentary rocks.

Microcracks ("crack porosity")

These are virtually two-dimensional objects mainly consisting of a narrow void separating two solid surfaces. The aspect ratio (ratio of the large dimension to the small dimension) is very high. These structures may react significantly to application of stress (e.g. by closing up) and the effect may vary considerably depending on the relative orientations of the object and the stress applied (anisotropy).

In rocks, these "microcracks" correspond to intergranular or intercrystalline joints. Figure 1-2.16 (p. 148) shows an example of this type of porosity encountered in a special facies of Fontainebleau sandstone. Each grain, enlarged by syntactic quartz cement, is separated from its neighbours by a very narrow interstice. The epoxy pore cast technique is ideal to study this type of structure and stereo photographs (§ 2-2.1.3, p. 330) can be used to estimate their thickness, which is very small. In this Fontainebleau sandstone, the origin of the intergranular microcracks is uncertain. The most likely cause is surface weathering.

Since stress has a relatively large effect on their dimensions, these "microcracks" may cause structural variations in the porous geometry, resulting in major consequences on the petrophysical properties. Note that the absolute value of "microcrack" type porosity is always low (less than 1%, i.e. 10^{-2} fractional), since otherwise the medium would disintegrate.

b) Normalisation by porosity variation

In practice, the shape of the relation PPn *vs.* ϕ is often more important than the special petrophysical values PPn associated with a set of samples. When studying the effects of stresses on a petrophysical property, it is therefore best to normalise this effect by taking into account the correlative variation in porosity.

We will take the example of a set of samples with a known relation between a special petrophysical property PPa and the porosity ϕ. If, in addition, on a sample from this set a stress variation (e.g. differential pressure variation) induces a variation $\Delta PPa_{(Pdiff)}$ in the petrophysical property studied as well as a simultaneous porosity variation $\Delta\phi$, we must first check the relative value of $\Delta PPa_{(Pdiff)}$ and $\Delta PPa_{(\Delta\phi)}$, the variation in PPa associated with a variation $\Delta\phi$, on the PPa *vs* ϕ relation established under room conditions.

If $\Delta PPa_{(\Delta\phi)} \approx \Delta PPa_{(Pdiff)}$, we may conclude that the differential pressure variation does not really disturb the geometric structure of the porous space. This is the situation to be expected in a "pore porosity" type medium.

However, if $\Delta PPa_{(\Delta\phi)} \ll \Delta PPa_{(Pdiff)}$, the porous geometry undergoes a radical structural change in the event of stress variation. This is the case of a "crack porosity" type medium.

This normalisation by porosity variation is essential if we are to gain a better understanding of the mechanisms involved in the application of stress variations. Unfortunately, normalisation is rarely possible since accurate measurement of porosity variations is often more difficult experimentally than measurement of the other petrophysical property. This may lead to ambiguous situations. We will take the example of the effect of stresses on the formation factor. We have a set of samples of porosity $\phi_{a1\ to\ n}$ and formation factor $F_{a1\ to\ n}$, measured under room conditions and leading to a "cementation" factor m_a (§ 1-3.1.1D, p. 203). Measurement (which is relatively inexpensive) of formation factors under stress $F_{(Pdiff)1\ to\ n}$ enables us to calculate a new exponent $m_{(Pdiff)}$. The problem is that for m_a to be truly significant, the formation factors $F_{(Pdiff)}$ must be related to the corresponding values of porosity measured under the same differential pressure. Although this should be obvious, not all laboratories bother to do this and some relate the formation factor values under pressure to "room condition" porosities ϕ_a and the variations then observed on m have no physical meaning.

B) Influence of temperature on petrophysical characteristics

As we mentioned above, it is essential that the effects of temperature on the various components of the "porous medium" assembly considered are analysed separately. As we also pointed out, we will only consider the temperature variations encountered in the first few kilometres of the Earth's crust (less than 200°C). In other words, the temperatures at which the usual minerals are not weathered if the pressures are high enough to keep water in liquid state.

We will therefore simply say that, in the order of magnitude of variations considered:

- The effect of temperature on the solid skeleton itself is negligible (expansion of the solid). The same therefore applies for the direct consequences on the "geometric" properties 'intrinsic permeability, formation factor).
- The effect on the properties of the saturating fluids: Viscosity, Resistivity, Nature of the phases may be very high but must not be considered as having an effect on the petrophysical properties as such. This is obvious as regards permeability. The situation is less intuitive if we consider the very marked effect which a change of phase of the saturating fluid may have on the seismic properties of the rocks. In our

opinion, however, this phenomenon should be studied in terms of the impact of the saturating fluid's mechanical characteristics on the wave velocities.

– The effect of temperature on wettability may be very high. This effect depends mainly on the fluid pair involved and may be largely governed by deposition/dissolution of chemicals (e.g. asphaltene) with temperature variations. It is difficult to accurately analyse the effect of wettability on petrophysical properties. Consequently, temperature variation is one of the main parameters to control in case of measurements strongly conditioned by wettability (relative permeability, capillarity).

C) Conclusion: discussion on the need for stress and/or temperature (PT) measurements in the laboratory

a) Relative cost of measurements under room and PT conditions

Obviously, the cost of measurements is not the determining parameter when making technical choices, but it could be worthwhile comparing the financial aspects of taking measurements under room as well as pressure and/or temperature conditions. This is more difficult than it would appear, since the price lists of servicing companies specialised in petrophysical measurements are semi-confidential and variations on the experimental protocol, which initially might appear relatively minor, may have a major impact on the difficulty and therefore the cost. We will simply make some very brief remarks. Compared with measurements under room conditions, working under P/T conditions involves two main modifications: use of much more sophisticated experimental equipment, with obvious consequences on amortisation, and often lengthy sample preparation times (cost of specialised labour) and pressure/temperature equilibrium times (experimental equipment time down).

The "simpler" the experiment, the greater the cost ratio between measurements under room and P/T conditions. In case of routine measurements (porosity, intrinsic permeability) offering equivalent quality (a non-negligible subtlety), the ratio may be 10 or more. This coefficient is probably even higher as regards seismic wave velocities. Even for "special" measurements (SCAL: special core analysis) such as relative permeabilities where data acquisition and processing represent a large proportion of the work, the ratio remains high. SCAL type measurements are rarely carried out under room conditions, however.

b) Main types of petrophysical measurement applied to reservoir characterisation

To continue with our analysis, we must consider the main applications of petrophysical measurements to reservoir characterisation studies:

Routine core analysis

Routine core analysis measurements are used in reservoir screening techniques. To be as representative as possible, 1" × 1" plugs (diameter x length) are sampled at regular intervals (generally once or twice every foot, cf. § 2-1.3.1D, p. 299). Horizontal axis plugs ("H" measurements) are always sampled and sometimes vertical axis plugs ("V" measurements) as well. The intrinsic permeability (which cannot be measured by log analysis) is the first data expected from these measurements. Porosity and matrix density, measured at the same time, are extremely useful when checking and calibrating log analyses and when carrying out accurate core depth matching (§2-1.3.1G, p. 309).

All these measurements are generally carried out under room conditions, primarily for financial reasons in view of the large number of measurements to be taken. Working under room conditions seems reasonable since the sensitivity of porosity to pressure, despite its major impact (compressibility), is not sufficient to have a serious effect on the estimation of porous volumes *in situ*, at least in a first study phase.

With intrinsic permeability the situation is slightly different, since this petrophysical parameter may be very sensitive to pressure in case of poorly permeable cracked media. This low permeability criterion provides a way of avoiding the problem: for samples of average to high permeability (e.g. > 10 mD), measurement under room conditions remains valid.

There are two exceptions to the use of room conditions for routine core analyses: low permeability samples, just mentioned, whose permeability must be measured under pressure, and unconsolidated or brittle samples which must always be measured under effective stress to estimate the porosity as accurately as possible. For routine analysis measurements, the true stresses *in situ* are not necessarily taken into account, isotropic confinement often being sufficient.

Capillary pressure measurements under perfect wettability conditions after thorough washing (centrifuging, "grouped" restored states, or mercury porosimetry) are taken on a proportion (often less than one tenth) of the routine samples. In this case as well, measurements are usually taken at room temperature and the pressure used generally depends on the experimental constraints.

In conclusion, note that apart from the few exceptions mentioned, taking Routine Core Analysis measurements under room conditions seems justified.

Measurements under strict reservoir conditions

Compared with routine core analysis measurements, Special Core Analysis (SCAL) measurements are located at the other end of the range of experimental possibilities. The aim is to study a limited number of samples subjected to perfectly recreated pressure and temperature conditions, to reproduce as accurately as possible the phenomena to be expected during reservoir production (pore compressibility, relative permeability, enhanced oil recovery (EOR)). These very costly and very lengthy experiments (the time factor is important since the results have to be available at least by the start of the modelling calculations) must be studied in the light of two different difficulties:

– The "geological" representativeness of the samples: for obvious financial reasons related to the high cost of these experiments, the number of samples studied is very limited (less than ten in most cases). The samples used must therefore be as representative as possible:

• In geological terms, i.e. the samples must be representative of the layer studied. Although this seems obvious, these samples are not always chosen carefully enough, in other words after a serious Rock Typing study (§ 2-1.4) carried out to check that the samples are indeed representative. This first selection must be

completed, however, by checking the position of the samples with respect to the oil/
water or gas/oil contact (OWC, GOC), to obtain better control over the wettability
(§ 1-2.2.2, p. 178).

- In terms of scale: obviously, experiments as complex as this must be conducted on
samples large enough to minimise the edge effects and the samples must be
"homogeneous" with respect to the experiments performed (§ 2-1.2.1). Full size
samples (diameter close to that of the core) are used to tend towards this objective,
although there is no guarantee of systematically reaching it.

– The true representativeness of the experimental conditions: this is obviously the most
difficult point. It has already been discussed in the paragraph on measurement of
relative permeabilities (§ 1-2.3.4B, p. 194). As a general rule, the reservoir
temperature and the fluid pressures are known fairly accurately. The types of fluids
and therefore their thermodynamic characteristics are also generally known. A
practical problem arises, however: reconstitution of the live oil, the oil in place with
its dissolved gases. Technically speaking, it is quite easy to reconstitute live oil from
dead oil and separator gases (or more simply an artificial mixture of various gases).
The use of live oil makes the experimental procedure much more complicated,
however (storage and handling of hot and compressed fluids). Note that reconstitution
of the live oil may not be perfect as regards asphaltene type molecules, which could
have precipitated out during decompression.

It is harder to determine the true stress state in the reservoir and therefore to recreate
it, but the main difficulty lies in accurately recreating the wettability state of the
samples. When studying this parameter (§ 1-2.2.), we saw its determining effect on
two-phase flow phenomena and the legitimate doubts as to the actual knowledge of
this state *in situ*. Despite all the care taken to restore the live oil for experiments under
reservoir conditions, there is a strong likelihood that this is where the biggest
problems will arise…

… The experiments carried out therefore are generally those not requiring perfect
reconstitution of *in situ* conditions:

Intermediate case: experiments under pressure and temperature conditions, with conventional simplifications

For the rigorous analyst, there simply is no intermediate between Routine Core Analysis and
Special Core Analysis under reservoir conditions since with SCAL, either we are under
reservoir conditions or we are not! In practice, such an extreme principle is rarely applied.
Core petrophysical analysis clearly shows the need for a compromise. Faced with the
enormous difficulties involved when conducting a study under rigorous "reservoir
conditions" and the omnipresent risk of forgetting one of the numerous parameters, it has
become common practice to make certain simplifications which, in the light of experience,
seems quite appropriate. These conventions include:

– Use of pseudo isotropic or diaxial stress state in Hassler type cells (see Fig. 1-2.4,
p. 130). Closed loop cells compensating for the lateral deformation can be used to create an oedometric state considered as more representative of reservoir conditions.

- Use of dead oil whose behaviour, as regards wettability effects, does not seem too different from that of live oil. The *in situ* viscosity must be simulated by adding a light cut of synthetic hydrocarbons (if the problem of asphaltenes does not arise).
- Use of "soft washing" methods (§ 1-2.2.2, p. 176) on native cores, to restore wettability. This means that samples from the water zone cannot be included in the experiment, even if they have the same rock type as their oil equivalents in a structurally higher position.

In practice these experimental conventions, which consist in "blocking" certain parameters in order to focus the study on the effect of other parameters assumed to be more important, are an absolute necessity. We must therefore always bear in mind that experiments described as being conducted "under reservoir conditions" always include conventional simplifications of this type, more or less clearly stated, and that since the reconstitution of true *in situ* conditions is somewhat utopic, perhaps there is no real need to refine the experimental conditions *ad infinitum*.

2-1.2 REPRESENTATIVE ELEMENTARY VOLUME, HOMOGENEITY, ISOTROPY

2-1.2.1 Definition of Representative Elementary Volume (REV)

Application of the macroscopic laws of physics to porous media assumes that these media are continuous, in other words that physical values (porosity, permeability, saturation) can be defined at each point as a differentiable function of the point considered. Discontinuity however is the fundamental characteristic of a porous medium since at microscopic scale a point is either in the solid or in the porous space, if we take the variable "porosity" as an example. The problem of discontinuities is quite common in physics, but what makes porous media so special is that the dimensions of the minimum volumes to be taken into account in order to include the effect of these discontinuities may vary considerably in the same medium, depending on the property considered.

Two different approaches are used to define local properties: firstly, the notion of Representative Elementary Volume (REV) which consists in assigning to a point in space the value of the petrophysical property measured over a certain volume surrounding this point. Secondly, the notion of Random Functions which consists in considering the porous medium as the result of a random phenomenon which will be characterised using statistical methods.

The REV approach has a number of disadvantages, one of the most obvious being that it is poorly suited to coping with macroscopic discontinuities (e.g. layer limits) since the REV method converts these discontinuities into continuous variations. This is the more intuitive approach, however and will therefore be the only one described, whilst emphasising the fact that the Random Function approach leads to much more productive developments, in geostatistics for example.

A) Definition of the representative elementary volume taking the example of porosity

a) Definition

To define this Representative Elementary Volume (sometimes also called the Minimum Homogenisation Volume, although this term seems to be becoming obsolete), we can take for example the case of porosity in a two-dimensional space, e.g. on a thin section.

We will consider an intergranular porous space whose average grain diameter d will be the measurement unit of length (Fig. 2-1.1). Starting from a point chosen at random, if we

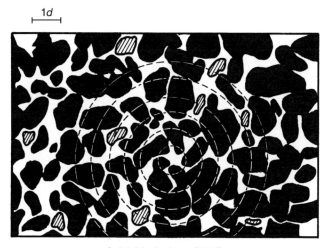

Solid: black phase (72%)
Pore: white and cross-hatched phase (28%)
The cross-hatched phase represents 10% of
the white phase (saturation)

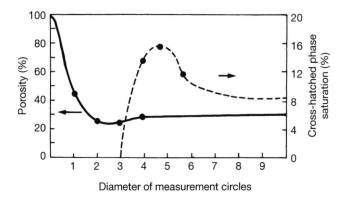

Figure 2-1.1 Definition of the Representative Elementary Volume (REV) for porosity and saturation, on an example of intergranular porous space (top diagram). The bottom graph shows the change in porosity (white phase) and in saturation (cross-hatched area) according to the diameters of the measurement circles.
[Bourbié *et al.*, 1987]

draw a number of concentric circles of increasing area, we can measure the change in porosity according to the circle diameter. On the example shown on Fig. 2-1.1, the centre is located in a pore, the initial porosity value is therefore 1. We reach the mean porosity value of the medium for circles of diameter $2d$ or $3d$. Using this method, we can statistically calculate the area required to obtain a stable value of the variable and thereby define the REV.

For the porosity variable, the order of magnitude of this diameter is 1 to 3 grain diameters in the well-sorted intergranular spaces. As soon as the porous structure becomes more complex (as in some limestone rocks, for example), the dimension of the REV may be very large compared with that of the petrological grain.

b) Specificity of REV with respect to a petrophysical property

Taking once again the example of the intergranular medium of Fig. 2-1.1, if we study a phase dispersed as bubbles representing low saturation (cross-hatched phase on the figure), we observe, using the same method of concentric circles, that the drawing is no longer large enough to define this saturation whose REV has a diameter greater than 6 or 8 d units.

For the same very simple medium, we therefore observe a clear variation in REV dimension for the "porosity" parameter and one of the "saturation" parameters. The same may be true for all the other petrophysical variables. For the mechanical parameters (moduli), we may *a priori* use a volume equivalent to that required for porosity as long as the material is not cracked. If the material is cracked, however, it is obviously the REV corresponding to the cracks which will have to be taken into account and its dimension may sometimes be very large.

Quantification is much more difficult for permeability, since the two-dimensional simplification is impossible. However, this explains why the dimension of the REV for permeability may be much larger than that corresponding to porosity.

The example shown on Fig. 2-1.2 [Greder *et al.*, 1996] illustrates the difficulties encountered in practice when defining an REV for permeability. The values are permeability measurements on cores from a vuggy dolomite reservoir, exhibiting two main vug distribution facies (A and B) at the pluricentimetric scale (Fig. 2-1.2 a). The measurements were taken on full size samples (Volume > 1 000 cm^3) which were then cut into 90 cm^3 and 12 cm^3 plugs. Measurements taken with a micropermeameter (Fig. 2-1.13) were used to investigate volumes of about 0.2 cm^3. The porosity measurements (Fig. 2-1.2 b) taken on the plugs (12 cm^3, 90 cm^3, 1 500 cm^3) show converging values indicating that it might be possible to define an REV at full size dimension. This is not the case for the permeabilities. Of the four samples shown on Fig. 2-1.2c, two demonstrate a tendency to converge (samples B1 and B3) which is not observed with the other two (A1 and B2). No REV can be defined at full size scale (in the meaning of the empirical definition we are using).

B) Characteristic lengths and representative samples

a) Various characteristic lengths: porous medium, physical phenomenon

We can understand the importance of the notion of REV dimension, which is a fundamental characteristic when studying a petrophysical property. We can also appreciate that a second

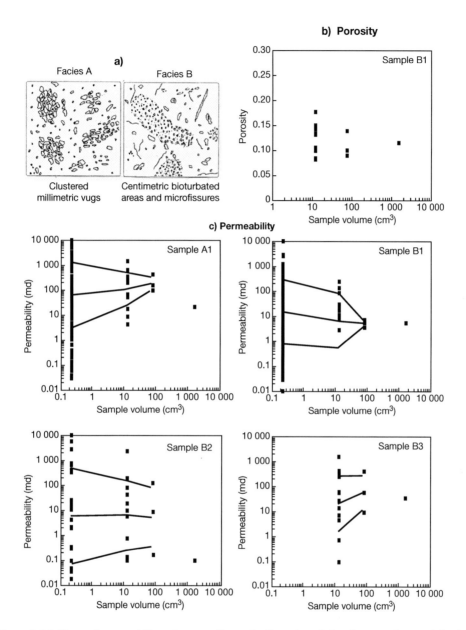

Figure 2-1.2 Change in permeability value according to the dimension of plugs in cores of vuggy dolomite
[Greder *et al.*, 1996]

a) Schematic diagram of two types of vuggy facies represented (A and B). The order of magnitude of the side of the squares is 5 cm.

b) Change in porosity according to plug dimension for the full size sample B1.

c) Change in permeability according to plug dimension for 4 full size samples. The central line connecting the various scale values corresponds to the values of the permeability geometric mean. The two lines on each side represent a difference of approximately one standard deviation around this mean.

type of characteristic dimension is involved: dimensions related the physical phenomenon itself. In the following lines we will use empirical definitions (which are therefore inaccurate and possibly incorrect from the specialist's point of view) to describe notions which would require a more rigorous approach. However, we consider that by so doing we will provide a few basic ideas, necessary in this field to correctly interpret the results of petrophysical experiments. Failure to respect the conditions concerning the characteristic lengths is the main cause of numerous experimental errors, leading to apparently inconsistent petrophysical behaviour.

The three types of dimension to be taken into account are schematised on Figure 2-1.3:

- The characteristic dimension of sample homogeneity for a property and in a given petrophysical state. The REV dimension gives an approximation of this dimension (d on the diagram). In practice, accurate measurement of this dimension involves the use of (geo)statistical methods (range of the variogram, for example) but we will not develop this point since we have not described the Random Functions method. We can therefore see the limits of our qualitative approach, but in fact it is often sufficient to detect the largest risks of error, as we will see in the example described below.
- The characteristic dimension of the physical phenomenon itself. The simplest example amongst the properties described in Section 1 is that of the wavelength Λ of elastic waves (§1-3.2.2, p. 228).
- The sample size C. In a given experiment, this characteristic length is always easy to find! However, it generates challenging practical problems when, in order to perform a rigorous analysis of the other characteristic dimensions, very large samples are required.

b) Notion of valid sample

Petrophysical measurements are taken for two main reasons:

- To characterise a geological formation by means of a limited number of samples, combining their individual properties in up-scaling processes to best define the layer parameters.
- To study the relations between various petrophysical properties in a particular geological formation (for example, to carry out rock typing, § 2-1.4) or more generally when carrying out research on petrophysical processes.

In both cases, the sample must satisfy certain validity conditions. Firstly, it must be homogeneous (further details concerning this notion will be provided below in § B) and implicitly its volume must be much larger than the REV for the phenomenon considered (in other words C ≫ d, using the notation given on Fig. 2-1.3).

The need for a homogeneous sample is quite obvious since, if we are studying the relations between the petrophysical properties A and B and the sample is the combination of two parts of properties A1, B1 and A2, B2, the measurement result will be a combination of relations A1B1 and A2B2 which is difficult to generalise!

The measurement must also have a physical meaning, in other words the REV dimension must be very small compared with the characteristic length of the physical phenomenon being studied.

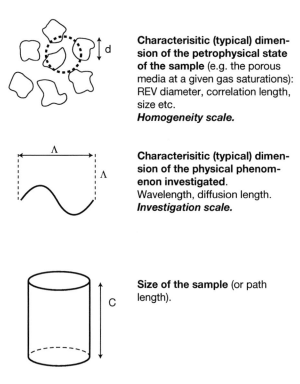

Characterisitic (typical) dimension of the petrophysical state of the sample (e.g. the porous media at a given gas saturations): REV diameter, correlation length, size etc.
Homogeneity scale.

Characterisitic (typical) dimension of the physical phenomenon investigated.
Wavelength, diffusion length.
Investigation scale.

Size of the sample (or path length).

Figure 2-1.3 Diagrammatic representation of various types of characteristic dimension in a petrophysical experiment

For experiments concerning elastic waves, the wavelength Λ must be compared with the REV. The cases studied are summarised in § 1-3.2.2A, p. 228:

- If Λ is significantly larger than the diameter of this volume, then the vibration will exhibit macroscopic homogeneous behaviour insensitive to microscopic discontinuities in the porous medium. Valid experimental results can be deduced from the characteristic values of the medium studied.
- If Λ has the same order of magnitude as this representative dimension, we observe scattering phenomena which radically change the behaviour of the wave in the porous medium.
- If Λ is significantly smaller, the porous medium "no longer exists" as such and the analysis must be repeated at a much smaller scale: the medium to be studied consists of individual grains and pores.

In conclusion, to perform good petrophysical experiments, it is important to respect certain basic validity conditions:

- the sample must be homogeneous;
- the relations of scale $C \gg d$ and $\Lambda \gg d$ must be respected.

Apart from practical considerations, there are *a priori* no special difficulties in satisfying these conditions, as long as the REV does not vary during the experiment. This is not always the case, however, and we may be faced with a difficult problem when varying a parameter such as saturation. We saw on the example of Figure 2-1.1 that low saturations may in some cases generate REV values much greater than that of the porosity as such: during an experiment, the REV may therefore vary to such an extent that the entire experiment could be invalidated.

An example of this type of situation is given below.

c) Variation in REV dimensions during an experiment: example of water/air saturation

Figure 2-1.4 [Cadoret *et al.*, 1995] represents water/air saturation maps obtained by CT scan (§ 2-2.3) on a 6 cm diameter rod of bioclastic limestone ($\phi = 0.28$, $K = 300$ mD). The saturation maps *a, b, c*, illustrate the same sample "slice" during drying. The most striking feature is the location of high gas (air) saturations in a single area on map *a* ($S_w = 0.96$) whose area increases on map *b* ($S_w = 0.92$) which shows a number of new occurrences. Still from the qualitative point of view, it is clear that the REV corresponds to the entire section (6 cm diameter) for map *a*. A detailed quantitative analysis would be required to determine whether the REV of section *b* is still equal to or slightly less than its diameter.

Maps *a* and *b* can be explained by considering that drying is carried out by drainage (§ 1-1.2.2, p. 56), the air (non-wetting fluid) expels the water (wetting fluid), a process during which the non-wetting fluid invades the porous space following the path determined by the series of the largest pore access radii. Air first enters the sample through the largest access radius intercepted by the sample surface, creating the unique region of map *a* (we can see on map *a* the first rare breakthroughs responsible for the other air regions on map *b*.

Saturation maps *a* and *b* are therefore representative of an "initial drainage" phenomenon for which the REV first corresponds to the entire sample, decreasing only progressively as the saturation in non-wetting fluid increases.

In contrast, map *c* ($S_w = 0.61$) shows that the air saturation distribution is virtually homogeneous. The characteristic dimension of REV is close to that corresponding to the porosity as such (i.e. about a few grain diameters). This is due to the fact that, at this stage in the drainage process, the non-wetting fluid has invaded nearly all the pores and the saturation variations "only" concern microscopic changes in the menisci.

In this drying experiment, therefore, the REV varies considerably from a dimension equivalent to the entire sample for low air saturations up to a dimension of several grain diameters at high saturations.

Map *d* illustrates another phenomenon, which is extremely important when interpreting petrophysical measurements. This map strictly corresponds to the same section as maps *a, b, c* and the water saturation ($S_w = 0.92$) is identical to that of map *b*. The distribution of air saturations is nevertheless radically different: the small quantity of air present in the rock is regularly distributed such that the REV dimension is a few grain diameters. This is due to the fact that the air was drawn into the sample in quite a different way, using a "depressurisation" process similar to imbibition. After the capillary rise of water in the dry sample (§ 1-1.2.7, p. 113), the air is progressively evacuated through a series of depressurisation phases under increasing vacuum, followed each time by repressurisation to atmospheric pressure.

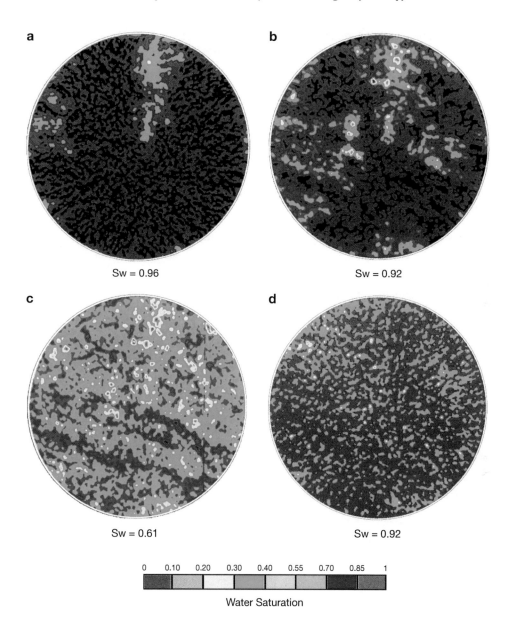

a

Sw = 0.96

b

Sw = 0.92

c

Sw = 0.61

d

Sw = 0.92

| 0 | 0.10 | 0.20 | 0.30 | 0.40 | 0.55 | 0.70 | 0.85 | 1 |

Water Saturation

Figure 2-1.4 Water/air saturation map for the same slice of bioclastic limestone.
Maps *a, b* and *c* correspond to water displacement using a drying process. For map *d*, water displacement occurred via a "depressurisation" process. [after Cadoret *et al.,* 1995].

This is a clear illustration of the principle whereby, in experiments involving multi-phase equilibria, the fluid displacement method may have extremely important consequences on the location of the fluids and therefore on the petrophysical characteristics. The quantitative saturation data alone is far from being sufficient to describe the state of a sample.

d) Consequences of REV variation during an experiment: case of seismic velocities of samples in water/gas saturation

The relations between the seismic wave velocities and the gas saturation are extremely valuable when interpreting reservoir seismics. Some laboratory results may appear to be contradictory, producing different types of V *vs.* S_w relations depending on the experiments and not corresponding to the values predicted by the Biot-Gassmann equation (§ 1-3.2.5, p. 236). This phenomenon illustrates the practical consequences of the previous remarks. Some data extracted from Cadoret *et al.* [1995] are given below.

The results shown on Figure 2-1.5 concern elastic wave velocities measured using the resonant bar technique applied to a large sample (length 100 cm, diameter 6 cm). This method [e.g. Bourbié *et al.*, 1987] consists in bringing the sample into resonance in extension or torsion mode. The resonance frequency is such that the half wavelength is equal to the bar length (or to an integer fraction of this length). Sonic frequency (in the kilohertz range) wave velocities can therefore be measured extremely accurately on rock volumes of pluridecimetric dimension.

In this experiment, the air saturation variation was obtained using the two methods described in the previous paragraph: drying and depressurisation. Note that the lowest water saturation which can be obtained easily by depressurisation is equal to the Hirschwald coefficient (§ 1-1.2.7, p. 114), which explains why the lowest saturations in the depressurisation series are approximately 0.65. The results are quite clear:

- Concerning the shear wave, we observe that the fluid displacement method used has no effect whatsoever on the velocity. The S-wave velocity (Vs on the graph) decreases slowly and regularly as the water saturation increases. In compliance with the theory (§ 1-3.2.5), the shear modulus is unaffected by the presence of fluid (and therefore the fluid displacement method). We only observe a density effect due to the higher water content.

- The situation is quite different for the E-wave velocities (Ve on the graph) similar at this stage of the analysis to compression waves. When the saturation variation is

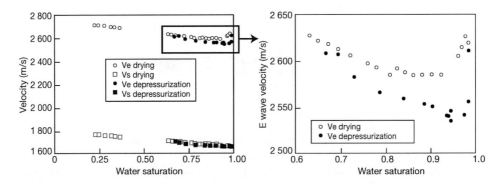

Figure 2-1.5 Variations in velocities Ve and Vs according to the water saturation, resonant bar method. The saturation variation is obtained by two different methods (drying, depressurisation). After Cadoret *et al.* [1995] © American Geophysical Union (reproduced by permission of American Geophysical Union)

obtained using the depressurisation method, the experimental result shows excellent agreement with the predictions of the Biot-Gassmann equation: as the water content increases, the E-velocity decreases regularly (density effect). It is only when the last air bubbles have disappeared (total saturation) that the sudden increase in saturating fluid incompressibility causes in a correlative increase in velocity. In contrast, when the saturation variation is obtained by drying, the velocities no longer follow the Biot-Gassmann equation and vary progressively between minimum values at around $S_w = 0.8$ and the maximum value at $S_w = 1$.

The saturation maps shown on Figure 2-1.4 explain this behaviour. When the saturation varies by depressurisation, the REV measuring a few grain diameters (millimetric, Fig. 2-1.4d) is very small compared with the wavelength (metric). The wave "sees" a homogeneous medium and the result of the experiment has an intrinsic physical value. When the saturation variation is obtained by drying, however, at high values of S_w, the REV is large (pluricentimetric, Fig. 2-1.4a and b) so the wave "sees" two media ($S_w = 1$ and $S_w < 1$) and the velocity measured is the result of a kind of average between the velocities of the two media. The experiment has no intrinsic value and its result cannot be upscaled to other situations.

The effect of REV variation on the result of an experiment is quite clear in this example concerning elastic waves but, obviously, it may be observed with all the other types of petrophysical experiments. It is a major cause of interpretation error.

2-1.2.2 Homogeneity and Isotropy

A) Definition

The notions of Homogeneity and Isotropy are widely used to describe petrophysical properties. Their strict definitions (especially for homogeneity) are greatly simplified by the use of concepts derived from Random Functions (e.g., stationarity). However, in view of our decision not to describe this approach, we will restrict ourselves to a pragmatic definition of these two notions.

Taking a medium on which we have defined a REV, we can consider the result of two types of operation (Figure 2-1.6):

- Make a translation of the REV "object" (Fig. 2-1.6a). If during this operation the petrophysical value considered does not vary, the medium is considered as "homogeneous" with respect to this property. Otherwise, it is "heterogeneous".
- Make a rotation of the REV "object" (Fig. 2-1.6b). By analogy with the previous case, if the value considered does not vary, the medium is "isotropic" and otherwise it is "anisotropic".

Using these criteria, we can therefore characterise a given volume of porous space (e.g. a sample), observing that:

- The notions of homogeneity and isotropy concern a porous medium **and** a petrophysical characteristic. A sample which is homogeneous as regards porosity may be heterogeneous as regards saturation (the example of Figure 2-1.4 could be studied from this angle).

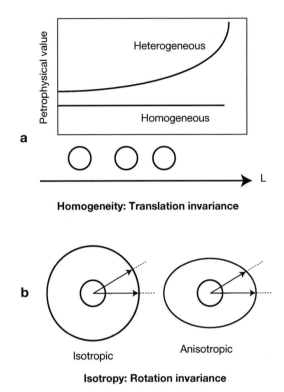

Figure 2-1.6 Definitions of homogeneity and isotropy

– Due to its very definition, the notion of isotropy is only meaningful for the properties likely to vary with direction (tensorial properties). Consequently, the concept of isotropy does not apply to scalar properties (e.g. all the contents: porosity, saturation, etc.). Qualifying porosity as "anisotropic" in finely bedded formations would therefore be incorrect and should be avoided.

B) Anisotropy in natural media

a) The anisotropy ellipsoid

We have seen that porosity and saturations may not be anisotropic. In contrast, the transport properties (permeability, formation factor, elastic wave velocity) depend on the direction considered and are therefore likely to exhibit anisotropic phenomena. The most conventional method of describing anisotropy is based on the tensorial expression of the values of the petrophysical property considered. This highly efficient approach soon involves complex mathematical equations and is unsuitable in view of the choices made for this book. Our description is therefore quite different, based on an analysis of experimental results. Although this method excludes calculations and therefore a quantitative expression of the results, it can be used to describe some points which we consider important, concerning measurement and REV problems.

Anisotropy can be simply illustrated by drawing a line, at a point in the medium, for each direction in space, of length proportional to the value of the property in the direction considered (it would be more appropriate to speak of vector, which would take us back to the tensorial expression of anisotropy). We can therefore build an anisotropy "ellipsoid" (some figures concerning elastic wave velocities, for example, are not ellipsoids). By observing the planes and axes of symmetry, according to a method familiar to mineralogists, we determine the symmetry system of the phenomenon studied.

In the simple case of permeability (like that of the formation factor), we obtain an ellipsoid generally exhibiting 3 orthogonal planes of symmetry (orthotropy). A special case frequently used to represent the permeability anisotropy of bedded sedimentary formations corresponds to an ellipsoid with a plane of symmetry (bedding plane) and an axis of symmetry normal to this plane (sedimentary "vertical"). We then speak of transverse isotropy.

b) Anisotropy measurement: example of permeability

In theory, to characterise anisotropy, we could simply measure the value of the property considered in a certain number of directions in space, the number of independent measurements depending on the symmetry system.

For permeability and the formation factor, this theoretical number is 6. In the more complex case of elastic waves, it may reach 21. In practice, however, in view of the uncertainty regarding the homogeneity of natural media, it is essential to take measurements in numerous directions and on the same rock volume, which is difficult for direct measurements as soon as we move away from cubic geometry in which only three directions can be measured.

Permeability probably has more technical proposals for anisotropy measurement than any other property (measurement on cube, on cylinder generator, etc.). We will summarise below a technique based on the use of CT Scan (§2-2.3) which avoids the difficulty regarding the volume investigated. Using this description, we will also develop a number of considerations regarding the causes of permeability anisotropy in sediments.

If at a point S (Fig. 2-1.7B) of the impermeabilised side of a porous medium (semi-infinite medium) we inject a liquid that is perfectly miscible with the liquid initially saturating the medium, ignoring diffusion, we show that the interface between the two liquids (the invasion front) describes a surface such that the distance from this surface to the injection point is proportional to the square root of the permeability in the direction under consideration (Fig. 2-1.7A) [Bieber *et al.*, 1996].

The interface between a liquid absorbing X-rays (e.g. potassium iodide solution) and a weakly absorbing miscible liquid of similar chemical nature (e.g. potassium chloride solution) can be easily followed using CT scanning. By producing seriated CT cuts (Fig. 2-1.7B), we can describe the three-dimensional shape of this interface and deduce the overall permeability anisotropy using a suitable inversion programme.

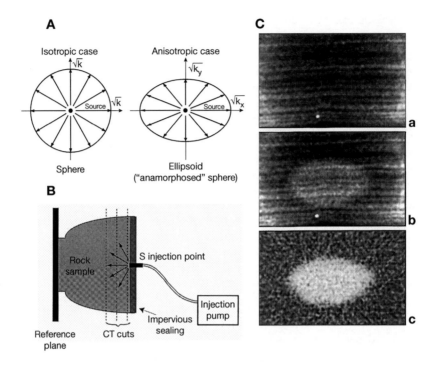

Figure 2-1.7 Measurement of permeability anisotropy by tracer injection and CT

A) Theoretical shape of the injection front Sphere (isotropy), Ellipsoid (anisotropy).

B) Experimental principle.

C) CT results. a): sample in initial state (KCl brine); b): after injection of KI brine; c): image subtraction revealing the shape of the invasion front; Triers Sandstone [Bieber *et al.,* 1996].

In practice, it is very rare to find surfaces resembling an ellipsoid (as in the case illustrated on Fig. 2-1.7Cc). We are more likely to observe interfaces such as those reproduced on Figure 2-1.8. The injection front is highly irregular. The position of the injection front is determined by the presence of microbeds of variable porosity/permeability clearly visible on the porosity map. The more compact (less permeable) beds block the progress of the invasion fronts. Overall, although we can detect that horizontal permeability is much greater than the vertical permeability, it is impossible to calculate a "true" permeability tensor.

Very clearly observed here at centimetric scale, this phenomenon illustrates the true situation encountered in reservoirs at metric and hectometric scales. The reservoir permeability anisotropy depends on the alternation of more or less permeable beds and the horizontal continuity of these beds. In practice, therefore, the strict physical definition of "anisotropy" is unsuitable since it cannot be applied to a homogeneous medium at the scale of definition of this petrophysical property. It is therefore better to keep to the usual definition, i.e. the ratio of horizontal permeability to vertical permeability: Kh/Kv.

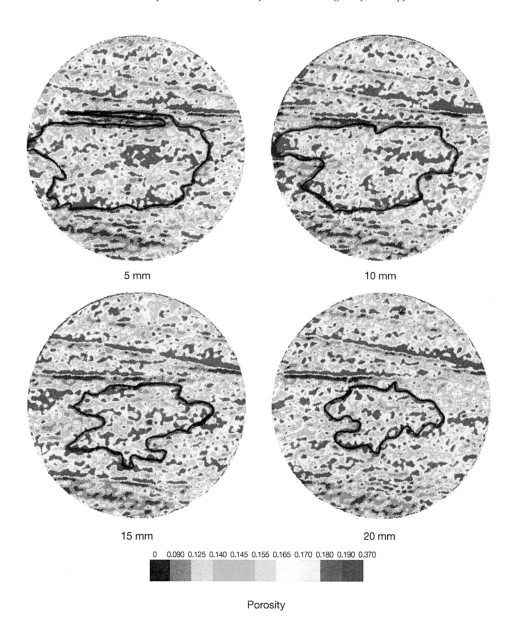

Figure 2-1.8 Invasion front superimposed on the porosity maps. Tomographic images 5, 10, 15 and 20 mm from the tracer injection point (see Fig. 2-1.7B). Sample diameter: 6 cm

c) Scale effects and plurality of anisotropy causes: Example of ultrasonic P-wave velocity anisotropy

We must emphasise that within a given natural medium and for a given petrophysical property, there may be numerous anisotropy causes related to textural features, contributing to the total anisotropy. However, since each structural feature can be expressed at a certain scale, the "additivity" of these anisotropies can only be expressed from the sampling dimension (or the characteristic dimension of the physical phenomenon) corresponding to the structure considered. Ultrasonic compressional wave velocity anisotropy (P waves, § 1-3.2.1, p. 222) in some sedimentary rocks can be used as an example. Concerning

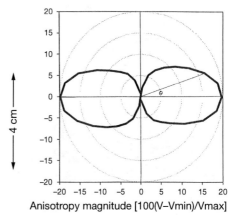

Anisotropy magnitude [100(V–Vmin)/Vmax]

Bedding related P-wave velocity anisotropy

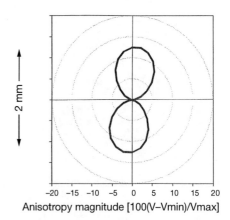

Anisotropy magnitude [100(V–Vmin)/Vmax]

Grain contacts (pressure-solution) related P-wave velocity anisotropy

Figure 2-1.9 Two main causes of P-wave velocity anisotropy in sedimentary rocks.
The parameter shown in polar coordinates is the difference between the velocity
in the direction considered and the minimum velocity, in percent

anisotropy in the propagation of elastic waves, note that shear waves (S waves) exhibit very specific anisotropy: on entering an anisotropic medium, the wave is split into two S waves with different velocities and perpendicular polarisation directions (shear wave splitting). This birefringence phenomenon, similar to that observed for light and behind the principle of the geologist's polarising microscope, will not be described in this book.

Two main causes of P-wave velocity anisotropy are shown on Fig. 2-1.9: bedding, which tends to induce maximum horizontal P velocities and preferential cementation of horizontal grain contacts, under the effect of compaction. This selective cementation favours maximum vertical velocities.

The cooperation (or competition in the case below) between these various causes can be observed after modifying the scale of the sample studied.

Figure. 2-1.10 shows the results of measurement in diametral planes of cylinders of smaller and smaller diameter (successive machining of the same sample). For the smallest diameter (6 cm) the measurement result is an almost perfect lemniscate indicating a maximum vertical velocity related to the effect of pressure-solution in contact with the grains of the oolitic limestone used for the experiment. The photograph on Figure 1-1.45b (p. 88) corresponding to an oolitic limestone of quite similar structure, gives an idea of the interpenetration of oolites by pressure-solution.

When the measurement diameter increases, the anisotropy figure undergoes deformation. For the largest diameter (18 cm), we observe the appearance of a clear high horizontal-velocity component caused by discrete bedding. It must be pointed out that the irregular anisotropy figures recorded for diameters above 6 cm cannot be strictly interpreted in terms of elastic anisotropy, therefore indicating a medium which is non-homogeneous as regards ultrasonic wave propagation. Although the phenomenon causing the maximum

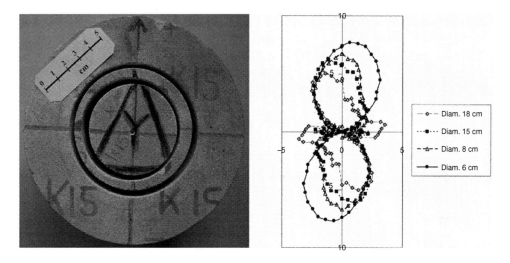

Figure 2-1.10 Measurement of ultrasonic P-wave velocity anisotropy on cylinders of different diameters machined from the same sample. As on Figure 2-1.9, the values shown in polar coordinates are the differences between the velocity in the direction considered and the minimum velocity, in percent

vertical velocity (grain joint) is at a small enough scale (compared with the wavelength and the sample used) to produce a "true" anisotropic medium under the experimental conditions (500 kHz), this is no longer the case for bedding. The bedding is expressed at a larger scale and waves of lower frequency would be required to return to a "physically" correct anisotropy figure.

This illustrates once again the need for extreme care when defining and criticising the notion of "homogenisation" in experiments concerning natural media.

2-1.2.3 Continuously variable REV: self similarity and fractal object

A) Empirical definition of self-similarity

In the empirical definition of the REV, we implicitly accepted that above a certain investigation radius, the value of the characteristic studied stabilises (at least over a dimensional interval) and that an REV can therefore be determined for the property concerned. Frequently, however, the values of some properties of natural media do not exhibit this stabilisation. When the investigation volume increases, the value of the characteristic varies continuously. It is no longer possible to define an REV, with the previous common meaning.

We will take the example shown on Figure 2-1.11. The photograph shows quite ordinary artificial outcrops (edge of a road) of limestone. Considerable natural fracturing can be seen, as is generally the case on the surface (compared with fractures of hydrocarbon reservoirs, we must always bear in mind that virtually all of these fractures disappear with depth, mainly under the effect of stress). If we want to measure the "unit block" (minimum volume of non fractured rock, which plays a central role in the matrix/fracture exchange calculations in oil recovery) we are quickly faced with a difficulty quite familiar to geologists studying natural fracture networks: when the observation scale used for the measurement is changed, we always obtain a larger (or smaller, depending on the case) block than for the previous scale.

It is as though during the scale change, the structure reproduced itself. This property is known as self-similarity. Note that the scale was deliberately omitted from Figure 2-1.11 and even an experienced observer would find it difficult to obtain an idea of the true scale of the photographs. This uncertainty is typical of self-similar objects. Images *a* and *c* are in fact enlargements of areas taken from photograph *b*, which is about 5 m high.

B) Self-similarity in natural media

This notion of self-similarity has important consequences since the media concerned have no easily definable characteristic dimension. For the observer, the measurement result depends directly on the dimension of the measurement instrument used. One famous example concerns measurement of the developed length of a highly broken coastline (e.g. the Brittany coast, [Mandelbrot, 1967]). We can easily understand that the length of this coastline measured on aerial photographs at different scales and on the ground with different

Figure 2-1.11 Limestone outcrops on the edge of a road.
Photographs *a* and *c* are enlargements of areas taken from *b*. Outcrop *b* is about 5 m high

measurement instruments (from the decametre to the decimetre for example) varies by several orders of magnitude.

Numerous cases of self-similarity have been described in nature. For example, topographic surfaces in rocky or mountainous regions. The easily identifiable self-similarity of hydrographic networks in poorly permeable environment is a special case of this situation. The surface of clouds is also self-similar (obviously over a finite range of distances, as already mentioned). Tree leaves and, more generally, forest surfaces also exhibit this trend.

Self-similarity is omnipresent in the human environment, for example in topographies, clouds, forest canopies, etc. This could explain the outstanding success – well outside scientific circles – of the notion of fractal object, after its popularisation by Mandelbrot [1977].

C) Overview of the notion of fractal object

a) Koch's snowflake

Although the notion of self-similarity is quite intuitive, its mathematical formulation is highly complex. Benoît Mandelbrot synthesised and popularised mathematical ideas dating back to the 19th century. One of the most unusual characteristics of the objects described is that they are both finite in some respects (e.g. surface, volume) and infinite in others (perimeter, specific area). The construction of these objects is based on iterative processes and self-similarity is one of their main features. We will restrict ourselves to a very simple example of fractal object: Koch's snowflake (Fig. 2-1.12).

To draw the snowflake, as with all fractal objects, the initial figure, the initiator, must be defined; in this case, it is an equilateral triangle. A generator defines the operation to be carried out on each iteration: a figure with 4 line segments replaces a single segment (Fig. 2-1.12a). On the Koch's snowflake, we can see the "mixture" of finite and infinite characterising fractal objects: as the number of iterations tends towards infinity, the perimeter of the snowflake also tends towards infinity although the area circumscribed by this perimeter is finite.

b) Notion of fractal dimension

Using this example, we can attempt to explain very simply (and therefore very roughly) the notion of fractal dimension. Although the notions of Euclidian dimension (dimension 0 for a point, 1 for a line, 2 for an area, etc.) seem so intuitive, defining the dimension of a space is an extremely complicated mathematical problem reserved for specialists, which we will therefore not discuss. The following explanation can be proposed: from a object of space dimension D, we will define a fragmentation rule such that at each successive application (iteration), the unit length of the new objects divided by the previous length is S^{-1}. We will examine the change in the multiplier N of the number of objects generated on each iteration.

In the example of Euclidian objects (line, area, etc.), we see that if $S = 2$, a segment gives after iteration $N = 2$ segments of length $1/2$, a square gives $N = 4$ squares of side $1/2$ and a cube $N = 8$ cubes of side $1/2$.

We therefore obtain a relation $N = S^D$.

What is the situation for Koch's snowflake? $S = 3$ since between each iteration, the length of the segment is divided by 3. If we study the change in the number of segments (NB) between each iteration, we obtain:

Iteration No.	0	1	2		n
NB	3	12	48		3×4^n

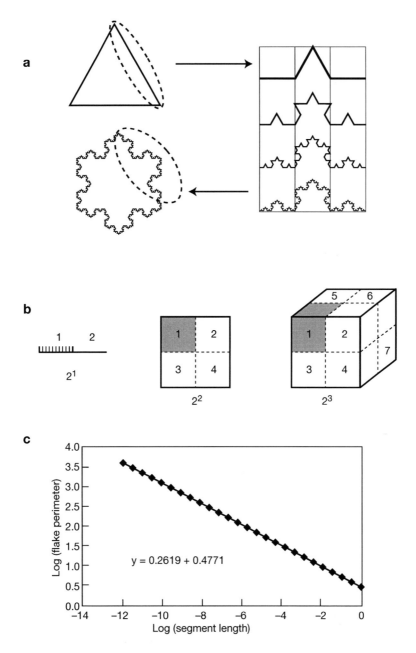

Figure 2-1.12 Example of fractal object and explanation of the notion of fractal dimension
a) Generation of Koch's snowflake.
b) Fragmentation of objects of Euclidian dimension.
c) Relation between perimeter and length of unit segment for Koch's snowflake.

Between each iteration therefore, the multiplier of the number of segments, N is

$$N = 3 \times 4^n/3 \times 4^{n-1} = 4$$

Returning to the definition $N = S^D$; $D = \log N/\log S = 1.2618\ldots$.

With Koch's snowflake, dimension D is not an integer (unlike the "Euclidian" examples). These non-integer dimensions are known as fractal dimensions.

Using the same example, still empirically, we will see how to obtain an approximate value of the fractal dimension experimentally. We will return to the definition of D. We can see that between two iterations, since the length of the segment l is multiplied by S^{-1} and the total number by N, the flake perimeter L is therefore multiplied by N/S.

Thus $L(n)/L(n-1) = N/S$ and $l(n)/l(n-1) = S$.

Writing these expressions in logarithmic format, we obtain:

$$[\log(Ln) - \log(Ln - 1)]/ [\log(ln) - \log(ln - 1)] = \log(N/S)/\log S.$$

Drawing a graph of $\log L$ *vs.* $\log l$ (Fig 2-1.12c), the gradient of the straight line is equal to $\log(N)/\log(S) - 1$.

By definition $D = \log(N)/\log(S)$, we therefore see that the gradient is equal to $D - 1$.

l, the length of the segment can be compared with the unit of the measurement instrument used to measure L. The log-log lines showing the change in measurement result according to the measurement unit can therefore be used to approximate the fractal dimension of a real object.

c) Conclusion

To conclude this very brief presentation of fractal objects, it seems important to emphasise a possible ambiguity with this approach:

- On the one hand, we must stress the fact that self-similarity is frequently observed in nature and that it cannot be ignored. This avoids falling into insoluble contradictions, such as those met when assigning, at all costs, representative values to typically self-similar objects (natural fracture network is a perfect example). We must nevertheless point out that the "nested structures" frequently found in nature are not necessarily self-similar: taking the porous medium of clays, for example (Fig.1-1.9), the aggregative structures are "packed" together but do not exhibit any morphological similarity.
- On the other hand, after detecting this self-similarity and defining the scale limits between which it can be observed – remember that this limitation is important – the "quantitative" characterisation may seem disappointing. The "fractal" dimension may correspond to a deceptive quantification since its value as such does not characterise an object any more than dimension 3 characterises a Euclidian volume.

Much research will probably still be required before this fractal dimension can be included in calculations for current applications (e.g. mechanics) but this promising avenue could lead to interesting innovations.

2-1.3 SCALE CHANGES

The physical data acquired during the discovery, evaluation and operation of a reservoir must be compared and summarised to allow rational analysis, especially if they are used to create a digital model designed to estimate the reservoir size and production output.

Several successive or simultaneous processing operations are required to produce this data summary. The information may be grouped pragmatically according to its supports and acquisition modes:

- cores and rock samples;
- well logging;
- well tests.

Note that this classification also establishes a hierarchy between the acquisition scales:

- The last group (well test) is close to the grid volume of the reservoir simulation digital model and it provides a global view of reservoir behaviour at the scale of interest to the production engineer. A similar approach is taken in hydrogeology.
- The second group (well log analysis) is by far the acquisition mode most frequently used in reservoir evaluation. This priority of well logging over direct acquisition on core is mainly due to the cost of coring (especially the time during which the drilling platform is immobilised). Note that in theory, and excluding the "production" logs: flowmeter, thermometry, etc., the intermediate log analysis scale would not be specifically required to obtain physical data.
- Lastly, the first group (core), whose scale seems far removed from the operational scale, is nevertheless the only one providing direct access to the basic physical properties: log analyses cannot be used to "directly" measure actual saturation, porosity, density and even less mineralogy; permeability cannot be accessed directly from well tests; etc.

The skill of the petrophysicist-log analyst is precisely to reconcile these scale changes imposed by the practical acquisition constraints. This chapter will focus more especially on the first link in this sequence of processes: processing of data acquired on cores. Note that direct access to the parameters and physical mechanisms involved in reservoir production is only possible at this scale. Transcription of information at well scale (log analyses) or at the scale of the reservoir itself (geophysics) in terms of petroleum physical properties and even in terms of reservoir architecture largely depends on the initial assumptions. Interpretation of the well test, despite the fact that it is the experiment closest to the production processes, itself depends on the choices made on the geometry of the volume investigated and its surroundings.

Lastly, note that some extremely important properties are still very difficult to access by core analysis. They include, in particular, identification of open fracture networks and their dynamic behaviour; without over-exaggerating, we could say that apart from the highly attractive "output" of modern processing software, many well logs in this domain (mainly well imagery) are actually not much better!

2-1.3.1 From plug to core and well logs

A) Overview; typology of data acquired on cores

There are various types of core data, acquired at several scales and which must be reconciled:

- *Quantitative data,* either petrophysical, discussed at length in the first section of this book, or "geological", which must be considered in the same way. To process these data, it is essential to correctly distinguish between "point" type quantitative data (identified by a number: porosity, permeability, mineral content, etc.) and "function" type data (characterised by a series of X, Y pairs: capillary pressure, resistivity index curve, granulometric curve, etc.). We must mention in particular the quantitative data derived from image analysis (microscopy, digital photographs, scanner, etc.). The result is often expressed in the same terms as the true physical or chemical measurement (e.g. mineralogical contents, porosity, saturation) although it is only an estimation. Forgetting this simplifying convention may lead to contradictions, which are discussed in § 1-1.1.5, p. 26.

- *Descriptive qualitative data* must be included in the process. This is even one of the strong points of the "core-property log", provided that a clear distinction is made between the strict intrinsic description of the material (petrography, lithology, oil impregnations, etc.) and the analysis (e.g. sedimentological environments) whose qualification may vary over time according to the experience acquired.

One of the key points when studying cores is the comparison with the logging results (even if only to calibrate them). This comparison is only justified if the various results are reconciled; we will refer to this point repeatedly. The most efficient way of reaching this objective is to draw up a "Core Petrophysical Log" which includes the various observations.

This petrophysical log will be mentioned on a number of occasions in the following pages. We must emphasise the difference between this document, which results from an analysis, and the various core logs described below. To avoid terminology confusions, we will use the expression "Core Interpolated Petrophysical Log" to designate the summary document.

Before describing the principles used to produce this log, we will discuss below the data acquisition techniques implemented: continuous core logging, sampling, database, etc.

B) Drilling data recorded during coring

These data do not concern the core itself but the drilling conditions observed during the coring operation. Systematically recorded by the driller, these parameters include: the weight on bit, the torque on the pipe string, the drilling rate of the core barrel in the formation and any mud losses in a high-permeability reservoir. These data taken routinely by the driller are only very rarely taken into account by geologists, despite the fact that they are extremely useful to explain some coring incidents.

For instance, crossing a highly brittle "super K", lost during coring, like that shown on Figure 1-2.18, p. 151, is generally clearly visible on the weight on bit recording and on the instantaneous drilling rate.

Another extremely useful data is the depth at which even slight mud losses occur. This information is critical to detect the presence of an "open" fracture, which often causes core dislocation, leading to jamming in the core barrel and therefore loss of the interesting zone.

The amount of gas dissolved in the drilling mud is often recorded. Once again, this data is extremely valuable.

We can never over-emphasise the importance of these drilling parameters when interpreting the core results.

C) Continuous core logging

Various devices which offer continuous measurement of some physical properties along the core, are grouped under the denomination of "core logs" (which may sometimes lead to confusion with "Core Interpolated Petrophysical Log"). In the core logs described below, the first two are (or should be) employed for routine acquisitions, use of the others being less widespread:

a) γ ray core logging

Care must be taken during acquisition as regards background radiation in order to pick up a meaningful signal, but processing this measurement is relatively straightforward. This technique is sometimes implemented purely to carry out core depth matching with respect to core logs (see § G below). Obviously, we must not attempt to compare the absolute values. It is in fact recommended to weight the measurement by a volume index for the core investigated, especially in case of poor core recovery.

Currently, spectral analysis (K, Th and U) is often carried out in addition to this measurement. It is an extremely useful technique provided, once again, that the absolute values of the data are not considered in terms of elementary content in the strict sense.

Considering the volume investigated, the acquisition step of this core log is decimetric.

b) Apparent density core logging

This electron density measurement is quite similar to that implemented in well logging. The only difference is its acquisition by "transparency", the emitter and the receiver being positioned on opposite sides of the core.

The gamma radiation attenuation measured depends on the electron density of the medium between the emitter and the receiver.

$$I = I_0 \, e^{-C\rho_e L}$$

where: I: intensity of the γ radiation measured by the receiver.

I_0: intensity of the γ radiation emitted by the source.

ρ_e: electron density (number of electrons per unit volume) encountered on the path between source and receiver.

L: distance between source and receiver.

C: constant which depends on the device geometry, the detector type and the energy of the γ rays.

Provided that calibration is carried out regularly to readjust the value of C (standard materials are positioned systematically at the start and end of the acquisition sequence), conversion of the electron density into apparent density is quite straightforward.

In order to use this log correctly, therefore, the thickness crossed must be measured. Acquired by mechanical or optical sensors, this correction is of little value when the portion of core analysed is too fragmented.

Apparent density core logs (as well as mini-sonic) must be carried out with the virtual emitter/receiver axis in the bedding plane (i.e. parallel to the strike, see Fig. 2-1.14) so that acquisition occurs in this plane. The X-ray radiographs must also be positioned in the same way to obtain the best possible resolution on the bedding limits (this precaution is not required if a CT-Scanner is used, although the acquisition process is much longer; in addition, many laboratories are only equipped with the much less expensive X-ray radiography apparatus).

This type of core logging can be carried out for qualitative purposes on unconsolidated cores in rubber sleeve or fibre glass tube. Quantification is made extremely difficult, more due to the side effect of the variable thickness of mud remaining between the core and the sleeve than to the sleeve itself, whose effect can be calibrated.

Since this type of core logging is generally carried out on rocks of variable saturation state (core more or less dry), the density measured must be considered as "apparent". As with the other types of core logging, the most important point is to measure the relative density variation depending on the core depth. The acquisition step is centimetric.

c) Mini-sonic core logging

As for the density log, mini-sonic measurement is acquired by transparency, on a diametral plane of the core (Fig. 2-1.13a) and the operating conditions to obtain meaningful data are similar. An additional precaution is required: since measurement is taken with no

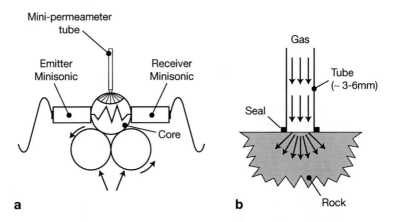

Figure 2-1.13 Schematic diagram of the mini-sonic and mini-permeameter devices
a) device simultaneously recording the two logs on several generators, using rollers to turn the core.
b) schematic diagram of the mini-permeameter, according to Jensen [1990].

confinement pressure, any cracks present have a considerable impact on the measurement. The sensors, and therefore the ultrasonic frequency implemented, are very similar to those used for acoustic measurements in laboratory, on sample (§ 1-3.2.3, p. 230). The most sophisticated devices can measure the compression and shear wave travel time. If the core is complete, acquisitions can be carried out along several generators.

The absolute values measured have little significance due to the uncertainty of the core saturation state but above all due to the possible effect of microcracks because the core is not under confinement. This type of logging, of centimetric acquisition step, is useful however:

- to carry out core depth matching (like the previous types of logging);
- to interpolate the isolated measurements acquired on samples (see § F below, core interpolated petrophysical log).

d) Mini-permeameter core logging

One or more permeability profiles are obtained along the core by a device (Fig. 2-1.13b) injecting gas (nitrogen) into the core.

The measurement is taken

- either by injecting nitrogen at constant rate:

$$Kg = \frac{2\eta M Pe}{G_o a \rho_g (Pe^2 - Patm^2)}$$

where: M: nitrogen mass flow rate,
      η: nitrogen viscosity,
      ρ_g: nitrogen density,
      a: probe radius,
      Patm: atmospheric pressure (pressure of the gas in the core),
      Pe: stabilised injection pressure,
      G_o: geometric factor. This geometric factor depends on the assumption made regarding the shape of the current lines in the sample.

- or by connecting the probe to a known volume of nitrogen of pressure P decreasing with time as it flows through the sample. This is the drawn-down method, suitable for low permeabilities (§ 1-2.1.2B, p. 135).

The measurements may be corrected for the Klinkenberg and Forchheimer effects (§ 1-2.1.2A, p. 133) and calibrated on known samples. In spite of these precautions, it would be most unwise to consider the values supplied as absolute values (core surface state, unknown liquid saturation state, etc.). Consequently, there seems to be little point in making the corrections for the Klinkenberg and Forchheimer effects. These profiles are nevertheless extremely valuable to detect permeability variations which may occur very rapidly and which measurements on plugs, of average sampling frequency 25 or 23 cm, are unable to capture.

On complete cores, acquisition is carried out on several generators. However, the most reliable data are acquired on the sawn face of the core (in this case, acquisition is carried out after slabbing). Working on the sawn face improves the contact between the probe and the rock. The optimum sampling step is semicentimetric. Note that this technique is also used to acquire permeability maps on the surface of full size samples intended for relative permeability measurements.

e) Other types of core logging

Other types of non-destructive physical measurements are sometimes implemented as continuous (or high-resolution) core logging. For instance, a Formation Factor longitudinal profile could be relevant in carbonate rocks but too many experimental difficulties remain, the most important being to obtain perfect core saturation in brine of known resistivity.

Implementation of magnetic susceptibility profiles is technically possible, but this method is rarely used. In case of quite specific clay phase mineralogy, this profile is an indirect indicator of the clay content.

Infrared spectrometry measurement (§ 2-2.4) can also be used to obtain information on the mineralogical assemblage. This adaptation is still in its infancy however.

Concerning the mechanical properties, automated tests based on the old scratch method are also implemented (measurement of scratch width at constant load on the stylus, or instead controlling the scratch depth and recording the force to be applied on the stylus). This profile correlates well with the rock mechanical resistance variation profile, of interest to drillers.

From the operational point of view, the most readily available logs to date are data resulting from the more or less sophisticated processing of core images:

- pseudo-density profiles obtained from processing of X-ray plates or CT Scan;
- hydrocarbon index profiles obtained from processing of core photographs under UV light;
- lastly, although to our knowledge not yet operational, texture profiles (discontinuities, etc.) obtained from processing photographs taken under natural light or X-ray radiographs. X-ray absorption logging deduced from CT Scan measurement is being increasingly used. The CT scan method is described in § 2-2.3.

D) Core sampling procedures

a) Sampling sequence and conditions

The core is a valuable asset shared by several specialists who may implement contradictory procedures. This is particularly obvious when irreversible operations (sampling, slabbing, etc.) are involved. Sedimentologists can only work correctly on slabbed cores. Petrophysicists wanting to acquire initial saturation or relative permeability data (§ 1-2.3.4, p. 195) will need a piece of core "preserved" against any mechanical aggression and carefully protected from the ambient air as soon as it is brought to the surface. Similarly, optimum acquisition of a density log or mini-sonic on core requires unslabbed cores, whereas mini-permeability logging is more meaningful if carried out on a sawing plane. The sampling phases, whatever the sampling type (petrophysical or geological; thin section, mineral or organic geochemistry, micropaleontology, etc.) are irreversible operations. They must be part of the core acquisition sequence, while maintaining a satisfactory compromise for the other operations.

Most acquisition types require precise orientation relative to the core diametral planes to obtain meaningful or highly accurate results: the strike of the bedding planes must be identified on each core or core section, either visually or by using an X-ray CT-Scan. By taking this preliminary precaution, there is no need to know the bed dip in advance or to allow for a possible deviation of the drilling angle from the vertical. A number of simple rules intended for the operators can therefore be defined:

- The core slabbing plane (observations of the lithology, sedimentary structures, etc. and photographs under natural and/or ultraviolet light) must be parallel to the dip (Fig. 2-1.14).
- The direction of the plug sampling axis must also be carefully chosen. By convention, "horizontal" plugs are sampled in the bedding plane, parallel to the strike. If we make the realistic assumption that the horizontal plane is virtually isotropic at the sample scale and with respect to the properties being investigated, applying this rule yields meaningful results irrespective of the dip of the reservoir horizons and the possible well deviation.
- The depth and well number should be marked on each plug to avoid losing any information (experience has shown the value of this apparently trivial remark!).

Apart from exceptions, most petrophysical properties must be acquired on these "horizontal" plugs. "Vertical" plugs (parallel to the core axis) are sometimes acquired in order to check for possible anisotropy. This is rarely justified.

We will take the example of routine permeability measurements. As illustrated above (§ 2-1.2.2Bb, p. 283), anisotropy is primarily a function of the scale and of the rock structure

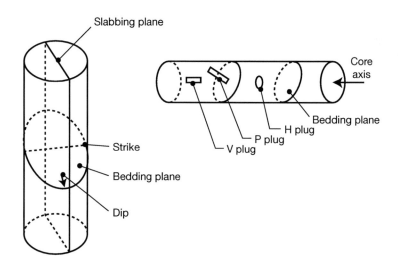

Figure 2-1.14 Orientation of core slabbing planes and plug axes
a) Slabbing plane parallel to the dip.
b) Orientation of various plugs: H parallel to the strike of the bedding plane; P perpendicular to the bedding plane; V parallel to the core axis (possible confusion with P). This terminology is not universally accepted and may lead to confusion.

organisation. We are more particularly interested in that concerning the scale of the model grid or of the volume investigated by the well test, …there is little likelihood of it depending on a possible "microscopic" anisotropy, meaningful at the scale of 1 inch plugs (oriented allochems, micas, etc.). This lack of meaning is further increased by another aspect: for practical reasons, "H" and "V" plugs are rarely sampled at the same depth (and spacing). Differences in measurement results are more often a sign of spatial variability in the property measured than true microscopic anisotropy. In addition, the notion of "vertical" is ambiguous: often taken to be the direction of the core axis, this direction may turn out to be oblique relative to the bedding, whether the well is deviated or the bed dip not horizontal. The sampling of "stratigraphically" vertical plugs must be carried out with extreme care.

In conclusion, as regards directional physical properties, the reference system used in practice is the local bedding plane. Measurements are usually taken in this plane. There are some exceptions:

- Compressibility measurements, when the experiment must be carried out under "œdometric" conditions (without lateral deformation, see § 1-1.1.6, p. 45) on samples taken along an axis parallel to the true geographic vertical.
- Relative permeability measurements whose results must be interpreted on a digital model (at least 1D), so that the vertical variability can be recreated along the laboratory sample.

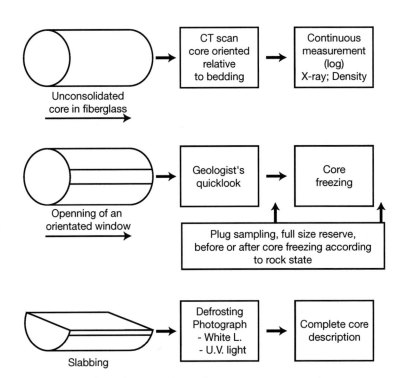

Figure 2-1.15 Example of unconsolidated reservoir core processing sequence

As an example, Figure 2-1.15 shows a preliminary unconsolidated reservoir core processing sequence, cores sampled and preserved in fibre glass or metal (steel, aluminium) tubes.

b) Sampling frequency for petrophysics

Full–size samples for special core analysis (SCAL)

A certain number of special measurements must be carried out on full diameter pieces ("full size"; typical cores have a diameter between 6 cm and 10 cm, reaching 12.5 cm for unconsolidated cores) which must be protected from ambient air to prevent drying and oxidation. Since the precise choice of the facies to be sampled is not determined in advance, and bearing in mind that apparently satisfactory pieces are often eliminated after the homogeneity study and that special measurements are often ordered at a late stage, or even during reservoir appraisal phases launched long after the drilling of the first appraisal wells cored, it is essential to anticipate by reserving enough full diameter pieces in order to protect them against future destructive operations: plugging, slabbing, etc. For information, a 30 cm long full size reserve should be made every metre in reservoir zones and every two or three metres elsewhere. As a precaution, these pieces must be "preserved", i.e. carefully wrapped (good quality plastic film, paraffin, etc., but not aluminium foil which reacts with the impregnated rocks).

When reserving full size samples, we must always remember that preserved samples should also be used for measurements other than conventional SCAL (measurements of rock mechanical properties, measurements on shale: porosity, density, formation factor, salt content).

Plugs for routine core analysis

Since the plugs intended for routine measurements must allow vertical "scanning" of the basic petrophysical properties, they must be sampled at a relatively high frequency: typically 3 to 4 samples per metre of core.

A priori, the operator samples at regular intervals (every foot, or 1/4 or 1/3 m). Initially chosen for the simplicity of control, this practice is often justified *a posteriori* by statistical considerations. Experience has shown however, that this sampling mode is often biased in practice. Although the identifying depth is determined regularly: nn.00 m; nn.25 m; nn.50 m; nn.75 m; etc., the sampling is physically more or less shifted depending on the mechanical quality of the core at the planned point, or even to avoid a level considered to be of no interest (e.g. shale level). This initially well-meaning intention disturbs the data processing, often making the final analysis too optimistic. It represents a source of inconsistency when producing the Core Interpolated Petrophysical Log.

To control rather than suffer this sampling bias, the sampling can be adjusted according to lithological variations and/or, if necessary, according to the oil impregnation, always observing an average frequency of 3 to 4 samples per metre, with two at the ends of the preserved full size samples. Each plug is assigned a vertical representativeness coefficient defined by the upper and lower limits of the layer that the plug is supposed to represent (Fig. 2-1.16).

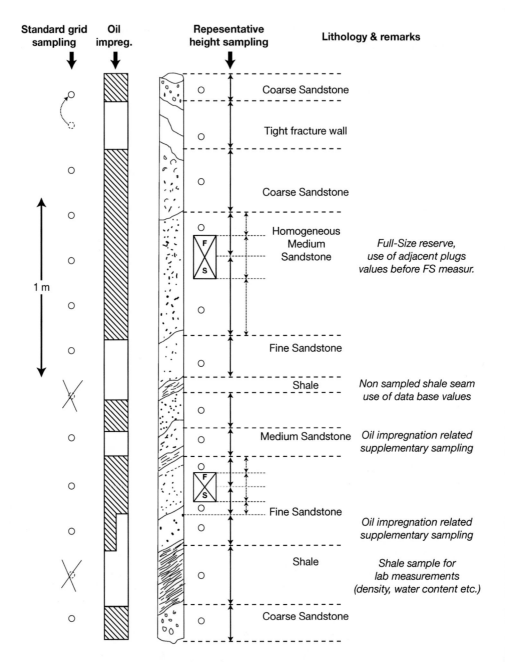

Figure 2-1.16 Routine sampling diagram, with estimation of a representative height
(and therefore a weighting coefficient) for each plug

Experience has shown that this process is not an operational handicap if included rationally in the organisation of the initial core processing sequences. On the contrary, it ensures that the maximum amount of information can be extracted from the data during future analysis.

Lastly, lithological, mineralogical, mineral and/or organic geochemical determinations must be executed on scraps adjacent to the plugs (or on the plug itself, after analysis). Studies of "physics/lithology" correspondence are therefore relevant.

E) Creation and verification of the "core" database

Rational use of data acquired on cores requires a "petrophysical summary" designed to:

- check, and possibly guarantee, the overall consistency of petrophysical and analytical data;
- check, and if necessary determine the consistency of these data with other information acquired at well scale (lithological and sedimentological descriptions, core logs and images, etc.);
- analyse these data to present them in an integrated form which is, if possible, at the scale of the future application: in other words, creation of the Core Interpolated Petrophysical Log.

Typically, users only receive their petrophysical data a little at a time, following their requests for services. The information is distributed over a very long period of time: from the conventional "porosity-permeability" measurements generally taken immediately after drilling the well, up to the heavy "relative permeability" measurements which are often taken shortly before the reservoir modelling studies.

Consequently, petrophysical analytical studies on the same field rarely refer to each other. This effect is even greater when the measurements are carried out by different service providers. This implies a risk of inconsistency between the petrophysical data themselves and, often, a lack of correspondence between the acquisition of petrophysical measurements and the other information such as the lithological and sedimentological analyses. In the end, the onus is on the end user to reconcile all these data, even though he should not really have to.

a) Data validation

The data validation process must be carried out as soon as possible after data acquisition and therefore in the laboratory taking the measurements. The laboratory must be able to provide information on: the resolution of the test considered, the scale and the volume investigated, the measurement quality (i.e.: the precision, accuracy and bias).

This "quality control" must take place before data integration in the analysis process. Nevertheless, it is always recommended to check this information *a posteriori* by crossed validation between the different sets of data. A typical example concerning solid density and mineralogy was given in the first section (§ 1-1.1.5, p. 26).

Disagreement does not necessarily indicate a laboratory error. The sampling may have been badly supervised or be incoherent between several disciplines. Equally, the material may simply be heterogeneous with respect to the property considered. This is when it becomes essential to know the scale investigated by the measurement.

It is always a good idea to remember one basic rule: the first validation, calibration processes, etc. must be carried out set by set (in the typological meaning defined at the start of the paragraph: a set corresponds to an acquisition mode). Comparison, or calibration of one set by the other (typical example of calibration of porosities, densities and clay contents analysed in well logging on core measurements) cannot be carried out directly without prior scaling. Unfortunately, we observe all too frequently log analysis calibration validated simply by a graphical presentation of a porosity (or density) graph resulting from log analysis involving a vertical seeding of more or less well adjusted "core" measurements (see Fig 2-1.20a). The problem with this practice is that it damages the credibility of core measurements - thereby justifying their non-use - and develops the feeling that log analysis is systematically accurate compared with the direct measurement, given the small size of the core sample.

Continuing with the example of the log/core comparison: in the same way as the raw log data must be corrected for environmental effects, depth matched, etc., before being interpreted in terms of petrophysical quantities, the core data must be validated by crossed exploratory analysis (see § 1-1.1.5), corrected and recalculated under confinement conditions, then scaled to match the resolution of the logging tools before being compared with the log data.

b) Construction of data consistency

This first essential step is the most tedious. The data must be obtained then processed so that they can be mutually analysed:

Compile a first version of the petrophysical and analytical database

The other core data (mainly upstream "geological" information) must be added to the other petrophysical data on routine sampling, clearly identifying the measurement/observation scales:

- Those which are at the same scale as the petrophysical sampling (e.g. lithological description of plugs, full-size samples; microlithological analyses on petrographic thin sections, mineralogical analyses, etc.).
- Those which are at the scale of the layers, sedimentary units (descriptive analysis of cores) and whose scale must be reduced to the volumes investigated by the logs.

Good control of this first scale change will help us to understand the scaling of the petrophysical parameters discussed below.

Process the data:

For example:

- if necessary, correct the petrophysical values to return them to "confinement conditions" (mainly the effect of stress and fluids);
- depth match the various data acquired on cores, which may have been shifted during the numerous handling operations.

A reference datum must be chosen for the core depth. The most reliable reference datum is the earliest information taken: this could be the initial photograph or the first "quick-look" of the sedimentologist or the well-site geologist. When reconciling the various results, the data depth must be precisely identified. Full-size samples intended for relative permeability tests are generally identified by the core depth of the top of the piece considered. The layers described by the petrographer or the well-site geologist are identified by the top and bottom depths with an additional continuity requirement between these intervals (except for coring loss). These apparently secondary considerations are essential when processing the core property log.

Note that, unlike the case of the "log sample", this depth identification is not sufficient. Several physical samples may be taken at the same depth. In addition, and still by comparison with the log analysis, an entire section of the data processing is independent of the depth: the search for the laws and relations between physical properties and lithological type; their determination is one of the bases underlying upscaling in reservoir geology (rock typing § 2-1.4).

A log analyst unfamiliar with processing core data could be surprised by this strict requirement, since it might appear obvious that the depth should be clearly indicated on all core data automatically. In practice, various specialists carry out a number of operations and transfers on the core pieces as soon as they are brought to the surface: from the usual cutting into sections of practical length, generally 1 m, up to the more or less coordinated sampling by all the analysts: petrophysicists, sedimentologists, geochemists, micropaleontologists, etc. The successive handling operations to photograph the core under natural or ultraviolet light, or by X-ray radiography, must also be added to the list. The numerous risks of positioning the pieces incorrectly or inverting them are accompanied by the inevitable expansion of the most fragile or fractured levels.

Carry out an Exploratory Data Analysis

This analysis may be limited to the petrophysical data alone. It consists in:

- checking the possibly atypical individuals which might not have been detected during the laboratory analysis;
- checking the first petrophysical "standard relations" such as the "Porosity-Permeability" correlation or the Leverett-J functions or other capillary pressure curve processing operations (§ 1-1.2.5, p. 94).

By introducing geological type data at this stage, it is possible to:

- improve the check (e.g. verify the compatibility between the mineralogy and the solid density);
- specify the standard petrophysical relations by geological units.

These methods were described in the first section of this book. For a simple study, this first phase alone may be sufficient.

F) Identification of the organisation along the well: the Core Interpolated Petrophysical Log

The purpose of the Core Interpolated Petrophysical Log is to describe the isolated (depth) measurements produced in the laboratory as a virtually continuous log of petrophysical properties, along the core, and representative of the confinement conditions. It is therefore an interpolation process based on an exhaustive analysis of all information available on the core. We have already mentioned the possible terminology confusion with core logs, discussed in paragraph C. Consequently, we will systematically use this complete denomination.

We therefore obtain a continuous high-resolution description of vertical petrophysical variations, allowing:

- easy core depth matching compared with well logs;
- calibration of the quantitative log analysis, provided that unbiased scaling is carried out;
- more rigorous comparison of the permeability measurements with the well test results (h. **K** and Anisotropy);
- critical analysis of the petrophysical measurements, identifying anomalic samples.

The basic principle when establishing the Core Interpolated Petrophysical Log is to interpolate the validated petrophysical measurement, based on a continuous datum (or a combination of continuous data) acquired on the core, ideally a physical log on core, otherwise the lithological description. There are generally two cases:

- Core logs are available and the vertical variation of one of them or of a combination of them correlates with the vertical variability of the petrophysical property to be interpolated.
- No relevant core log is available, but the plug "representative heights" were carefully indicated during sampling. In addition, the geologist's description is available and is as detailed as possible, at a sufficient resolution (in the worst case, to within $^1/_2$ decimetre).

a) First configuration: one or more core logs are available

The method is based on the principle of kriging with external drift. Kriging (named after the South African statistician D. G. Krige) consists of interpolation which takes into account the spatial structure specific to the variable considered. It is one of the Random Function approximation applications we decided not to discuss in paragraph 2-1.2. From a property (micropermeability in the example of Fig. 2-1.17a) for which numerous measurement points are available and whose variation correlates with the property studied (intrinsic permeability on Fig. 2-1.17b), we define the spatial variability (in our example the external drift) to calculate a "petrophysical log" going through the validated/corrected measurement points and between these points fitting the shape of the support log.

The input variables must be preprocessed in order to use this procedure:

The measurements on plugs are:

- validated by the procedure described above and recalculated under confinement conditions, if possible;
- weighted, in case of obvious heterogeneity of the core section corresponding to where the plug was taken (nodules, pluricentimetric vugs, etc.);
- completed when significant levels have not been sampled (typical example of levels which are too thin or measurements considered to be not significant: shale, evaporite, etc.) by values from a database built specifically for the reservoir.

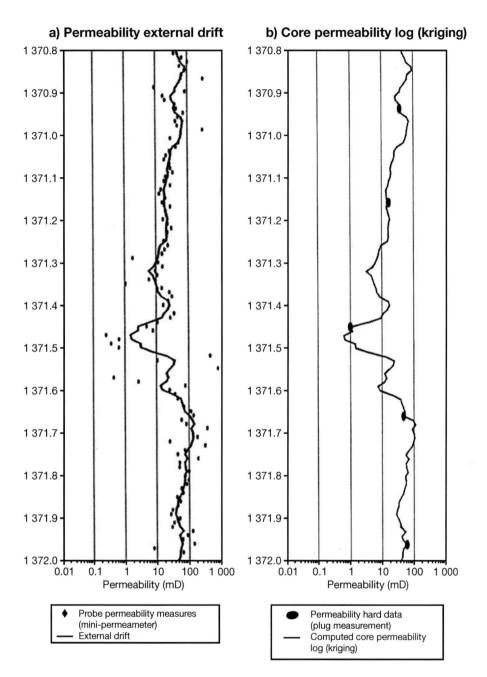

Figure 2-1.17 Example showing the construction of a core permeability log (**b**),
by kriging, from the permeability spatial structure information (external shift)
obtained using a minipermeameter (**a**). *In* [Greder *et al.*, 1994]

Similarly, core logs are also:

– validated/cleaned (artefacts, lack of material, etc.);
– possibly completed from more or less sophisticated processing of images acquired on cores (typical and simple example of X-ray plates processed in terms of density index log);
– recalculated with a centimetric resolution, either by resampling (natural radioactivity log) or by smoothing (minipermeameter of semicentimetric acquisition frequency).

Obviously, all these data are depth matched on the basis of well identified reference depth.

b) Second configuration: no relevant core logs are available (Figure 2-1.18)

The method is based on the plug representative heights defined during sampling (Fig. 2-1.16), after processing the measurements on plugs as previously.

The segmentation by representative heights is sometimes too basic and the sedimentological analysis is often too approximate at the required scale (centimetric

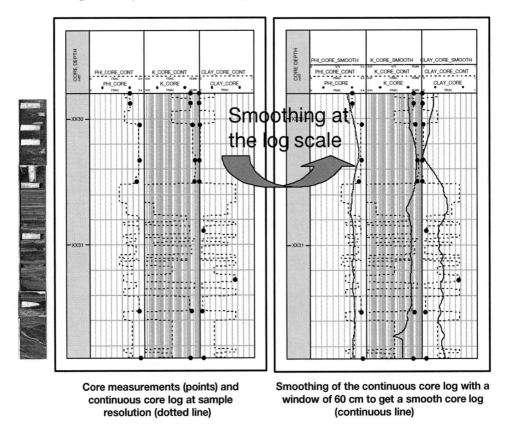

Core measurements (points) and
continuous core log at sample
resolution (dotted line)

Smoothing of the continuous core log with a
window of 60 cm to get a smooth core log
(continuous line)

Figure 2-1.18 Example showing the construction of a Core Interpolated Petrophysical Log from the core description, in the absence of any relevant core logs. *In* [Levallois, 2000]

resolution). It must then be repeated by direct observation on core or, if the core is inaccessible, by analysing all the photographs available: natural light, UV light (the impregnation variations are sensitive indicators) and X-rays.

The concerned parameter value of the plug is assigned to each level identified in this way. The missing values are rebuilt as above. In this case, the data base is systematically created for each well. We therefore obtain a petrophysical log by high resolution step. Clearly, the better the coring (recovery) quality, the better the accuracy.

G) Depth matching

This obviously essential process must be carried out very carefully. Too often, this operation is carried out on an individual basis and repeated for the same well as the core data becomes available, without too much discussion between the independent disciplines: geologists sedimentologists on the one hand, petrophysicists or log analysts on the other. At the end of the studies, it is not uncommon to observe a number of conflicting depth matching analyses, leading to errors and lost time when building the reservoir models. This operation is carried out in two stages:

a) Harmonisation of the various core depths

The same process must be applied to the core data and the various well log runs. By convention in log analysis, one run is chosen to be the "depth" reference. Typically, it could be the "natural radioactivity/neutron log" combination, acquired almost systematically. A natural radioactivity probe (γ-ray) is then associated with each tool combination.

If necessary, this depth is corrected for side effects (catching, partial sticking, etc.), using the accelerometers connected to the tool combinations.

Equally, for the cores, a reference datum must be chosen as indicated above since, as soon as they are brought to the surface, the cores undergo a number of operations and manipulations which inevitably disturb their sequencing.

b) Harmonisation of core/log depths

Depth matching consists in adjusting these two references: the logger depth on the one hand and the driller depth on the other hand. For the latter, the best support is the Core Interpolated Petrophysical Log defined above, which offers the fundamental advantage of being able to display the parameters recalculated at the scale of the volume investigated by the logs, including simulating their resolution.

The Core Interpolated Petrophysical Log is therefore produced at two scales, one at high resolution to refine the physical and geological interpretations and the other, where the information is degraded to match that of the logs, intended for calibration of the log analyses.

We must remember that choosing, at the end, the "logger depth" as the "true depth" is a pragmatic convention based on the fact that not all wells are cored. The logger depth is no more "true" than the driller depth.

On the contrary, experience has shown that in a correctly cored well, the driller depth may be closer to the true value, since the length of the pipe string can be measured directly, the coring losses are taken into account knowing the core barrel dimensions. The comparison between well images and core sections must be strictly calibrated on the "driller depth" of each end of the core.

c) Terminology remark on depths in deviated wells

With deviated wells, various terminological conventions must be respected to unambiguously characterise the bed depths and thicknesses. Some definitions are schematised in Figure 2-1.19, the standard abbreviation Z representing the depths:

ZKB or ZRT Apparent depth (in fact, the well length) measured from the Kelly Bushing or the rotary table (there may be a difference of 1 or 2 feet between the two). It is the depth driller in the common meaning.

ZTVD: True Vertical Depth. Vertical depth, calculated from the well trajectory logs, relative to a "geographical" reference, generally mean sea level (TVDMSL) (TVDSS, for sub sea).

TVDT: TVD Thickness: Bed thickness calculated by subtracting the ZTVD values of the bed top and bottom (note that this thickness is far from being "true").

TVT: True Vertical Thickness. Vertical thickness, measured vertically through the wellbore entry point.

TST: True Stratigraphic Thickness. Thickness calculated perpendicular to the bed. The TST is the closest to the "geological" thickness.

DT: Drilled Thickness: Bed length measured along the well.

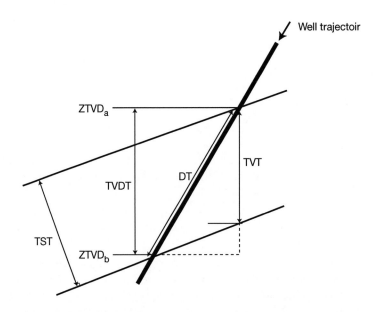

Figure 2-1.19 Diagrammatic representation of the terminological conventions used when defining bed depths and thicknesses in deviated wells

H) Conclusion: comparison of the "core log" and the well logs

One of the main applications of the Core Interpolated Petrophysical Log, after reintegration of missing data (shale levels are systematically omitted during standard procedures) is calibration of the quantitative log analysis.

To compare the two data types, the Core Interpolated Petrophysical Log must be brought to the scale of the well log. Initially, to simplify matters, this consists in calculating a sliding mean, more or less weighted, on a vertical window whose height is of the order of magnitude of the log depth resolution. Experience has shown that the resolution of the logs interpreted is rarely less than 0.5 m and even much larger, e.g. for saturations interpreted using resistivity logs.

The type of mean to be used depends on the petrophysical property considered. The arithmetic mean is always used for porosity, the solid density and mineral contents. For the total saturation, an intermediate variable ($\phi . S_w$) is required to homogenise the fluid volume with respect to the total volume. An arithmetic mean of this parameter is then used. For permeability, although the problem of comparison with a log does not arise, it may be worthwhile producing a log at the same scale. To allow for the fact that fluids mainly flow "horizontally" towards the well, the arithmetic mean is generally used in stratified formations. With "non stratified" rocks (non-sedimentary rocks but also reef limestone, etc.), it is best to use the geometric mean.

For comparison, the log analysis must also be calculated in variables of the same type as those of the Core Interpolated Petrophysical Log, e.g.: total porosity, total saturation, mineral contents by weight, etc. If all these precautions are taken, the log/core comparison is a valuable tool. Figure 2-1.20 illustrates the benefit of this method: while a crude comparison of log results and isolated plug data (Fig. a) is extremely disappointing, comparison of the well log and the "core petrophysical" log is excellent.

Ideally, the Core Interpolated Petrophysical Log should be "completed" with the corresponding core rock type. However, this sorting operation is difficult and is described in § 2-1.4, p. 317.

2-1.3.2 From core to well test ("bulk" and "matrix" property)

A) Notion of matrix and bulk property

In practice, the only scale to be considered to approximate reservoir characteristics is that of the grid used for the production computer model. Modellers ensure that this grid represents the actual reservoir as closely as possible, but despite constant progress, limits are nevertheless imposed by the calculation constraints. A typical dimension is the decametre for thickness and the hectometre for lateral extension.

The only data acquired directly at this scale concern seismic characterisation (but very little petrophysical information can be extracted from these data) and the production tests on wells (well tests) which are of paramount importance.

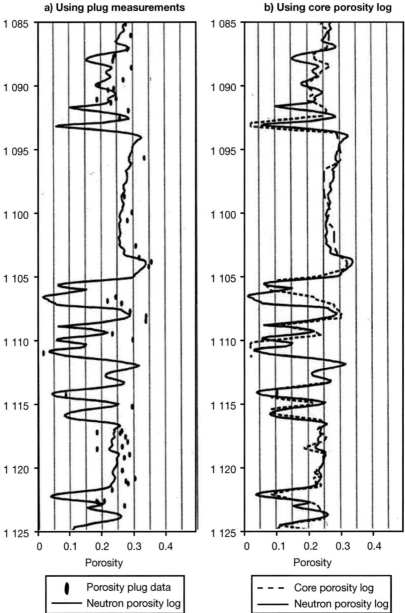

Figure 2-1.20 Example of comparing core log and well log results;
a) using the raw results on plug; **b)** using a Core Interpolated Petrophysical Log. *In* [Greder *et al.*, 1994]

The other core or Log measurements represent much shorter characteristic lengths and all the work carried out in reservoir characterisation must be aimed at upscaling these data.

In petroleum practice, these operations are often considered as being outside the conventional field of petrophysics. Two other technical fields are more specifically concerned:

- Geology, mainly through its quantitative disciplines (geostatistics) to extend between the wells the precise petrographic description obtained using cores and logs.
- Reservoir modelling (fluid mechanics in porous media), firstly to interpret the well tests and secondly to calculate the grid equivalent permeability using the information supplied by geological extrapolation.

Our description of this highly important aspect of reservoir characterisation will therefore be brief. We will restrict ourselves to describing the notion of matrix property as opposed to that of bulk property, emphasising the classical example of permeability.

The methods described in the previous paragraphs outline the petrophysical properties at the scale of a decimetric sample. By convention, the characteristics obtained in this way are defined as "matrix properties", as opposed to those measured at reservoir scale, called "bulk properties".

For porosity, which is a volume ratio, we should not expect to see much difference between the two types of value. This can be easily demonstrated by considering that a karst network consisting of 10 m diameter spherical cavities located according to a grid of side 100 m in a 20 m thick limestone bed generates less than $3 . 10^{-3}$ fractional porosity. Similarly, a network of 10 parallel fractures per metre, 1 mm thick, only induces 0.01 porosity. In practice, we often rightly consider that the total porosity is very close to the arithmetic mean of the matrix porosities.

The situation is not the same for permeability since a drain (or a barrier) of negligible volume, can totally modify the bulk permeability of the block, in a given direction.

B) "Core permeability" and "build-up permeability"

a) Parameter h . K and the h . K-core

In sedimentary reservoirs, fluids generally flow towards the well along roughly horizontal lines of current. The parameter to be taken into account when evaluating the production of a bed is therefore its horizontal permeability. More precisely, the product of this permeability by the bed thickness is directly related to the flow for a fixed head loss. The parameter h . K, usually measured in darcy. metres (or darcy. feet), is therefore used to characterise a bed.

If a core permeability petrophysical log is available (see above), it is easy, by integration, to obtain the value of h . K over the cored interval and to define the matrix permeability of the formation.

With this simplified calculation of h . K we are assuming that the permeabilities are homogeneous in the horizontal plane (significant stratification, homogeneous bed). If this is not the case, depending on the geological model chosen, a more complicated calculation method must be adopted, involving locally geometric means, for example.

b) The h. K-build-up (well test)

The term h. K-build-up is traditionally used, but h. K-well test would be more appropriate. The principle of the well test consists in measuring the build-up (hence the term used for h. K) of reservoir pressure in the well after a production phase. The rate of this pressure build-up depends on how easily the well is resupplied with fluid. We can therefore understand intuitively that the reservoir permeability (and incidentally the permeability immediately surrounding the well: the skin effect) is a major parameter in this pressure build-up.

Correct interpretation of well tests is extremely difficult and represents one of the main fields in reservoir engineering [see for example Bourdarot, 1998]. The principle consists in matching the time/pressure function observed (or more precisely a transformation of this curve, for example the derivative) with that calculated from a model assumed, *a priori*, to be as close as possible to the case studied. This interpretation, including the effects of plugging around the well (positive skin factor) or an increase in permeability (stimulation, negative skin factor) can be used to estimate the bulk permeability of the reservoir as a value of h.K and decide whether the reservoir porous medium is single or double (fractured reservoir). Still regarding the assumptions made on the reservoir geometry, the distance between the well and a permeability barrier (e.g. a non-conductive fault) can be estimated from the interpretation.

The results of this interpretation primarily depend on the assumptions made regarding the reservoir geometry and, while considering them as determining for the final interpretation and therefore taking priority over petrophysical considerations, we must always bear in mind that any particular well test interpretation is only one choice out of numerous configurations which could just as easily account for the data logged. It is the h. K-build-up parameter which is of most use to the petrophysicist.

c) The h. K-core/h. K-build-up comparison

When good quality h. K data are available (and when they concern the same sections of the reservoir), the permeability can be estimated at two different scales. What matters is the comparison of the results. If we consider the ratio (h. K-build-up)/(h. K-core), there are three possible cases. We will focus on the last one:

- A build-up/core ratio of about 1 indicates a "matrix" reservoir, i.e. one in which, as regards permeability, there is no significant intermediate scale between the core and the layer. This frequently observed case is the simplest to process, since reservoir petrophysics depends mainly on the sedimentological and diagenetic evolution of the formation: maps can then be drawn by interpolating between the wells with a certain degree of confidence and even by extrapolating using a geological logic.
- If the build-up/core ratio is much less than 1, and ignoring interpretation uncertainties (in particular incorrect consideration of a highly positive skin factor: damaged or plugged well, etc.), and if the single-phase approach adopted in the interpretation is justified, the simplest interpretation is the existence at a certain distance from the well of reduced permeability areas, such as plugged faults or fractures, channel banks, etc. Note, however, that a careful interpretation of the well test should be able to confirm the assumption of an individualised permeability barrier.

– A build-up/core ratio much larger than 1 indicates that there is a highly permeable drain in the well or its immediate vicinity. It generally consists of a conductive fracture, but we must always bear in mind that conductive fractures are not the only type of drain that can be found in a well. We have already mentioned the case of thin but highly permeable dolomitic drains ("super-K": § 1-2.1.4.Ba, Fig. 1-2.18, p. 151) which are practically never recovered by coring and which therefore do not participate in the h. K-core. We must also mention the paleokarst networks present in some reservoirs or even stylolithic joints which have undergone dissolution processes and which could locally induce significant horizontal permeability. Generally, however, they are fracture networks.

d) Brief remarks concerning the effect of fractures on reservoir bulk permeability

Fracture networks play a central role in the permeability of numerous oil reservoirs or aquifers. The study of fractures is an important field in reservoir geology or hydrogeology, which is outside the scope of this book. We will simply mention the importance of these networks when defining reservoir bulk properties and use them as an example of a potential difficulty due to the scale change.

Fractures play a major role in non-sedimentary reservoirs (e.g. crystalline rocks) where they are generally the only cause of effective permeability. In weakly tectonised sedimentary formations, which is the case with most oil and gas reservoirs, and simplifying matters to the extreme, we observe two broad types of fracture network:

– **"Diffuse" fractures**, extremely common in well bedded formations (an example is given on Figure 2-1.11). They are generally perpendicular to the beds and exhibit no relative displacement between the two blocks separated by the fracture (joint). Within a bed, we generally observe a fairly clear relation between the spacing between two fractures and the bed thickness, provided that the notion of bed is associated with a meaning which is more mechanical (homogeneity/continuity of mechanical properties) than sedimentological.

This relation between mechanical bed thickness and the spacing of "diffuse" fractures helps explain the scale variations observed in these networks and the correlative difficulty encountered when defining an REV, discussed in § 2-1.2.3A. The high fracture spacing variability (and the self-similarity observed) reflects the high layer thickness variability in the sedimentary series (from the millimetre to the hectometre). Diffuse fracturing, often quite spectacular at outcrops, drops drastically with depth, which might lead to question the presence of conductive fractures (open fractures, or more precisely partially open fractures) in deep reservoirs. Conductive diffuse fractures are probably relatively rare and the characteristic lengths to be considered (irrespective of the self-similarity aspect) are probably greater than one might expect in a first analysis. It is therefore difficult to include this type of fracturing in a homogenisation process, even when taking into account large REVs such as the modelling grid.

– **Fracture swarms.** An example of these highly specific geological objects is given on Figure 2-1.21. These swarms which may affect tabular (and therefore *a priori* weakly tectonised) zones are characterised by the presence over a relatively short width of a number of vertical fractures with very low spacing and splitting a thick, rigid bed with no displacement. In this case, the relation between bed thickness and fracture spacing, observed in case of diffuse fracturing, is no longer true. These types of swarm can sometimes be followed for several kilometres.

Figure 2-1.21 Example of fracture swarm (or fracture corridor) affecting the Dogger limestone and dolomite bar at the Pas de l'Escalette (Hérault, France). In this highly characteristic example, vertical fractures sometimes less than one metre apart split a layer several decametres thick, with no vertical displacement whatsoever. These types of swarm can sometimes be followed for several kilometres. If some fractures remain partially open in reservoir conditions, these swarms completely modify the reservoir permeability at kilometric scale

Recently, the presence of such objects has been clearly identified in oil reservoirs where, by drilling horizontal wells several kilometres long, it has been possible to intersect some of these swarms, easy to reveal on well imagery (continuous "photography" of electric resistance or ultrasound impedance properties of the borehole walls). As for diffuse fracturing, the effect of these fractures obviously depends on their opening state *in situ*. Fractures are frequently cemented across their entire plane of intersection with the reservoir, but it may happen that the plugging is only partial. In this case, the fractures radically modify the reservoir permeability at kilometric scale. For bulk permeability, the REV must be considered to be the entire reservoir.

This brief description of the fracture network effect focused only on bulk permeability. The fracture effect is in fact most apparent in this field. We must nevertheless mention that the consequences of open fractures on reservoir recovery extend far beyond this aspect. All the recovery mechanisms are directly affected (oil/water/gas exchanges between fractures and "matrix"). But this is another story…

2-1.4 ROCK TYPING

2-1.4.1 Rock Typing, description and terminology difficulties

Discussing "Rock Types" and "Rock Typing" is complicated by a major terminology ambiguity. Until the 90's, the term "rock type" was reserved for reservoir modelling operations: a rock type corresponded to a "KrPc" (relative permeability, capillary pressure) definition characterising a model grid.

This term then became increasingly used in reservoir characterisation in the broader sense (including "seismic rock type", "geomechanical rock type", etc.) such an extent that it sometimes corresponded to quite different practices. The operators all have their own rock typing protocol and therefore their own definitions. In a large oil company, these definitions may even vary from one subsidiary to another.

Any attempt to produce a summary is stymied by the absence of publications. The absence of publications is more due to the changing and therefore constantly incomplete nature of this process than to a possible desire to keep information confidential.

It would be best to respect a certain degree of terminological rigour, using instead the term "facies" and its traditional qualifiers:

- **Lithofacies, depositional facies**, widely used in geology.
- **Electrofacies**, defining all beds exhibiting similar logging responses.
- **Petrophysical facies**, designating all layers exhibiting strong petrophysical similarities (porosity, permeability).

The term rock type would then be reserved to define a particular petrophysical facies, focused on the KrPc aspect, for the construction of reservoir models.

Apart from these terminology problems, we can say at least that all rock-typings are pragmatic activities intended for an application on an oil field. It is therefore best to tackle rock typing from the point of view of its purposes. We will present below two rock typing approaches which, although highly entwined in practice, must be separated to simplify the description:

- Core Rock Typing, whose purpose is to prepare the synthetic description of the core analysis, select representative samples intended for special petrophysical measurements and, eventually, do away with the need for these special measurements through a better understanding of the relations between the various petrophysical properties. This is an essential step before Log Rock Typing.
- Log Rock Typing, from which it should ideally be possible to define, using log data, the petrophysical characteristics of the entire reservoir, to help building the reservoir model.

2-1.4.2 Core Rock Typing

A) Petrophysical identification of geological facies.

This is simply the formalisation of the traditional reservoir geology method.

It consists in identifying, facies by facies (geological), the petrophysical parameters (mean value, dispersion) and building the usual relations, e.g. K *vs.* ϕ and Leverett functions. Then, possibly, in grouping geological facies with similar characteristics. Positioning this approach in the process described here serves as a reminder that we must always handle processed petrophysical data and not "raw" data (see above). This approach is sometimes relevant but often results in the definition of "geologico-petrophysical" facies which cover extensive ranges of petrophysical values and are therefore poorly discriminating in this respect. This approach implicitly assumes the existence of a kind of one-to-one relation between lithofacies and reservoir properties. In practice, this is very rarely observed.

This explains why the definition of "petrophysical facies" (rock type) is an efficient alternative.

B) Core rock typing as such

a) Principles

The aim is to define in the reservoir a limited number of "rock types", like "boxes" in which all the samples can be stored. A Reservoir-Rock-Type is a rock category with "reservoir" properties (RP), i.e. with capillary equilibrium and fluid transport, similar enough to be characterised by unique RP = f[ϕ, (Pc)] relations. The porosity ϕ is the essential normative parameter. Ideally, in order to best integrate the capillarity and wettability information, we should also be able to include a parameter corresponding to a structural datum: The distance of the sample from the zero capillary pressure zone (Free Water Level, § 1-1.2.6, p. 106).

The first rock typing approach is straightforward: plot the R.P. *vs* ϕ relations to define rock type zones (petrophysical facies would be more appropriate) in two dimensional spaces, as schematised on Figure 2-1.22a. The most common example is that of the K *vs.* φ relation (numerous types were shown in § 1-2.1.4, p. 144) which is the basis of the traditional method discussed above. Quite frequently, in fact, routine rock typing is restricted to this permeability approach.

One major difficulty soon arises, however: although for simple porous media (e.g. intergranular), rock types based on one petrophysical property (e.g. permeability) frequently overlap those based on another (e.g. S_{wi}) (Fig. 2-1.22b1), as soon as the porous media become more complex there is no reason why we should continue to observe this favourable situation. The highly schematised case shown on Figure 2-1.22b2 is much more frequent than one might have expected. This is due to the fact that the various petrophysical properties can be governed by different geometric characteristics of the porous medium. This point was often mentioned in the first part of this book when describing the various petrophysical properties. Capillary properties represent the clearest example of this situation: during drainage they are governed by the absolute value of the pore access radius, whereas during imbibition the determining parameter is the pore radius/access radius ratio. This observation shows just how difficult it will be to define rock types including both a property during drainage (S_{wi}) and during imbibition (S_{or}) (§ 1-2.3.3, p. 186).

In practice, several solutions are available to bypass this major difficulty. The simplest is just to ignore the problem: in numerous cases, rock typing is based more or less explicitly on permeability. This may lead to disastrous results if this method is used to choose samples intended for expensive special core analysis measurements in reservoir conditions.

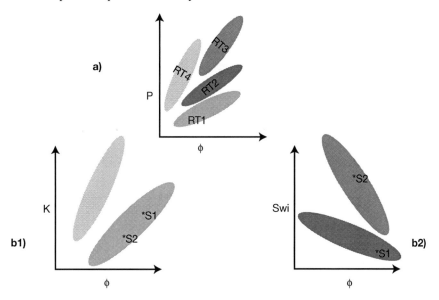

Figure 2-1.22 Simplified definitions concerning the Reservoir-Rock-Types (RRT)
a) Definition of RRT.
b) Diagrammatic example showing the possible dependence of the notion of rock typing with respect to a given petrophysical property. Samples S1 and S2 belong to the same K-Rock Type (b1) and to two different Swi-Rock Types (b2).

A better approach is to include the petrological analysis results. In carbonate reservoirs, the intersection of a simplified rock typing, based on permeability, and sedimentological characterisation of samples (e.g. Dunham classification: mudstone, wackestone, etc.) frequently produces excellent results.

To make further progress, a statistical analysis of all the data available is required.

b) Variables available for statistical analysis, perspective

The "core database" built according to the principles defined above (§ 2-1.3.1.E) is obviously the most important statistical analysis tool; it may be extremely useful, however, to work on petrophysical information based on special parameters which are both similar in essence to the characteristics studied in the laboratory and cheap enough to measure so that data can be obtained on a large number of samples. In practice, these parameters must be economically measurable on the entire series of 1 inch plugs sampled traditionally at a frequency of 3 or 4 per metre (§ 2-1.3.1D) for ϕ/K characterisation, so that they can be included directly in the core database. In addition, using this method avoids the cost of sampling and washing and minimises damage to the cores.

We can look for these "simplified" petrophysical parameters which would bear some resemblance to the traditional parameters, without having the physical "quality", but which would be obtained much more easily, along several axes:

- Firstly, through some basic characterisation methods, proposed in the early stages of petrophysics and since abandoned, one of the reasons being that they produced only poor reproducibility results. Two examples come to mind:
  - Measurement of drying kinetics to define a notion similar to irreducible saturation.
  - The spontaneous uptake of perfectly wetting fluid against air to measure the Hirschwald coefficient, sometimes wrongly disregarded by petroleum petrophysicists (§ 1-1.2.7, p. 113). In some cases, this test is reinstated by systematic measurement of the residual gas saturation (Sgr) in oil/air system. This data can be used to rough out the residual saturation problems. Note that these methods can easily be applied to imperfectly washed full-size samples (using solvent as the wetting fluid).
- Recent physical methods such as Nuclear Magnetic Resonance (NMR) can also be employed for empirical characterisation. This method, which has undergone numerous developments (§1-3.3) can be applied in a simplified form to a large number of samples, thereby proving highly suitable for the objective in question.
- Adapting automatic analysis of porous medium images to this practical problem may yield interesting results. This technique was first applied to petrophysics more than 30 years ago, with the exceptional studies conducted by Delfiner [in French, 1971]. Since then, all the technical visualisation and calculation problems have been ironed out. However, the results as regards practical applications in rock typing do not appear to have kept pace with technological progress. This relative failure may be due to the fact that most attempts have consisted in tackling the problem head-on: the basic idea was to measure N geometric characteristics which would be compared, using a sophisticated statistical instrument with the M petrophysical characteristics. This approach often proved disappointing. A much more limited and pragmatic approach

could nevertheless help solve the rock typing problem. We must define the two or three "geometric" parameters most important for the problems of multi-phase flow (already more or less known, see the corresponding paragraphs in Section 1), visualizing them roughly in the porous spaces using selective resin injections (§ 2-2.2.2, p. 336) and measure the result on thin section. Once adapted to processing a large number of samples, the method should not be very expensive.

c) Overview on statistical analysis

Statistical partitioning methods can be used if there are enough data. Numerous digital techniques are available. One proven method is to proceed step by step, first processing the most abundant data (porosity, permeability, clay content and, if necessary, the formation factor or rather the "cementation" exponent m), not forgetting a datum not strictly useful for reservoir evaluation but systematically acquired: the grain density (rhoma; § 1-1.1.5, p. 26) which will subsequently simplify the correspondence with the geological data, used when extrapolating the interpretation. This method allows for integration of the smaller number of physical data obtained, alternating between:

- the classification methods used to define the "boxes" which will store the samples. These are known as clustering methods (k-means Clustering, Agglomerative Hierarchical Clustering) or neural network;
- and the classification techniques: (Discriminatory Analysis) to decide which is the most suitable class for a new sample.

These methods, some being highly developed, are now included in the range of increasingly user-friendly petrophysical interpretation software programs. Programs dedicated to core analysis should be used rather than those which are inserted in the log analysis software packages. Otherwise, two precautions are required:

- Ensure that there is no "overlap" between the core data sets and the log data sets analysed. In logging software, there is a real risk (and temptation) to merge these data too quickly, at the expense of a rigorous interpretation.
- Check that the core data expressed as curves, such as capillary pressures, granulometries and resistivity indices, are clearly identified in the internal databases of these programs. A practice derived from log acquisition habits based on the organisation of vertical variations considers these data as "arrays", in other words as a series of numbers which represents the ordinates of the curve concerned. In this case, the abscissa values are implicit and regularly spaced. This is not true for many petrophysical results. Numerical tricks are available in the software to overcome this problem of irregular step… These may lead to confusion.

Lastly, in spite of the ease and speed now offered by numerical processing, we must not forget that these tools remain blind and are not intrinsically explanatory.

d) Scale effects and rock typing

The previous analysis deals exclusively with the matrix properties which can be properly analysed at plug scale. However, we must not forget those matrix properties whose minimum representative volume is decimetric or pluridecimetric (i.e. larger than the full-size sample). Examples could include, in the carbonates, the reef facies and the low porosity facies with networks of microcracks enlarged by dissolution. These cases, relatively frequent in practice, are a nightmare for the petrophysicist who must then rely entirely on the geologist's experience to select full-size samples, which may not always be representative. Developing a new, more objective Rock-Typing method would probably be highly worthwhile, but the avenues to be taken are not obvious. One might think that a systematic analysis of cores by CT scanning (§ 2-2.3) should be generalised by the development of automatic image analysis and statistical characterisation methods. CT scanning, although now widely used in petrophysics to obtain a "qualitative" image of the cores or an in-depth description of the multi-phase saturations on a limited number of samples, seem to be rarely used as a systematic semi-quantitative characterisation tool. This is probably due to the fact that it is impossible to obtain porosity maps on "raw" samples. Despite this technical difficulty (financially difficult to overcome?), trying to systematise the interpretation of raw absorption maps could be extremely beneficial.

2-1.4.3 Log Rock Typing

Log Rock Typing consists, using the wells for which complete information is available, in selecting a limited number of rock-types (as in the previous paragraph, core rock typing being an essential preliminary step). The logging signature and/or the sedimento-diagenetic definition of these rock-types must be determined in order to identify them in non- or poorly- cored wells (and therefore devoid of any immediate "reservoir" data). The result of this rock typing is a "reservoir property" (RP) log of all wells in the field. This basic information is one of the most important items of data in the various well interpolation methods (geostatistical, sedimentological, seismic, etc.) leading to identification of the flow-units in the reservoir model.

The single-phase permeability (K) is very easy to measure on core. In a cored well, reliable values of K are routinely available, every foot or 6 inches. Determination of K by log analysis is much more difficult, however. The numerous attempts made since the emergence of log analysis techniques have not led to a overall solution, hence the practical interest in *permeability* Reservoir Rock Typing (RRT), aimed at determining the various K *vs.* ϕ relations in a reservoir. It is currently the only RRT performed on a routine basis. Once this RRT has been determined, the mean values of the various other "reservoir properties" are assigned to each "K Rock Type", according to the (rare) measurements available on the K Rock Type considered.

The method implemented to perform this RRT in K-ϕ depends directly on practical problems: data types available, time available, etc. For example, with fast studies it is generally impossible to reconsider the core data. The methodology must therefore be highly empirical. It is based on two broad approaches:

- *Log analysis*, which is always performed, to identify electrofacies, some of which will correspond to K-Rock Type signatures. Statistical analysis software (in the broad sense, including the neuronal networks) associating log values and core values (ϕ, K), is especially useful for this type of study.

– *Sedimento-petrographic analysis*, focused on porous space geometries, is carried out less frequently since it implies access to cores and thin sections. It is also more time-consuming and is greatly dependent on the technician's skills. It is nevertheless essential to achieve high quality K-rock typing.

We already pointed out the major problem with this type of routine rock typing when discussing core rock typing: there is no guarantee that properties other than K will be included in the subdivisions so defined. Experience tends to show that, with "simple" reservoirs such as relatively clean sandstones and some micritic carbonates, the ϕ-K classes may be quite representative of the other petrophysical properties. We can be certain, however, that as soon as a reservoir exhibits a certain level of complexity (shaly sandstones, arkosic sandstones, numerous carbonates), this is no longer the case and we are faced with the problem of the "other" petrophysical properties.

As a conclusion: Rock Typing and multi-phase flows

We must emphasize one point which currently represents the main difficulty in practical applications concerning oil reservoirs: the necessary and difficult adaptation of the Rock Typing methods to multi-phase flows: We must remain extremely cautious when using Rock-Types in reservoir modelling, even consolidated in their multi-phase aspect by addition of parameters such as capillary pressures (or initial saturations S_{wi}) and residual gas saturations S_{gr} (or the Hirschwald coefficient).

In oil reservoirs, as soon as we no longer have strong wettability conditions (in other words, as soon as we move away from a situation of perfect water wettability), which is by far the most frequent case, determination of two-phase flow rock types must include wettability indicators. Taking wettability into account is an absolute necessity. As it is, the wettability indicators are neither *a priori* related to the morphology or structure of the porous media, nor are they generally strongly "geologically dependent" [Hamon & Bennes, 2004]. In addition, the two-phases flow rock types should in practice, depend on the recovery process implemented…

CHAPTER 2-2

Porous Network Observation Techniques

2-2.1 THIN SECTIONS AND EPOXY PORE CASTS

2-2.1.1 Impregnation of porous networks by resins

The porous space of samples under room conditions (after cleaning and drying) is "saturated" with air. The first experimental operation required before any visualisation attempt is to replace this air (difficult to see!) by a material easy to identify by its colour (optical observation) or its physico-chemical characteristics (observation under electronic microscope).

After being placed under vacuum, the samples are saturated with resin loaded with coloured and/or fluorescent pigments (fluorescent for examination under microscope illuminated with UV light). The traditional preparations can be carried out once the resin has polymerised under the effect of time and/or heat: polished sections (Fig.2-2.1), thin sections, epoxy pore casts. Two colours are generally used: deep red, which allows the resin to be clearly seen against the often very light rock background, and blue, which although more difficult to see can be used with the red dolomite characterisation stains (alizarin).

Rock can be impregnated with numerous different types of resin. They are purchased from various sources, for example the companies supplying thin section machining equipment. Most are epoxy resins, but some laboratories use methyl methacrylate (Plexiglas, Lucite) which despite very low viscosity is difficult to implement. When choosing the resin a compromise has to be made between a number of factors: viscosity during implementation, duration in liquid phase before polymerisation, volume to be processed. Concerning the volume to be processed, it is important to mention that since polymerisation is exothermic, the kinetics of this reaction may largely depend on the volume of resin used. Above a certain volume, polymerisation may become almost explosive. Some resins, easy to use but purely designed to consolidate the sample mechanically before machining the thin sections or polished sections, should be avoided. These viscous resins do not penetrate the microporosity and therefore cannot be used to observe the porous networks.

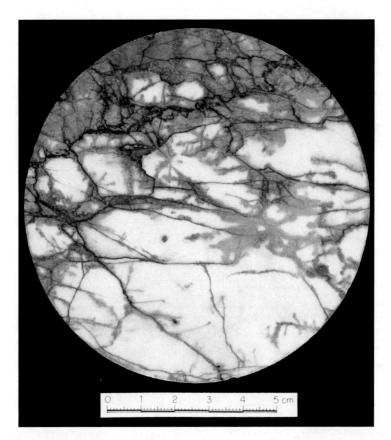

Figure 2-2.1 Section of full size core impregnated with red epoxy resin. In this low porosity limestone the resin clearly shows a dense network of microcracks (almost certainly closed under confinement conditions). We can also see coloured matrix areas: these are areas of very fine microporosity. Despite the care taken during injection, the resin will probably not have reached all these very poorly permeable areas. When diagnosing on the presence of microporosity at a given point, therefore, the distance to a microcrack must be taken into consideration

A) Physical parameters affecting the impregnation quality

In order to fully appreciate the precautions required during resin impregnation, we must consider the relative importance of the various physical parameters in this operation. The simple straight capillary model already mentioned for capillary rise (§ 1-1.2.7, p. 113) can be used again. Using the Washburn equation we can calculate the depth of penetration h of a resin of viscosity η, at time t, in a straight capillary of radius R under a total pressure P. This total pressure is the sum of the capillary pressure (P_c) and the injection pressure (P_i, at least atmospheric pressure, when the vacuum is released). Since the resins are generally perfectly wetting, the capillary pressure is a major driving force in the phenomenon. The Washburn equation therefore gives: $h = A\sqrt{(R^2 P)}.\sqrt{(1/\eta)}.\sqrt{t}$. Taking into account only the capillary pressure ($P_c = 2\sigma \cos \tau/R$), we obtain:

$h = B\sqrt{R}.\sqrt{(1/\eta)}.\sqrt{t}$ where A and B are constants.

If we add an injection pressure P_i simplification on R is no longer possible, but an approximation in the form of a coefficient of $\sqrt{(1+0.15P_iR)}$ to be applied to h can be used, pressure being measured in MPa and the radius in μm.

We can therefore consider that the efficiency of an injection will be proportional to the square root of the time (i.e. the duration of the liquid phase before polymerisation) and inversely proportional to the square root of the viscosity. If the only driving force is the capillary pressure, the radius of the capillaries acts as its square root (we can therefore very roughly consider that the efficiency is proportional to the fourth root of the permeability). If we add an injection pressure, it still acts as its square root (approximately). We therefore see that when choosing the physical values to be implemented (and therefore when choosing the resins and injection cells), the various factors must be carefully analysed and weighted. A resin of low viscosity (but fast polymerisation) is not necessarily better than a more viscous resin of longer lifetime (and one which can generally be used in large quantities).

Similarly, applying a high injection pressure is less useful than one might think, since pressure only acts as its square root. Concerning the possible injection pressure, we must also mention one point which may have practical consequences. The injection pressure is often transmitted to the resin by a nitrogen gas plenum; however, under high pressure, nitrogen diffuses very rapidly in the resins, causing bubbling when depressurisation takes place. This is another reason for not overestimating the relative utility of high injection pressures.

B) Checking the injection efficiency

When there is a strong colour contrast between the material and the resin (see Fig. 2-2.1), the resin penetration is easy to check at macroscopic scale. There may be some doubt, however, as to the impregnation efficiency at microscopic scale. Since resin is wetting, it tends to entirely cover the solid phase, guaranteeing high-quality microscopic impregnation. The best way to check this point is to observe the shape of the resin moulding a very fine geological object of well-known structure. The example of Figure 2-2.2 concerns chalk coccoliths, but all pore casts confirm this observation: the wetting resin impregnates all the spaces up to the finest intercrystalline contacts (e.g. marbles) provided that the kinetic conditions described above are respected. In marbles with crystalline joints, the penetration depth does not exceed a few fractions of a millimetre.

2-2.1.2 Observation of thin sections under optical microscope

The (polarising) optical microscope is the most traditional instrument used to observe rocks in very thin and therefore virtually transparent sections (thin sections). Measuring about 30 μm thick, these sections are obtained by machining down a piece of rock glued onto a glass support. The top side is covered with a glass slide or, preferably, very finely polished. Thin sections are the petrographer's preferred observation method and are therefore manufactured in large quantities for the description of geological samples.

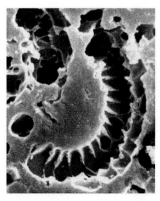

Figure 2-2.2 Checking the resin impregnation efficiency by comparing the image of a geological object (in this case, a coccolithophoridae of chalk, on the left) and the image of resin moulding a similar object (pore cast technique, see below, the calcite has been dissolved with HCl) on the right. SEM photographs

If the colouring of the impregnation resin is strong enough to remain visible on a 30 μm thick section, the pores will be very easy to identify (numerous photographs of thin sections are included in this book).

We must nevertheless point out that only pores present throughout the thickness of the section, i.e. of at least 30 μm "diameter" will be easy to observe. The smaller pores will be intimately intermingled with the rock matrix and will only be observable under special conditions which we will specify below.

Transmitted light and diffused light (Fig. 2-2.3)

The method generally used to illuminate the preparation under polarising microscope is transmitted light crossing through the section and producing an image by transparency. Under these conditions, pores of diameter greater than 30 μm stand out very clearly due to their colour. On Figure 2-2.3 (left photograph), the intergranular pores saturated with red resin are highly visible. What about the black (opaque) areas inside oolites? To describe these areas, the top surface of the preparation must be illuminated obliquely and the diffused light observed (photograph on the right). The oolites are coloured pink: they consist of microcrystalline calcite (< 10 μm) exhibiting highly abundant intercrystalline microporosity. Examination of epoxy pore casts under scanning electron microscope (see below) confirms this analysis.

Although individual micropores cannot be seen under diffused light, the microporous areas are easy to identify during conventional reservoir rock descriptions under polarising microscope.

This observation, which should be carried out on a routine basis, is extremely important in special cases where the most useful porous space corresponds to pores of intermediate dimension (maximum radius 50 μm, for example in leached limestones).

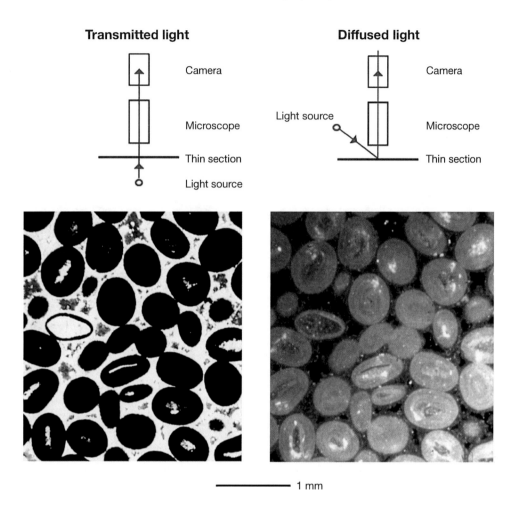

Figure 2-2.3 Thin section observed under transmitted light and under diffused light

The left photograph (transmitted light), clearly shows the intergranular porosity (red), the oolites with opaque cortex (black) and the rim cement surrounding the oolites (white).

The right photograph shows the same area of the section observed under diffused light (oblique lighting). The pink cortex of the oolites stands out clearly. The pink colour corresponds to the mixture of resin and microcalcite forming the cortex.

An example showing the clear advantage of observation under diffused light is given on Figure 2-2.4. It consists of a limestone whose micritic matrix (limestone mud) present between the grains (foraminifers) has undergone weathering, leading to the creation of intermediate porosity (dimension between micro and macro). The method involving two resins by centrifuging clearly identifies this porosity (in red, equivalent throat radius greater than 1 μm) but the distribution of this porosity in the overall rock structure can only be immediately revealed by observation under incident light.

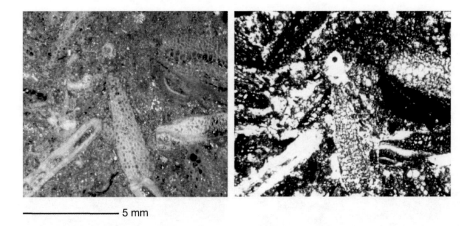

———————————— 5 mm

Figure 2-2.4 Leached foraminiferal pack/wackestone. The sample has undergone double-resin preparation
by centrifuging (§ 2-2.2). The blue resin fills the porous space accessible by radii of less than 1 μm,
the red resin by radii greater than this value. On the right, observation under transmitted light,
on the left observation under oblique incident light

2-2.1.3 Observation under Scanning Electron Microscope (SEM)

Conventional electron microscopy implements two broad types of imaging: backscattered
electron imaging, revealing in particular the composition variations of the material studied,
and secondary electron imaging, mainly sensitive to the topographic variations of the
surface considered. The latter case corresponds to the more traditional SEM images.

A) Polished sections (Backscattered Electron Imaging)

This type of imaging is ideal when studying polished sections of samples impregnated with
resin (Fig. 2-2.5). The contrast in elementary composition between resin and minerals
produces high-quality images from 20× magnification up to very high magnifications (note,
however, that the definition obtained with this technique is not as good as that with
secondary electron images). It represents an excellent way of obtaining digital images of
porous networks for automatic geometric analysis applications.

B) Pore casts (Secondary electron imaging)

Widely used for petrographic descriptions, conventional secondary electron imaging on rock
fractures is unsuitable for investigations of porous networks. On a fracture, it is always very
difficult distinguishing between "true" porosity and grain tear out. It is therefore preferable
to image the porosity itself, observing only the impregnation resin, after using acid to
destroy the entire mineral phase. This is known as the replica or epoxy pore cast technique.

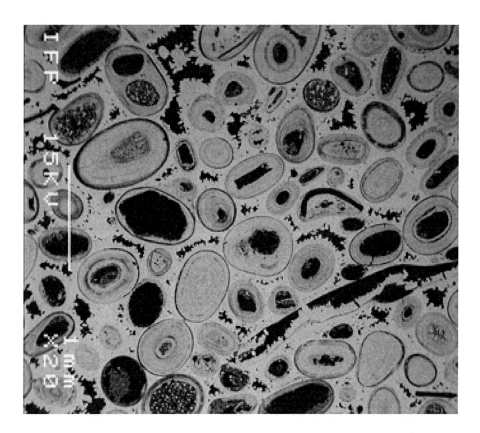

Figure 2-2.5 Example of backscattered electron image of a polished section (IFP document, E. Rosenberg). The scale bar inside the photograph represents 1 mm. The image shows the same type of oomoldic limestone as that illustrated on Figures 2-2.11 (p. 338) and 2-2.20 (p. 349)

a) Epoxy pore cast production principle

A "thick" section (compared with thin sections) (e.g. 1 mm) of sample impregnated with resin is first glued onto a 1 to 2 cm diameter plastic support and its visible surface then polished to obtain a reference plane to study the pore geometry. This sample fragment is generally cut from the same small block as that used for the thin section. The "thick" section corresponds to the top photograph of Figure 2-2.6.

The mineral fraction contained in the thick section is completely destroyed by acid attack. The process consists of two phases:

- Firstly, attack with hydrochloric acid (HCl) to destroy the carbonates. For routine studies when working on carbonate reservoirs (calcite and dolomite), hydrochloric acid attack may be sufficient, thereby simplifying the preparation.
- Secondly, attack with hydrofluoric acid (HF) dissolves the silicates. We must emphasise that working with hydrofluoric acid is extremely dangerous for inexperienced operators. The substance should be handled with extreme care (risk of very serious

1 mm

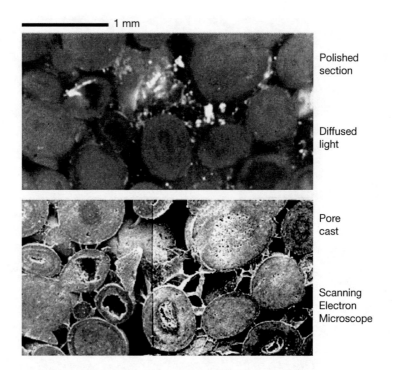

Polished
section

Diffused
light

Pore
cast

Scanning
Electron
Microscope

Figure 2-2.6 Principle used to produce epoxy pore casts

Top photograph: polished "thick section", optical microscope, incident light. We can see the oolites, col-oured red since they contain a high proportion of resin (intraoolitic microporosity), and the vitreous reflec-tions of the large calcite crystals (intergranular sparite).

Bottom photograph: pore cast, same sample observed under SEM after total destruction of the mineral phase by acid attack. The oolite "ghosts", formed solely of resin, are clearly visible. Between the oolites, we also observe the presence of resin films corresponding to the very fine intercrystalline voids which existed between the calcite crystals.

burns). Calcic silicates are a special problem, and one for which we can offer no prac-tical solution. Since the calcium fluoride produced by the reaction is very poorly solu-ble, it is deposited directly on the preparation, making it unusable. This problem is often encountered when studying some clays or arkosic sandstones rich in calcic feld-spars.

Apart from the case of calcic silicates, visualisation by epoxy pore casts is extremely easy, over the entire range of magnifications possible under SEM (from 20x). Numerous epoxy pore casts are shown in this book. Figure 2-2.7 shows the benefit of this observation method for non-calcic clay sandstones.

b) Stereo photographs of epoxy pore casts

Stereo photographs of epoxy pore casts under SEM are very easy to produce. The epoxy pore cast is simply photographed twice, changing the inclination of the preparation under the

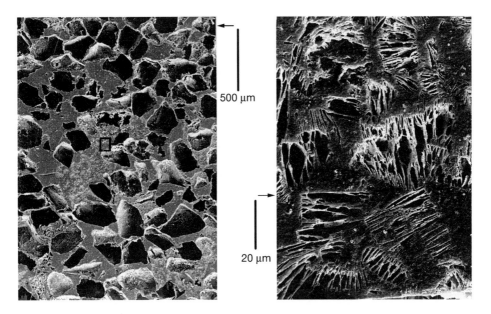

Figure 2-2.7 Epoxy pore cast of a kaolinitic sandstone

The left photograph (low magnification) shows the large areas of resin corresponding to intergranular porosity (quartz grains completely dissolved by HF and rare grains of weathered microporous feldspar – e.g. bottom left corner –). We also see that these large areas sometimes contain elements easy to identify at high magnification on the right photograph: kaolinite crystals generating an original microporous network. Similar crystals are shown on thin section (Figure 1-1.16, p. 32) and under SEM on rock fracture (Figure 1-1.46, p. 90). [Bourbié *et al.*, 1987].

microscope by a few degrees (e.g. 10°). The relief image is obtained by observing the two photographs through a stereo viewer.

Apart from the fact that the method provides a better understanding of the porous geometries, the possibility of observing a relief image is extremely useful when taking semi-quantitative measurements. On Figure 2-2.8, for example, we can estimate the thickness (very small) of intercrystalline films.

This simple and inexpensive observation method seems to be rarely used. This is probably due to the fact that only field geographers and geologists are familiar with stereo photographs.

2-2.2 VISUALISATION OF FLUID LOCATION IN PORES

It is extremely important to know exactly where the fluids are located in the pores to obtain a better insight into capillary equilibria and, ideally, solve wettability problems directly. Inversely, once the rules concerning the location of these phases have been understood, this

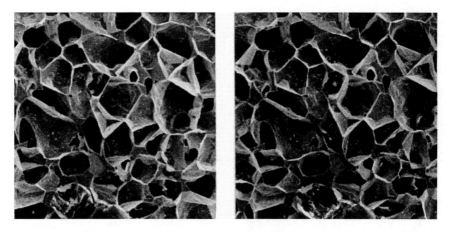

Stereoscopic set up

Figure 2-2.8 Stereo photographs of an epoxy pore cast of a very low porosity Fontainebleau sandstone
($\phi < 0.03$). The photograph is 1.2 mm long. Once the quartz has been completely dissolved by HF,
we observe the "microcracks", which are in fact the very fine porous films remaining between the grains,
as well as the rare intergranular pores. Figure 1-2.16, p. 148, shows a thin section photograph
of a very similar sample. [Bourbié *et al.*, 1987]

visualisation represents a powerful tool to evaluate the complex geometries of true porous
media, using two-dimensional observations. The principle of this visualisation is to "freeze" a
microscopic equilibrium of phases (at least one of them being liquid) so that it can be observed
at pore scale. Two broad experimental directions can be used. Either a capillary equilibrium is
frozen in a "natural" water-oil state, for example, and then observed under electron
microscope, or liquids which are easy to solidify by polymerisation or cooling are used, for
observation on thin sections or polished sections. A large number of articles are available in
the literature, describing fluid location visualisation experiments. We may mention, in
particular, due to its historical interest, the film of the Stanolind Oil and Gas company (1952).

2-2.2.1 Direct observations of natural fluids: Cryo-scanning electron microscopy

This is an application to reservoir rocks of the methods used in biology for observation of
tissues and microorganisms under electron microscope.

Erwin Sutanto [1991], for example, gives a good description of the method, summarised
below:

– A small sample (6 mm diameter cylinder, 20 mm to 30 mm long) is cut from the sample to be studied, taking all necessary precautions regarding the choice of coring liquid (brine, oil, etc.) depending on the type of experiment envisaged.
– The sample, which may be given special preparation (saturation centrifuging,
flushing, etc.) is placed on a small metal support and frozen in pasty nitrogen or liquid
propane. Although the sample is small, freezing does not occur fast enough to vitrify
the liquids present (brine, oil). We will therefore observe some volume variation

effects (water expansion, oil contraction) which could disturb the shapes slightly, but without modifying the relative positions of the oil and water accumulations.

– Placed under vacuum in a microscope side chamber, the sample is fractured using a micromechanical device. Note that at the temperature of liquid nitrogen, there is virtually no water or oil sublimation. A good quality vacuum can therefore be maintained in the chamber.

– The fracture is then covered with a very fine layer of chromium to provide electrical conductivity and observed according to the usual methods.

– When the accumulations are large enough (micron), it is relatively easy to differentiate between oil (dark colour) and water (lighter grey), as shown on Figure 2-2.9. If the accumulations are less visible, it is still possible to differentiate between the frozen liquids by performing an X-ray spectrometry analysis of the characteristic elements (such as chlorine for brine).

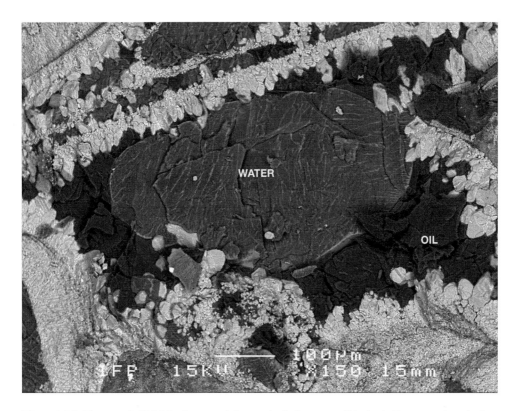

Figure 2-2.9 Photograph (SEM, backscattered electrons) of a fragment of bioclastic limestone viewed using cryo-scanning electron microscopy: Water appears light grey, oil dark grey and the calcitic matrix white. Scale bar: 100 μm. NB. This excellent example of strong oil wettability corresponds to a situation which is quite rarely encountered in oil reservoirs IFP document (E. Rosenberg, A. Moctezuma)

Figure 2-2.9 shows an example of applying the method. It concerns a limestone made strongly non water-wet. A water globule (light grey) can be clearly seen in an intergranular macropore, surrounded by oil (dark grey) in contact with the pore walls. In this case, cryo-scanning electron microscopy immediately reveals the undeniable oil wettability.

Since cryo-scanning electron microscopy allows "direct" observation of the relative positions of the fluids, this technique is highly efficient in wettability state diagnosis. It is a vital tool in this field. The main difficulty with this technique is the need for specialised apparatus, which increases the cost of the experiment, but in particular the need for special handling of the core samples.

2-2.2.2 Indirect observations using polymerisable liquids or fusible metal

A) Epoxy resins

a) Overview on epoxy resins

The epoxy resins used for these experiments must satisfy two essential conditions: polymerisation with no significant volume change or gas release (bubbling), and viscosity as low as possible. For practicality and efficiency reasons, the following qualities are also required: ease of handling, possibility of bulk preparation (several kilos), long lifetime before gelification, etc. In addition, the resin/hardener pair must not chemically react with the deep dyes (e.g. Organol) used to deeply colour the resin, of very low thickness in thin sections.

In the past, we have used the mixture CY192-HY918, still sold by the company Vantico. Its viscosity at room temperature is very high (300 mPa . s) but, at 70°C, since the viscosity remains less than 20 mPa. s for over ten hours, capillary equilibria can therefore be reached, even in low permeability rocks. More information in French can be found in Zinszner and Meynot [1982].

b) Visualisation of equilibrium during drainage: centrifuge

Method

The principle consists in conducting a drainage experiment by centrifuge (see § 1-1.2.4B, p. 74) using an epoxy mixture as wetting fluid and air as the non-wetting fluid. There is no doubt that the rock is perfectly epoxy resin-wet. A rock sample injected under vacuum with coloured resin undergoes centrifuging, which drains the resin. The acceleration is chosen so as to be:

- Either in the transition zone (see § 1-1.2.2, p. 58) where there is a relation between saturation in wetting fluid and drainage pressure, in which case the resin occupies the porous volumes whose access radius is smaller than a radius which depends on the imposed acceleration ("cutoff radius").
- Or in the so-called "initial@Pc_{max}" saturation state, in which case the resin occupies the same sites as the "initial" water after the drainage tests.

The mixture is then polymerised (ideally leaving the sample under acceleration, but usually in practice after removing the sample from the centrifuge).

Checking the result

It is relatively easy to check the result of the experiment for drainage in the transition zone, since a cutoff radius has been calculated, beyond which the epoxy has been expelled from the porous spaces. By direct comparison between the result of mercury porosimetry (see § 1-1.2.4D, p. 79) on an untreated sample and on an adjacent treated sample, we can immediately measure the effects of centrifuging, the mercury-resin wettability angle being virtually the same as the mercury/rock wettability angle. This comparison yields convincing results (Fig. 2-2.10).

Checking the result for centrifuging in "irreducible" saturation is based on a comparison of the final saturation after centrifuging the resin and that after a conventional drainage experiment (restored states, centrifuge). Irreducible saturation in epoxy resin generally corresponds to the "Water S_{wi}" obtained under capillary pressures lower than that involved by centrifuging the resin (0.3 MPa *vs.* 0.6 MPa for example). This is probably due to the fact that the viscosity of resin is much higher than that of water, and to the differences in interfacial tensions.

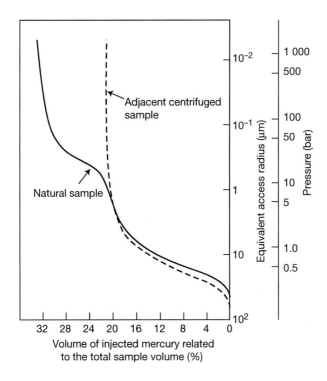

Figure 2-2.10 Checking the result of the epoxy resin drainage experiment in transition zone. Mercury porosimetry on a natural sample and an adjacent sample centrifuged under acceleration such that the equivalent capillary cutoff radius (R_c) is about 1 μm

Preparation examples

The example on Figure 2-2.11 shows the result of standard preparation by centrifuge. It represents an oomoldic limestone, whose porosity is mainly located inside dissolved oolites (intragranular macroporosity) and only connected to the outside through the microporosity present in the remaining layers of cortex (intragranular microporosity). The result of centrifuging with a 1 μm cutoff radius is clear: only some of the intergranular macroporosity is well connected, some small intergranular pores being poorly connected together due to thick layer of rim cement surrounding the grains. We can easily appreciate that the permeability of this exceptionally ($\phi = 0.38$) porous oolitic limestone is relatively low (15 mD). The situation shown in Figure 2-2.11 is so obvious for geologists used to observing thin sections that one could almost question the need for such preparations.

In numerous less simple cases, however, these sections (only two or three times more expensive than ordinary sections) are extremely useful, either for qualitative observations or for automatic measurement of porous geometry characteristics, which is considerably simplified by these preparations.

Figure 2-2.12 shows a different type of example, consisting of a more complicated preparation made possible by the wide range of porosities of the bioclastic limestone investigated. Three different centrifugings have been carried out with increasing cutoff radii. After polymerisation of the first resin centrifuged, the sample has been completely resaturated with a different coloured resin, and so on. These preparations are rarely used

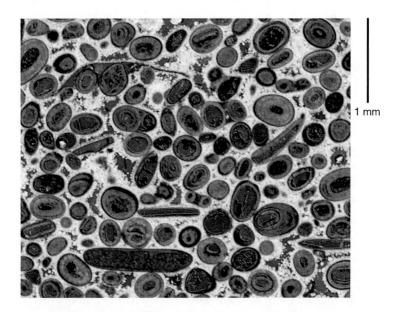

Figure 2-2.11 Oomoldic limestone (Brauvillers, $\phi = 0.38$; $K = 15$ mD)

Centrifuge, 1 μm cutoff radius. The blue resin corresponds to pores accessible via equivalent radii less than 1 μm.

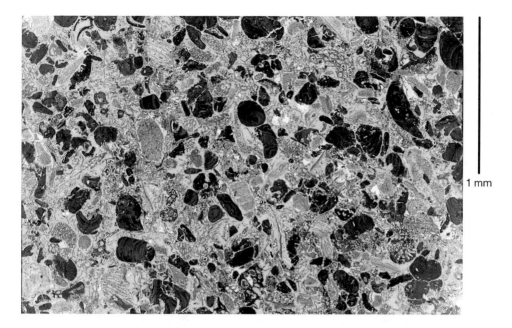

Figure 2-2.12 Bioclastic limestone from the Lubéron ($\phi = 0.37$; $K = 1.5$ D).
Three successive centrifugings corresponding to cutoff radii of 1 μm, 5 μm and 10 μm. The increasing equivalent access radii correspond to the following colour order: royal blue, red, yellow and light green

since the extra information obtained compared with a single centrifuging does not generally justify the more complicated experimental procedure. On photograph 2-2.12, we can see qualitatively that the characteristic dimension (§ 2-1.2.1, p. 274) corresponding to the best connected fraction of the porous space (light green fraction) is much larger than that of the other types of porosity.

An example of centrifuging in "terminal" saturation is shown for information on Figure 2-2.13. The sample investigated was first subjected to centrifuging at very high acceleration. Spontaneous imbibition was then carried out after polymerising the resin. Since the resin present is mainly located in the microporosity, where it is difficult to observe using conventional thin section methods, this type of preparation is often only of limited use if the experiment is only carried out for observation purposes. If imbibition is carried out afterwards, however, the mechanism involved may be substantially modified by this thin layer of resin. Unfortunately, the true effect is poorly understood and is a subject of debate.

c) Visualisation of equilibrium during spontaneous imbibition: capillary rise

Method

Capillary rise is carried out to visualise the relative position of the wetting/non-wetting fluids after imbibition (see § 1-1.2.7, p. 113). The bottom of the sample is placed in the epoxy mixture which, expelling some of the air, progressively invades the sample, due to the sole effect of the capillary forces. We have seen that the final result does not depend, to the

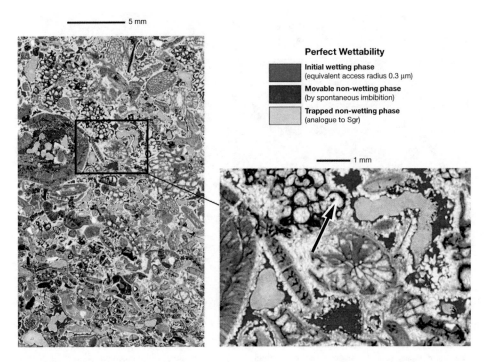

Figure 2-2.13 Bioclastic limestone from the Lubéron ($\phi = 0.2$; $K = 0.7$ D) which has first undergone centrifuging in "initial@Pc$_{max}$" saturation followed by spontaneous imbibition by capillary rise

The red (saturation 0.36) and yellow (saturation 0.39) resins correspond respectively to the displaceable and trapped fractions of the non-wetting fluid taken up by drainage. Note the highly characteristic shape of the trapped non-wetting "globules" which occupy the central regions of the largest pores.

Considerable caution is required when extrapolating similar models, but we may consider that this photograph gives a fairly reliable representation of the distribution of fluids in a gas reservoir (or an oil reservoir, in case of strong water wettability) which has undergone waterflooding. The red and yellow regions represent the hydrocarbon present initially, the red fraction corresponds to the water which has replaced the hydrocarbon produced, the yellow fraction represents the hydrocarbon trapped by capillary phenomena (residual saturation § 1-2.3.3B, p. 188); recovery from this fraction may only be possible through the use of physico-chemical phenomena "breaking" this capillarity (tertiary recovery, [e.g. Latil *et al.* 1980]).

first order, on the fluid pair under investigation (provided that the wettability contrast is good). We therefore obtain a capillary equilibrium corresponding to that reached at the end of total imbibition.

The resin is polymerised when the capillary rise has stopped. To avoid volume variations due to expansion of the trapped air, the temperature is only significantly increased after gelification. After complete polymerisation, thin sections are cut from the samples and, to improve observation, it may be worthwhile colouring the pores not saturated with resin (i.e. saturated with air, non-wetting phase) with resin of a different colour. In this special case, this second injection raises a practical problem: since imbibition has blocked all the accesses, the samples are impermeabilised (the relative permeability of the non-wetting phase is zero, § 1-2.3.1, p. 181) and the traditional technique by injection under vacuum is ineffective. The following method can be applied: one side of the rock section is polished ready for gluing onto the object holder. The sample then undergoes injection under vacuum followed by thorough wiping of the polished side, so that after polymerisation only light polishing is required before the sample is glued onto the object holder.

If a large number of samples are going to be studied, the method can be simplified. Pieces of rock 1 cm thick are used, and immersed directly in resin. After polymerisation, they can be observed directly on uncovered thin section. The fractions of the non-saturated porous space can be observed under reflected light. The result is generally similar to that obtained during an experiment conducted according to the complete protocol.

Checking the result

The validity of the experiment relies mainly on the fact that the result of capillary rise (spontaneous imbibition) is practically independent of the fluid pair used, provided that there is a clear wettability contrast. This point has been clearly demonstrated (see § 1-1.2.7, p. 113) and is true for the epoxy resins, despite their relatively high viscosity (Fig. 2-2.14).

An example of preparation obtained by spontaneous imbibition (on a sample previously centrifuged in "irreducible" saturation) is given on Figure 2-2.13. Details concerning this sample are provided on the figure legend. We will simply emphasise the advantage of this type of preparation (rather inexpensive) when studying porous geometries using automatic image analysis methods. Identification of the fraction of porous space corresponding to the trapped non-wetting fluid is obviously important for oil reservoir studies, but inversely it also provides a way of identifying the zones where the pore radius/access radius ratio is very high, to obtain an indirect approximation of the third dimension which considerably simplifies automated descriptions of the porous space.

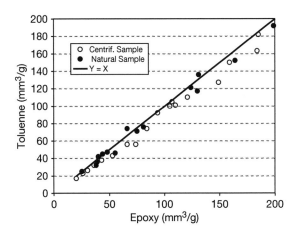

Figure 2-2.14 Comparison of toluene and epoxy resin uptakes
during standardised capillary rise experiments

The blue resin (similar to "S_{wi}") represents a saturation of about 0.25, calculated on the entire sample. It is difficult to see on the photograph, being mostly located in the microporosity inside the bioclasts (in beige on the photograph). It can only be observed in rare intrabioclastic, very poorly connected pores (white arrow on the zoom).

B) Wood's metal or similar

Wood's metal is a tin/bismuth alloy of melting point about 70°C. Above 120°C, its capillary behaviour is very similar to that of mercury at room temperature. Experiments identical to mercury porosimetry (§ 1-1.2.4D, p. 79) are therefore conducted with molten metal then, by rapid cooling, the capillary equilibrium obtained is frozen [Dullien and Dhawan, 1974; Swanson, 1979]. The Wood's metal is easy to see in polished sections and thin sections prepared from samples treated in this way, thereby indicating the zones in the porous space accessible by an "access radius" (average curvature) which depends on the pressure exerted. In practice, rather than Wood's metal, it is better to use an alloy with a melting point of 137°C, readily available off-the-shelf and exhibiting better capillary and mechanical (polishing) characteristics than Wood's metal.

This method is not suited for fast preparation of numerous samples. However, it offers the advantage of allowing observation at various scales, from decimetric to microscopic, making it very useful for studies on large models of porous media.

The block shown on Figure 2-2.15 consists of ground, calibrated glass, packed in a glass tube about a ten centimetres diameter and about 50 cm long. The glass tube (removed when sawing the sides of the block) was placed in a highly inclined position (Fig. 2-2.15 a) so as to observe a long drainage front during invasion of the block, previously placed under vacuum, by the fusible metal. The drainage front can be studied at decimetric scale (Fig. 2-2.15 b1), centimetric scale (Fig. 2-2.15 b2 and b3) and submillimetric scale (Fig. 2-2.15c, thin section). We can therefore observe the characteristics of a starting drainage front (characteristic dimension equal to the sample dimension) and the simultaneous appearance of the smallest heterogeneities. The discontinuity shown on photograph 2-2.15 b3 actually corresponds to a very slight variation in the packing of the ground glass.

2-2.3 X-RAY TOMOGRAPHY
(also known as CT or CAT Scan, computerised or computerised axial tomography)

X-ray tomography for medical purposes developed rapidly after its invention by G. Hounsfield in the early 1970's. The first tests applied to petrophysics date back to the 1980's [e.g. Wellington & Vinegar, 1987]. It is now used routinely in the geosciences [Mees *et al.*, 2003]. Used amongst other things to accurately describe the location of fluids inside samples during experiments, X-ray tomography is an extremely useful tool in the field of petrophysics.

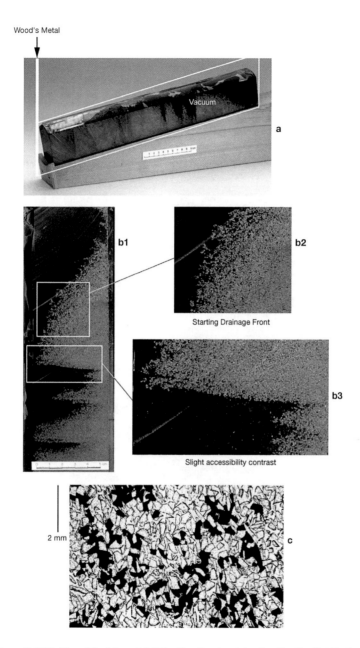

Figure 2-2.15 Use of fusible metal (Wood's alloy type) to visualise the distribution of wetting and non-wetting phases

Block of ground glass subject to drainage of the "vacuum" phase (metal vapour) by the molten alloy, under the effect of gravitational forces. After the alloy has solidified, the zones not invaded by metal have been saturated with red epoxy resin. On the polished sides (a and b), the metal is silvery in appearance and the glass resin mixture is very dark red. On the thin section (c), the metal is black, the resin very pale pink and the ground glass grains lighter.

2-2.3.1 The medical scanner

A) Principle

The principle of all tomographic reconstructions is to calculate, for a given object and for a given physical characteristic (e.g. X-ray absorption, acoustic impedance), the geometric distribution of this property "justifying", as accurately as possible, the modifications undergone by a set of rays (e.g. X-rays, acoustic beams) crossing this object along different paths. We can appreciate intuitively that the greater the number of paths, the better calculation result will be.

The principle of the medical scanner (Fig. 2-2.16) is to rotate a frame supporting an X-ray emitter and a large number (e.g. 512) of receivers around the patient. The attenuation of the X-rays crossing the body along millions of different paths can therefore be recorded. For each pixel of the slice considered, a tomographic reconstruction program recalculates the most likely X-ray attenuation.

The tomographic definition depends on the X-ray beam collimation accuracy and the linear density of the receivers. The voxel (3D pixel) definitions of current standard scanners are about 0.2 mm × 0.2 mm × 1 mm.

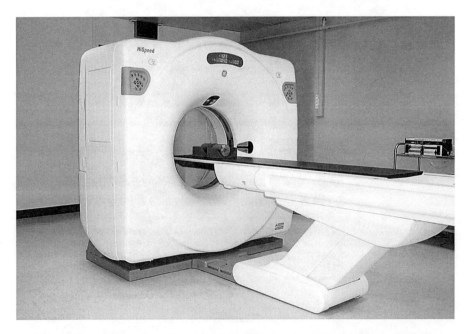

Figure 2-2.16 Example of medical scanner used for petrophysical applications. The diameter of the core placed on the examination bed is about 12 cm. The partially torus-shaped housing, measuring about 2 m diameter, surrounding the end of the examination bed contains the rotating frame supporting the X-ray emitter and the receivers. By translating the bed, the core is accurately positioned in the vertical X-ray acquisition plane. IFP photograph

Before recording the CT scans (and in particular to precisely locate them), a conventional X-ray radiography is taken, moving the object under examination through the stationary X-ray machine (translation of the bed).

On tomographic reconstruction the radiodensity (transparency to X-rays) unit used, the hounsfield, is adapted for medical purposes: this unit is –1 000 in air, 0 in water (and about +1 000 in compact bone). Radio density is generally always over 1 000 hounsfields for petrophysical objects. It depends on the energy of the X-rays used and the chemical nature of the materials. In a first analysis, the most important parameter is the specific gravity; consequently, in a monomineral rock in single-phase saturation, the tomographic image is similar to a porosity image.

The X-ray absorption by some elements such as iodine and barium is extremely high, and their salts can be used as markers in experiments on displacement through porous media (see Fig. 2-1.8, p. 285)

B) Porosity and saturation maps

To measure porosity and saturation precisely, the images acquired in various saturation states of the core studied must be processed by subtraction and division. The principle is summarised on Figure 2-2.17. A porosity map is obtained by subtracting, pixel by pixel, the

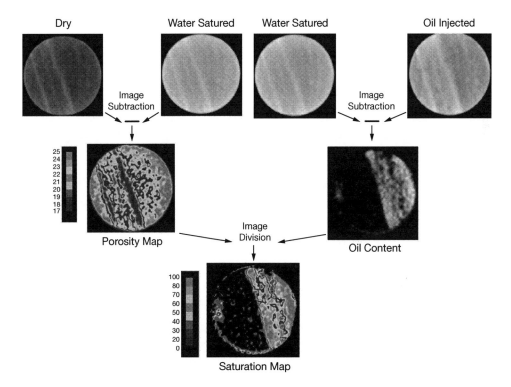

Figure 2-2.17 Porosity and saturation map calculation principle

hounsfield values of the CT scan in dry state from those in saturated state (saturated with a given liquid) and by applying to the result a proportionality coefficient which depends on the absorption in the liquid. Similarly, by subtracting the CT scan results in two different saturation states (e.g. after oil injection on Figure 2-2.17), we obtain a map of the content in a given fluid which, when divided by the porosity, produces a saturation map.

The main practical difficulty lies in precisely repositioning the sample in the X-ray acquisition coordinate system, since it will be impossible to obtain the expected result if there is the slightest shift between two acquisitions. For SCAL experiments, fluid displacement is carried out "under scanner", i.e. without moving the experimental set-up, but at the cost of immobilising the machine throughout the duration of the experiment. As an alternative, mechanical systems for highly accurate positioning of the cells may be available. If cells are used, the translation shifts can be corrected using specific software. Rotations cannot be corrected, however, and for routine measurements, repositioning entails serious practical difficulties when a large number of samples have to be analysed.

C) Example of use

The numerous petrophysical applications of X-ray tomography can be divided according to two main themes.

a) Routine core description and plug selection

Cores, often under protective wrapping ("preserved core", § 1-2.2.2, p. 175; § 2-1.3.1D, p. 298), must be described continuously to produce a first "core log" and simplify subsequent description but also, and perhaps above all, to choose the best plug sampling locations. X-ray radiography and CT scans are taken on a routine basis (Fig. 2-2.18) along the length of core available (often several tens of metres). The spacing between each tomographic cut (5 cm on Fig. 2-2.18) is a compromise between the required accuracy and the cost. Note, however, that acquisition times are becoming shorter and shorter and that there is a trend to reduce the spacing between each slice.

Current CT scan slice acquisition times are very short. In contrast, quantitative processing of the mass of data accumulated would probably be a problem. Many users restrict themselves to a qualitative analysis of the images obtained, however.

For samples intended for SCAL measurements, a very fast tomographic examination can be used to assess their homogeneity and therefore their representativity for the experiment. This preliminary characterisation of the samples is absolutely vital.

b) Location of fluids in multi-phase flow experiments

Although the number of CT scan slices recorded in this context is probably less, the most spectacular contributions of X-ray tomography are observed in the field of fluid location. Figure 2-1.4, p. 279 shows an example of a routine application concerning water-air saturation maps. The liquid-gas densitometric contrast is such that these saturation maps can be produced without any special precautions. Liquid-liquid saturation maps can be produced easily with "marker" elements such as iodine and barium. A "qualitative" example is given

RADIOGRAPHY   COMPUTED TOMOGRAPHY (CT)

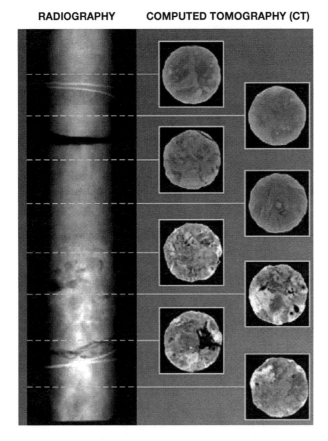

Figure 2-2.18 Example of routine CT Scan analysis on preserved core.
X-ray radiography and CT scan slices at 5 cm intervals

on Figure 2-1.7C, p. 284, where it is easy to distinguish between potassium iodide brine and potassium chloride brine.

The possibility of accurately locating fluids inside the sample during experiments represents a major breakthrough which only those who experienced the "pre-scanner" period can truly appreciate.

2-2.3.2 Microtomography devices

The medical scanner does not have sufficient definition to approach pore size and generally, porous geometries are never identified as such. In order to investigate porous geometry, definition of the order of the micrometre is required (definition of the optical microscope). Devices capable of approaching or reaching this definition are available, using the CT scan principle with different types of X-ray beam.

A) Synchrotron X-ray microscanner

The principle is based on a beam of X-rays from a synchrotron [Sham & Rivers, 2002]. The beams offer the advantage of being parallel and monochromatic. Monochromatic beams avoid the artefacts related to beam hardening (X-ray filtering in the object under investigation). A CCD camera adapted to detection of X-ray photons acts as 2D detector. The sample crossed by the X-ray beam is rotated to acquire a large number of attenuation projections.

Images with a definition close to the micrometre can be obtained by tomographic inversion of these data (Fig.2-2.19). Due to the high beam strength, acquisitions can be taken fairly quickly (a few minutes) on samples measuring up to 1 cm diameter. The main disadvantage with this method is the difficulty in obtaining access to a synchrotron installation.

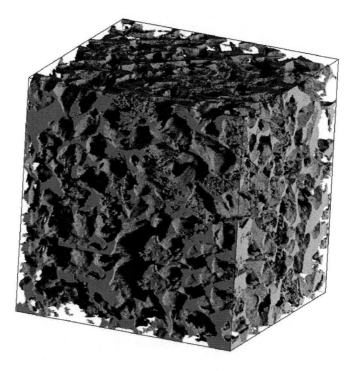

Figure 2-2.19 Example of synchrotron imaging; Castlegate sandstone, $512 . 10^6$ voxels, resolution: 1.67 µm. Courtesy J. Friedrich *et al.* [2006]

B) Industrial microscanner (cone beam)

Much smaller devices adapted to tomographic description of small industrial parts are commercially available and making extremely rapid technical progress (image definition, object dimension). They can be implemented to describe porous media with an increasingly fine definition (Fig. 2-2.20). The definition is currently a few micrometres.

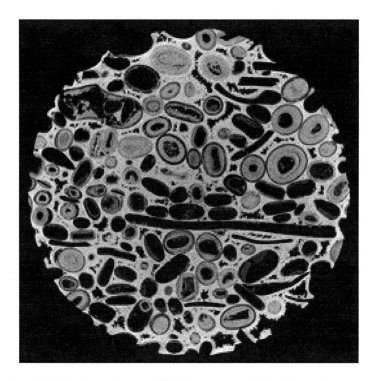

Figure 2-2.20 CT scan of an oomoldic limestone sample. Cone beam microscanner.
The rock cylinder has a diameter of 5 mm. A photograph of a thin section of the same limestone
is shown on Figure 2-2.11 (IFP document, E. Rosenberg)

The beam used is generated by a point source emitting a conical beam (hence the name of this type of scanner [Wang *et al.*, 1994]. The object under investigation rotates in the X-ray beam and, as with the synchrotron device, a camera records a 2D image of the attenuation projections. Rock samples measuring several centimetres in diameter can therefore be studied, despite long acquisition times (several hours).

In the very near future, devices of this type will probably be able to acquire 3D images of porous media at submicron definitions. This will radically change the practical data acquisition conditions for automatic analysis of porous geometries.

2-2.4 MINERALOGICAL ANALYSIS APPLIED TO PETROPHYSICS: X-RAY DIFFRACTION AND FLUORESCENCE

In § 1-1.1.5 (p. 26) we demonstrated the importance of quantitative mineralogical analysis in the progress of a petrophysical synthesis during reservoir evaluation.

- This analysis is one of the key features of the interdisciplinary consistency analysis as regards data acquisition.

- It acts as support for extrapolation/interpolation of important parameters whose acquisition is more complex or more expensive (Cation Exchange Capacity, Magnetic Susceptibility, etc.), or which are simply calculated (Hydrogen Index).
- It also represents one of the basic calibration parameters in quantitative log analysis.

We can appreciate that an accurate and precise quantitative analysis is complex and involves several complementary techniques. A less precise quantification may often prove sufficient for petrophysical applications, however, if easy to implement and cheap enough to systematise to a large number of samples. Nevertheless, it is essential to correctly quantify the major minerals, as well as the more rarely encountered minerals which have a determining impact on the physical properties considered. The amorphous phases must be identified and quantified. The first step of a mineralogical analysis is an exhaustive identification of the mineral assemblage composing the material analysed.

2-2.4.1 Some reminders about mineralogy/crystallography

Mineralogy lies outside the scope of this book and interested readers can refer to the literature on the subject [e.g. Deer *et al.* 1970] as well as to the numerous web sites targeting a wide audience, from the collector to the university specialist [for an exhaustive list see: http://www.uhh.hawaii.edu/~kenhon/mineralogy/links/default.htm]. We will give a few definitions which will be useful in the remainder of the discussion.

A mineral is defined both by its chemical composition and its crystallographic characteristics. The latter play a central role in the determination.

A crystal is characterised by a distribution of elementary atoms in a regular, periodic pattern forming its three-dimensional lattice. The basic unit is the unit cell defined by the angles made between the cell edges and the length of these edges (Fig.2-2.21a).

The same chemical species may exist in different crystal structures: this is known as pleomorphism or polymorphism. Two typical and common examples are often given: firstly silica SiO_2 which can crystallise in the hexagonal structure (quartz), the tetragonal structure (cristobalite) and the triclinic structure (tridymite), and secondly calcium carbonate $CaCO_3$ which is observed in the rhombohedric or trigonal structure (calcite), the orthorhombic structure (aragonite) and more rarely in the hexagonal structure (vaterite).

Minerals of identical crystalline structure with similar chemical and physical properties may syncrystallise, forming solid solutions. This is known as isomorphism. A common example is the series of plagioclases which extends between a sodic end-member, albite (frequently found in sedimentary rocks), and a calcic end-member, anorthite. Some types of clay also exhibit this feature, observed in numerous mineral families: amphiboles, pyroxenes, etc.

Note also that "isotype" minerals have the same crystalline structure but contain limited quantities of ions different from those forming the mineral basis. By undergoing slight deformations, crystalline structures can accommodate atoms of different size but similar chemism. This explains the existence of magnesian calcites, calcic or ferriferous dolomites, slightly sodic orthoses, etc. When the content of "foreign" ions exceeds a certain threshold, however, the structure splits into separate minerals: calcite/dolomite, orthose/albite, etc.

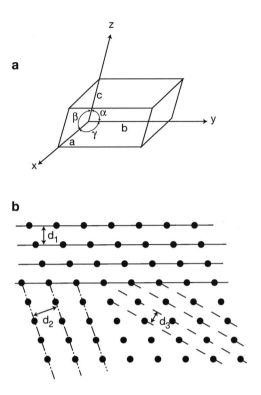

Figure 2-2.21 Crystallographic elements
a) unit cell of a crystalline structure.
b) examples of reticular lines (two-dimensional expression of reticular planes).

One major consequence of the distribution of elementary atoms in the regular, periodic structure is that in the crystal, the atomic density is extremely high in some planes. These are known are reticular planes (sometimes called atomic planes and even cleavage planes) (Fig.2-2.21b) which form the basis of the X-ray diffraction method. The distance (d) between planes in the same family (often simply referred to as the "d-spacing") and the relative orientation of these planes are functions defined by the unit cell.

2-2.4.2 Laboratory techniques

A) Mineral identification under polarising optical microscope

The method is based on observation of thin sections (§ 2-2.1) under polarised light. A mineral is characterised by its extinction angle and its polarisation colour ([MacKenzie and Guilford, 1980] for example, give numerous illustrations). Quantification is possible with a point counter which moves the thin section at a constant step over the surface of the microscope stage. The proportion of each mineral in the rock and, consequently, the quantitative mineralogical composition can be obtained by mineralogical determination at each randomly selected point. Although this is only a surface quantification (2D), volume extrapolation (3D) is justified provided that the thin section analysed is representative (for example, only sections cut in a vertical plane should be used if the rock exhibits pronounced horizontal bedding). We must also bear in mind that the microporosity inside the phase observed, of dimension less than the resolution of the optical microscope, is included in this mineral phase (§ 1-1.1.5B p. 33).

B) X-ray diffraction

This phenomenon was discovered by M von Laue (1912) and applied to the description of crystal structures by WH and WL Bragg (father and son – 1913).

Since the wavelength of the X-rays is the same order of magnitude as the interatomic distances (a few angstroms), the interference of the scattered rays is alternately constructive or destructive. Depending on the direction of the space, the flow of X-ray photons will therefore be very high or very low. These variations depending on the direction result in X-ray diffraction. Application to the determination of minerals is known as the Debye-Scherrer method.

When a beam of X-rays of wavelength Λ "falls" on a network of reticular planes of spacing d at an angle θ, the beam is reflected (diffracted) by the crystal if

$n\,\Lambda = 2\,d\,\sin\theta$: this is the "Bragg condition"

If a beam interacts with a powder consisting of crystal fragments with no relation in their relative orientation (reticular planes arranged equally in all directions in space), all reticular planes respecting the Bragg condition ($\sin\theta = n\Lambda/2d$) (Fig. 2-2.22) form cones of centre corresponding to the location of the power and of axis parallel to the incident beam and making an angle θ with this axis. The diffracted rays therefore form cones of angle 2θ. Since there are families of different d-spacing d_1, d_2, d_3... d_m in the mineral, we therefore observe as many angles respecting the Bragg condition θ_1, θ_2, θ_3.... θ_m and as many cones of reflected X-rays. The intersection of these cones with a circle whose incident beam is a ray will produce a set of points forming the diffractogram.

Originally, the points (lines in fact) corresponding to the various reticular planes were recorded on a strip of photographic paper forming a circle around the powder sample (Debye-Scherrer diagram). The current devices (schematised on Fig. 2-2.22) use an electronic X-ray measurement device (scintillometer) mounted on a goniometer allowing rotation around the sample to record the X-ray intensity according to the angle θ and therefore the d-spacing, applying Bragg's equation, since the wavelength of the emitted X-rays is known. A typical recording is shown on Figure 2-2.23.

The rock analysed is reduced to a powder of granulometry as homogeneous as possible (isogranulometric grinding) and completely disoriented to obtain statistically representative reticular plane angles.

The mineral mixtures produce complex spectra which can be deconvoluted using the current automatic algorithms. The XRD method can be used to identify the crystallographic nature of minerals, irrespective of their chemical composition. It offers the advantage of

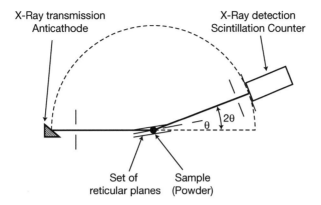

Figure 2-2.22 Diagram of X-ray diffractometer

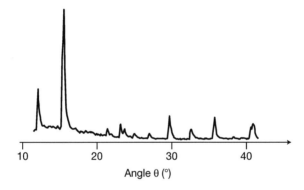

Figure 2-2.23 Example of X-ray diffractogram. Quartz powder. The peak height depends on the X-ray intensity recorded for a given angle θ (abscissa) and therefore for a d-spacing given by the Bragg equation. Cobalt anticathode ($\Lambda = 1.78892$ Å)

distinguishing between the various crystallisation types of a given compound (e.g. for silica, mentioned above). Although it is unable to detect the amorphous phases, it is currently a necessary step in all mineralogical analyses.

To optimise the analysis, attempts were made at a very early stage to interpret the diffractogram in quantitative terms. Interpretation is based on the intensity (the area) of the diffraction peaks of each mineral species, correcting the response for the numerous artefacts, some inherent to preparation of the powder analysed (granulometry, compaction, etc.) and some to the equipment (including the ageing of the X-ray generator), others inherent to the nature of the material itself (absorption, etc.). In the past, the laboratories often supplied results with no further correction, producing in this case a semi-quantitative type analysis.

The most rigorous quantitative method is to introduce an "internal standard" (a known quantity of a pure, stable mineral) into the powder to be analysed. Another method, this time numerical, is to consider that all species present are measurable and measured. This additional equation (sum of contents equal to 1) can be used to solve the problem. Powerful numerical resolution software applications are now available.

Sources of error still exist however. We will mention two of the most frequent errors in common rocks.

- The existence of amorphous phases, in particular the "non-extractable organic matter" which always forms a minimum background of a few tenths of a percent in sedimentary rocks. Detecting and quantifying the mineral amorphous phases is important since, in shallow sedimentary rocks, they are frequently hydrated (silica, alumina, ferric hydroxides, etc.). Neglecting them results in overestimation of the water saturations, during the quantitative well log analysis.
- The existence of clay phases impossible to identify on disoriented powder, apart from well-crystallised species (micas, dickites, some chlorites, etc.). Clay families are identified on oriented pastes to amplify the response of the main reticular planes and to process these pastes in order to identify any variations in the interlayer distances (heating, swelling).

The result of the purely numerical interpretation of the diffractogram must therefore be compared and if necessary be calibrated with independent methods.

The results of the other analyses conducted must be taken into consideration. We have seen that one of them is essential: measurement of the non-extractable organic carbon, carried out by automated oxidation systems allowing routine acquisitions on large series. Rock decarbonation is essential with this method, to avoid including the mineral carbon.

C) X-ray fluorescence spectrometry

Another complementary method providing good results is now conducted on a routine basis: X-ray fluorescence spectrometry. It consists of an elementary chemical analysis method based on the fact that the X-ray spectrum emitted by matter when struck by electrons is characteristic of the sample composition. The elementary composition, i.e. the weight concentrations of the elements, can be determined by analysing this spectrum. Note that the same physical principle is used to produce X-rays of known wavelength, using anodes of precise chemical nature (copper, cobalt, etc.).

Analysis can be carried out on powder, the same as that used in X-ray diffraction. Similar precautions are required to ensure that the quantitative interpretation is correct. Although the lightest elements are difficult to detect, this methods allows elementary chemical analysis of the rock.

We are faced with the known case of normative resolution of the mineralogical analysis which can be solved by automatic iteration methods, knowing:

- the exhaustive analysis of the mineralogical phases using X-ray diffraction, completed by some additional analyses (at least, non-extractable organic carbon);
- the elementary chemical composition of the total rock by X-ray fluorescence;
- the elementary composition table of the minerals encountered, on the basis of the reference table, possibly adjusted according to a few micro-analyses, or more cheaply on the basis of the detailed examination of a few diffraction spectra (e.g. slight shifts

from the main peak of calcite may indicate magnesian calcites; this is also the case for the plagioclases).

The X-ray diffraction/X-ray fluorescence combination therefore represents an efficient and reliable quantitative mineralogical analysis tool, which can be automated to supply results on large sample series similar to those of routine petrophysical measurements. Some analytic laboratories of oil operating companies have acquired the necessary equipment and developed extensive know-how.

Accuracy is further improved if the sampling has been carried out rationally (§ 2-1.3.1.D, p. 298). It is in fact easy to calculate the density of a mineral species of known elementary composition. In addition, since the total dry density of the sample is always acquired during the porosity measurement (§ 1-1.1.5.A, p. 26), this represents an additional numerical constraint when solving the previous equations by iteration.

D) Infra-Red Spectroscopy (IRS)

If the wavelength of an infrared radiation is equal to the resonance frequency of one of the chemical bonds of the body crossed by the IR ray, the ray is absorbed by the chemical component. We therefore see that a chemical analysis can be conducted by recording the transmittance of the infrared rays through a body, depending on their wavenumber (in IRS, the parameter used to characterise the radiation is the wavenumber, equal to the reciprocal of the wavelength in centimetres). Use of the Fourier transform method, which allows direct analysis of the various frequencies (FTIR), has led to significant improvements in routine analytical methods and application to rocks [e.g. Matteson and Herron1993].

The problem regarding quantitative mineral interpretation is quite similar to that with X-ray fluorescence. IRS offers the advantage of being easier to implement. When used alone, however, this method does not allow the analyst to benefit from the exhaustive crystallographic balance inherent to X-ray diffraction, which is often extremely useful to solve doubtful cases.

References

Aase NE, Bjorkum PA, Nadeau PH. 1996. The effect of grain-coating microquartz on preservation of reservoir porosity. V. 80, N° 10, pp. 1654-1673.

Al-Kaabi AU, Mimoune K, Al-Yousef HY. 1997. Effect of Hysteresis on the Archie Saturation Exponent. SPE Paper 37738.

American Petroleum Institute. 1952. Recommended practice for determining permeability of porous media.

Amott E. 1959. Observations relating to the wettability of porous rocks. Trans. AIME, 216.

Amthor JE, Okkerman J.1998. Influence of early diagenesis on reservoir quality of Rotliegende Sandstones, Northern Netherlands: AAPG. Bull. V 82, N° 12, pp. 2246-2265.

Amyx JW, Bass DM Jr, Whiting RL. 1960. Petroleum reservoir engineering – Physical properties. McGraw-Hill. New York.

Archie GE.1942. The electrical resistivity log as an aid in determining some reservoir characteristics. Trans. AIME.

Anderson WG. 1986-87. Wettability litterature survey 1986-87, Six part JPT vol 32.2 N° 11, 12, 13; vol 39.2 N° 10, 11, 12.

Archie GE.1952. Classification of carbonate reservoir rocks and petrophysical considerations. AAPG Bulletin V. 36, N° 2, pp. 278-298.

Atkinson CD, McGowen JH, Bloch S, Lundell LL, Trumbly P N. 1990. Braidplain and deltaic reservoir, Prudhoe Bay Field, Alaska: in Sandstone Petroleum Reservoirs, Barwis JH et al. Eds, Springer-Verlag, pp. 205-224.

Baize D. 2004. Lexique de pédologie. INRA Editions. Versailles France.

Banton O, Bangoy M.1999. Hydrogéologie – Multiscience environnementale des eaux souterraines. AUPELF, Montreal Canada.

Bieber MT, Rasolofosaon P, Zinsner B, Zamora M.1996. Measurement and overall characterization of permeability anisotropy by tracer injection. Rev. IFP, V 51, N° 3.

Biot MA. 1941. General theory of three-dimensional consolidation. J. Applied Phys. V12, pp. 155-164.

Birch F. 1960. The velocity of compressional waves in rocks to 10 kilobars, part I- J. Geophys. Res., 65, pp.1083-1102; part II- J. Geophys. Res., 66, pp. 2199-2224.

Bloch S. 1991., Empirical prediction of porosity and permeability in sandstones: AAPG. Bull. V 75, N° 7, p. 1145-1160.

Bourbié T, Coussy O, Zinszner B. 1987. Acoustics of Porous Media. Technip. Paris.

Bourbié T, Zinszner B. 1985. Hydraulic and acoustic properties as a function of porosity in Fontainebleau sandstone. GJR 90. pp. 11524-11532.

Bourdarot G. 1998. Well testing: interpretation methods. Editions Technip. Paris.

Bousquié P. 1979. Texture et porosité de roches calcaires. Thesis, Paris School of Mines.

Bouteca M. 1992. Elements of poro-elasticity for reservoir engineering. Rev. IFP, V. 47, N° 4, pp. 479-490.

Bouvier L, Maquignon S. 1991. Reconciliation of log and laboratory derived irreducible water saturation in a double porosity reservoir. Adv. in C. Evaluation II/Res. Apr. pp. 275-292. Gordon and Breach Sc. Pub.

Brady NC, Weil R. 1999. Nature and Properties of Soils. Prentice Hall.

Brie A, Pampuri F, Marsala AF, Meazza O. 1995. Shear sonic interpretation in gas bearing sands. SPE 360595.

Brown RJS, Fatt I. 1956. Measurements of fractional wettability of oilfield rocks by the nuclear magnetic relaxation method. Pet. Trans. AIME 207, pp. 262-264.

Cadoret T. 1993. Effet de la saturation eau/gaz sur les proprietes acoustiques des roches. Thesis, Université Paris VII.

Cadoret T, Marion D, Zinszner B.1995. Influence of frequency and fluid distribution on elastic wave velocities in partially saturated limestones. J. Geophys. Res., 100, N° B6.

Cargill G S.1984. Radial distribution functions and microgeometry of dense random packings of hard spheres, in Physics and Chemistry of porous media, Johnson DL and Sen PN Edts, API, New York.

Carman PC.1937. Fluid flow through granular beds. Trans. Inst. Chem. Eng. 15: 150-166.

Castagna JP, Batzle ML, Kan TK. 1993. Rock Physics, the link between rock properties and AVO response, in: offset dependent reflectivity – theory and practise of AVO analysis. SEG, Tulsa.

Chardaire-Riviere C, Roussel JC. 1992. Principle and potential of nuclear magnetic resonance applied to the study of fluids in porous media. Rev. Inst. Français du Pétrole. 47, 4, pp. 503-523.

Choquette PW, Pray LC.1970. Geologic nomenclature and classification of porosity in sedimentary carbonates. AAPG Bull., V. 54, N° 2, pp. 207-250.

Clavier C, Coates G, Dumanoir J.1984. Theoretical and experimental basis for the dual water model for interpretation of shaly sands. SPE Journal, Paper N° 6859PA.

Corey AT. 1954.The interrelation between gas and oil relative permeabilities. Producers Monthly, Nov., 38-41.

Coussy O. 2004. Poromechanics. John Wiley and Sons, New York.

Christiansen R L. 2001. Two-Phase Flow through Porous Media. Colorado School of Mines. rchristi@mines.edu.

Cuiec L. 1975. Restoration of the natural state of core samples. SPE paper 5634.

Dake L P.1978. Fundamentals of reservoir engineering. Elsevier science, Amsterdam.

Darcy H.1856. Les fontaines publiques de la ville de Dijon. Ed Dalmont. Paris.

Deer WA, Howie RA, Zussman J. 1970. An introduction to the rock-forming minerals. Longmans.

Delage P, Cui YJ, Yahia-Aïssa M, De Laure E. 1998. On the unsaturated hydraulic conductivity of a dense compacted bentonite. Proc. of Unsat'98, Beijing, V 1, pp. 344-349.

Delfiner P. 1971. Etude morphologique des milieux poreux et automatisation des mesures. en plaques minces. PhD thesis, University of Nancy. France.

De Waal JA, Smits RMM. 1988. Prediction of Reservoir Compaction and Surface Subsidence: Field Application of a New Model. SPE14214.

Diedrix KM.1982. Anomalous relationships between resistivity index and water saturations in the Rotliegend sandstone, the Netherlands. 23th SPWLA Annual Logging Symp.

Dixon JR, Marek BF. 1990. The effect of bimodal pore size distribution on electrical properties of some Middle East Limestones. SPE 20601.

Domenico SN. 1977. Elastic properties of unconsolidated porous sand reservoir. Geophysics V 42, pp. 1339-1368.

Donaldson EC, Thomas RD, Lorenz PB.1969. Wettability determination and its effect on recovery efficiency. SPE Journal. March.

Donaldson EC, Siddiqui TK. 1989. Relation between Archie saturation exponent and wettability. Soc. Petr. Eng., Formation evaluation, pp. 359-362.

Dubesset M. 2000. Le manuel du système international d'unités. Ed. Technip. Paris.

Dullien FAL, Dhawan GK. 1974. Characterization of pore structure by a combination of quantitative photomicrography and mercury injection. J. Colloid Interface Sci.

Dullien FAL. 1977. Porous Media: Fluid Transport and Pore Structure. Academic Press.

Dunn KJ, Bergmann DJ, Latorraca GA. 2002. Nuclear Magnetic Resonance; Petrophysical and Logging applications. Handbook of geophysical exploration. Pergamon.

Fatt I. 1956. The network model of porous media, Part I to III, Pet. Trans. AIME, 207.

Fatt I. 1958. Compressibility of sandstones at low to moderate pressure. AAPG Bulletin; August; v. 42; N° 8; pp. 1924-1957.

Faust LY.1951. Seismic velocity as a function of depth and geologic time. Geophysics 16, pp. 192-206.

Ferfera FMR, Sarda JP, Bouteca, M, Vincke O. 1997. Experimental study of monophasic permeability changes under various stress paths. International J. of Rock Mechanics. V34, Number 3, April, pp. 413-413.

Fjær E, Holt R, Horsrud P, Raaen A M, Risnes R. 1992. Petroleum related rock mechanics. Dev. Pet. Sci. vol. 33.A., pp. 1-325. Elsevier, Amsterdam.

Fleury M. 2002. Resistivity in Carbonates: New Insights. SPE paper 77719.

Forbes P. 1997. Centrifuge data analysis techniques: an SCA survey on the calculation of drainage capillary pressure curves from centrifuge measurements. SCA-9714.

Forchheimer P. 1901. Wasserbewegung Durch Boden. Z Ver Deutsch Ing, V 45, pp. 1782-1788.

Fredrich JT, DiGiovanni AA, Noble DR. 2006. Predicting Macroscopic Transport Properties Using Microscale Image Data. J. Geophys. Res. Vol. 111, B03201.

Gardner GHF, Gardner LW, Gregory AR.1974. Formation velocity and density. The diagnostic basis for porosity traps. Geophysics 39, pp. 770-780.

Gassmann F. 1951. Uber die elastizitatporosen medien. Vierteljahrsschrift der Naturforschenden Gesellshaft in Zurich, 96, pp. 1-23.

Giouse H. 1987. Propriétés pétrophysiques et résistivités des grès argileux pétrolifères. Ph. D. dissertation. School of Mines. Paris.

Godefroy S, Korb JP, Fleury M, Bryant RG. 2001. Surface nuclear magnetic relaxation and dynamics of water and oil in macroporous media. Physical Review E, V64, 021605.

Gomez-Rivero O. 1976. A practical method for determining cementation exponent etc. The Log Analyst.V. 17, N° 5 pp. 8-24.

Gonsalves 1999. http://www.physics.buffalo.edu/gonsalves/Java/Percolation.html

Graton LC, Fraser H. 1935. Journal of geology, 43, 785.

Greder HN, Marion DP, Pellerin F M.1994. Petrophysical logs on cores: a new tool for core-log calibration and reservoir characterization. Paper Q. SPWLA 35th Annual Logging Symposium.

Greder HN, Biver PY, Danquigny J, Pellerin FM. 1996. Determination of permeability distribution at log scale in vuggy carbonates. 37th Annual SPWLA Symposium, New Orleans. USA.

Guillot D. 1982. Propriétés physiques et texturales d'un modèle artificiel de matériau à porosité intergranulaire. Thèse, Mines de Paris.

Guyon E, Oger L, Plona TJ. 1987. Transport properties in sintered porous media composed of two particle sizes. J. Phys. D: Appl. Phys. V 20, 12, pp. 1637-1644.

Hall HN. 1953. Compressibility of Reservoir Rocks. Petroleum Transactions, AIME.

Hamon G. 2000. Field-Wide Variations of Wettability. SPE paper 63144.

Hamon G, Suzanne K, Billiotte J, Trocme V. 2001. Field-Wide Variations of Trapped Gas Saturation in Heterogeneous Sandstone Reservoirs. SPE paper 71524.

Hamon G, Bennes M. 2004. Two-phase flow rock typing: another approach Petrophysics, vol. 45, N° 5.

Harari Z, Wang ST, Saner S. 1995. Pore-Compressibility Study of Arabian Carbonate Reservoir Rocks. SPE, Paper Number 27625.

Hirschwald J. 1908. Die Prüfungder natürlichen Bausteine auf ihreWetterbeständigkeit. W. Ernst & Sohn. Berlin.

Jackson PD, Smith DT, Stanford PN. 1978. Resistivity-Porosity-Particle Shape relationships for marine sands. Geophysics, v. 43, N° 6, pp. 1250-1268.

Jacquin Ch. 1964. Corrélations entre la perméabilité et les caractéristiques géométriques du grès de Fontainebleau. Rev. IFP, XIX. pp. 921-937.

Jadhunandan PP, Morrow NR. 1991. Effect of wettability on waterflood recovery for crude-oil/brine/rock systems. SPE paper 22597.

Jaeger JC, Cook NGW. 1969. Fundamentals of rock mechanics. Methuen and Co Ltd, London.

Jeannette D, Hammecker Cl. 1992. In Philippon J, Jeannette D, Lefevre RA. La conservation de la pierre monumentale en France. Presses du CNRS.

Jensen JL.1990. A model for small scale permeability measurement with applications to reservoir characterisation. Paper SPE/DOE 20265. SPE/DOE Symp. on Enhanced Oil Recovery, Tulsa.

Jones FO, Owens WW. 1980. A Laboratory Study of Low-Permeability Gas Sands SPE, Paper Number 7551.

Johnson EF, Bossler DP, Naumann VO. 1959. Calculation of relative permeability from displacement experiments. Trans. AIME, V. 216, p. 370.

Johnson PA, Rasolofosaon PNJ. 1996. Nonlinear elasticity and stress-induced anisotropy in rocks. J. Geophys. Res., 101, B2, pp. 3113-3124.

Jouniaux L, Bernard ML, Pozzi J P, Zamora M. 2000. Streaming potential in volcanic rocks from Mount Pelée, J. G Res., 105, pp. 8391-8401.

Juhasz I.1981. Normalised Q, the key to shaly sand evaluation using the Waxman-Smits equation in the absence of core data. Trans. SPWLA 22nd Ann. Log. Symp., Z1-36.

Keighin CW, Law BE, Pollastro RM. 1989. Petrology and reservoir characteristics of the Almond Formation, Greater Green River Basin, Wyoming: in Petrogenesis and petrophysics of selected sandstone reservoirs of the Rocky Mountain region, Coalson EB *et al.*, eds., Rocky Mountain Association of Geologists, Denver, Colorado, p. 281-298.

Klinkenberg. 1941. The permeability of porous media to liquids and gases. Drill. Prod. Pract. pp. 200- 213.

Lambe TW, Whitman RV. 1979. Soil mechanics, SI version. Wiley, New York.

Land CS. 1971. Comparison of calculated with experimental imbibition relative permeability, SPE 3360.

Langford RP, Grigsby JD, Howard WE, Hall JD, Maguregui J. 1990. Sedimentary facies and petrophysical characteristics of cores from the lower Vicksburg gas reservoirs, McAllen Ranch Field, Hidalgo County, Texas: Transactions Gulf Coast Association of Geological Societies, v. 40, pp. 439-450.

Laroche C, Vizika O. 2005. Two-Phase Flow properties prediction from small-scale data using Pore-Network Modeling. Transport in Porous Media 61. pp. 77-91.

Latil M, Bardon C, Burger J. 1980. Enhanced oil recovery. Editions Technip Paris.

Lauterbur PC. 1973. Image formation by induced local interactions: examples employing Nuclear Magnetic Resonance. Nature 242, 190-191.

Lenormand R. 2003. Interpretation of mercury injection curves to derive pore size distribution. SCA Meeting Pau, France. SCA2003-52.

Levallois B. 2000. Core petrophysical measurements on unconsolidated sands in deep water reservoir. SCA Meeting Abu Dhabi. SCA2000-34.

Leverett MC. 1941. Capillary behaviour in porous solids. Trans AIME, Vol 142, pp. 152-169.

Lindquist SJ. 1988. Practical characterization of eolian reservoirs for development: Nugget Sandstone, Utah-Wyoming thrust belt. Sedimentary Geology, 56, 315-339.

Looyestijn W, Hofman J. 2006. Wettability-Index determination by Nuclear Magnetic Resonance. SPE paper 93624.

Lucet N, Zinszner B. 1992. Effects of heterogeneities and anisotropy on sonic and ultrasonic attenuation in rocks. Geophysics 57, 8, pp. 1018-1028.

Lucia FJ.1995. Rock-fabric/petrophysical classification of carbonate pore space for reservoir characterization. AAPG Bull. v. 79; n° 9, pp. 1275-1300.

Magara K. 1978. Compaction and fluid migration: Practical Petroleum Geology, Dev. Pet. Sci., 9. Elsevier, Amsterdam.

Maidment RD. 1993. Handbook of Hydrology. McGraw-Hill. New York.

Mandelbrot BB. 1977. Fractals: form, chance, and dimension. Freeman. San Francisco.

Mandelbrot BB. 1967. How long is the coast of Britain? Statistical Self-Similarity and Fractional Dimension. Science, V.156, N° 3775, pp. 636-638.

Marle C. 1981. Multiphase flow in porous media Ed. Technip Paris.

Marle CM.2006. Henry Darcy et les écoulements de fluides en milieu poreux.
Oil & Gas Sc. and Tech. Rev. IFP. Vol. 61, n° 05, pp. 599-609.

Marsily G de. 1986. Quantitative hydrogeology. Groundwater hydrology for engineers. Academic Press, San Diego.

Marzano MS. 1988. Controls on permeability for unconsolidated sands from conventional core data, offshore Gulf of Mexico. Transactions Gulf Coast Association of Geological Societies, v. 38, pp. 113-120.

Matheron G. 1967. Éléments pour une théorie des milieux poreux. Masson, Paris.

Matteson A, Herron MM. 1993. Quantitative mineral analysis by Fourier transform infrared spectroscopy, SCA, Annual Technical Conference. Paper SCA-9308.

Mavko G, Mukerji T, Dvorkin J.1998. Rock Physics Handbook. Cambridge University Press.

McDougall SR, Sorbie KS. 1997. The application of network modelling techniques to multiphase flow in porous media. Petroleum Geosci, 3, 161.

MacKenzie WS, Guilford C. 1980. Atlas of rock-forming minerals in thin section. Longman New York.

Mees F, Swennen R, Van Geet M, Jacobs P. 2003. Applications of X-ray computed tomography in the geosciences. Geological Society special publication 215.

Meyer FO, Price RC, Al-Raimi SM. 2000. Stratigraphic and petrophysical characteristics of cored Arab-D super-K interval, Hawiyah Area, Ghawar Field Saudi Arabia. GeoArabia, Vol 5, N° 3.

Monicard R. 1980. Properties of reservoir rocks: Core analysis. Technip, Paris.

Morriss CE, Freedman R, Straley C, Johnston M, Vinegar HJ, Tutunjian PN. 1997. Hydrocarbon saturation and viscosity estimation from NMR Logging in the Belridge diatomite. The Log Analyst, 1997.

Morrow NM. 1970. Physics and thermodynamics of capillary. Ind. and Eng. Chemistry, vol. 62, N° 6, pp. 32-56.

Morrow NM. 1971. The retention of connate water in hydrocarbon reservoirs. Jour. Can. Pet. Tech. pp. 38, 55, Jan-March.

Muller MM, Coalson EB. 1989. Diagenetic and petrophysical variations of the Dakota Sandstone, Henry Field, Green River Basin, Wyoming. In Petrogenesis and petrophysics of selected sandstone reservoirs of the Rocky Mountain region. Coalson EB *et al.*, Eds, Rocky Mountain Association of Geologists, Denver, Colorado, pp. 149-158.

Nafe HE, Drake CL. 1963. Physical properties of marine sediments, in: The Sea, Vol 3, pp. 794-815. Interscience.

Neasham JW. 1977. The morphology of dispersed clay in sandstone reservoirs and its effect on sandstone shaliness, pore space and fluid flow properties. SPE6858.

Nelson P H, Kibler J E. 2003. A Catalog of Porosity and Permeability from core plugs in siliciclastic rocks. USGS Open-file Report 03-420.

Nes OM, Holt RM, Fjaer E. 2000. The reliability of core data as imput to seismic reservoir monitoring studies. SPE 65180.

Newman GH. 1973. Pore-Volume Compressibility of consolidated, friable, and unconsolidated reservoir rocks under hydrostatic loading. SPE, Paper Number 3835.

Nur A. 1992. Critical porosity and the seismics velocities in rocks. EOS, Transactions A.G.U. 73, pp. 43-66.

Owens WW, Archer DL. 1971. The effect of rock wettability on oil-water relative permeability relationships. AIME 251, 1, 873-878.

Pickell JJ, Swanson BF, Hickman WB. 1966. Application of air-mercury and oil-air capillary pressure data in the study of pore structure and fluid distribution. SPE 1227.

Pickett GR. 1963. Acoustic character logs and their applications in formation evaluation. J. Petrol. Technol. 15, pp. 650-667.

Poupon A, Loy ME, Tixier MP. 1954. A contribution to electrical log interpretation in shaly sands. Transaction of AIME.

Purcell WR. 1949. Capillary pressures-their measurement using mercury and the calculation of permeability therefrom. Trans. AIME, v. 186, pp. 39-48.

Raiga-Clemenceau J, Fraisse C, Grosjean Y. 1984. The dual porosity model, a newly developed interpretation method for shaly sands. SPWLA 25th Ann. Log. Symp.

Rasolofosaon P, Zinszner B. 2002. Vérification expérimentale de la formule de Gassmann dans les calcaires poreux. Rev. IFP, Vol 57 N° 2.

Raymer LL, Hunt ER, Gardner JS 1980. An improved sonic transit time-to-porosity transform. SPWLA, 21st Ann. Logg. Symp.

Reedy GK, Pepper CF. 1996. SPE Annual Conference, Omega Volume, Paper N° 36506, pp. 119-127.

Robin M. 2001. Interfacial phenomena: reservoir wettability in oil recovery. Oil & Gas Sc. and Tech. Rev. IFP, V. 52, N° 1; pp. 55-62.

Schlumberger Co. 1989. Log interpretation principles. An updated electronic version is available on internet at http://www.oilfield.slb.com

Sen PN. 1997. Resistivity of partially saturated carbonate rocks with microporosity. Geophysics, Vol 62, N° 2, pp. 415-421.

Sham TK, Rivers ML. 2002. A brief overview of synchrotron radiation. In: Applications of synchrotron radiation in low-temperature geochemistry and environmental science. Reviews in Mineralogy and Geochemistry v. 49. Fenter PA *et al.* Eds. Geochemical Society and Mineralogical Society of Americ, pp. 117-147.

Slichter C P. 1990. Principles of Magnetic Resonance Springer-Verlag.

Stenger BA. 1999. Regional temperature gradient: A key to tilted OOWC. SPE Paper 53197.

Stanolind Oil and Gas Co. (then AMOCO). 1952. Fluid distributions in porous systems. A preview of the motion picture.

Sutanto E. 1991. Liquid distribution in porous media. PhD Thesis. University of Minnesota.

Swanson BF.1979.Visualizing pores and nonwetting phase in porous rocks. JPT Paper N° 6857-PA.

Swanson B F. 1981. A simple correlation between permeabilities and mercury capillary pressures. Journal of Petroleum Technology, pp. 2498-2504.

Taylor TR, Soule CH. 1993. Reservoir characterization and diagenesis of the Oligocene 64-zone Sandstone, North Belridge Field, Kern County, California. AAPG. Bull. v. 77, n° 9, pp. 1549-1566.

Terzaghi K. 1943. Theoretical Soil Mechanics. John Wiley and Sons, New York.

Thomsen L., 1990.Poisson was not a geophysicist! TLE v. 9; n° 12; p. 27-29.

Thomeer JHM. 1960. Introduction of a pore geometrical factor defined by the capillary pressure curve. J. Pet. Tech. March 1, pp. 73-77.

Thomson A, Stancliffe RJ. 1990. Diagenetic controls on reservoir quality, eolian Norphlet Formation, South State Line Field, Mississippi. In Sandstone Petroleum Reservoirs, Barwis JH, *et al.* eds. pp. 205-224. Springer-Verlag.

Tiab D, Donaldson EC. 1996. Petrophysics. Gulf Publishing Company. Houston, Texas.

Unesco/RILEM, 1978. International Symposium. Deterioration and Protection of Stone Monuments.

Walsh JB, Brace WF. 1966. Cracks and pores in rocks. Proc. 1st Congress Int. Soc. of Rock Mechanics. Lisbon, vol 1, pp. 643-646.

Wang G, Lin TH, Shinozaki DM, Kim HG, Cheng PC. 1994. Cone beam X-ray microtomograph. In Multi-dimensional microscopy. Cheng P C *et al.* Eds. Springer Verlag. New York.

Wardlaw NC, Cassan JP. 1979. Oil recovery efficiency and the rock pore properties of some sandstone reservoirs. Bull. Can. Pet. Geol. V. 27.

Waxman HM, Smits LJM. 1968. Electrical conductivities in oil bearing shaly sands. SPE Journal, June.

Waxman HM, Thomas E C. 1974. Electrical conductivities in shaly sands.

I - The relation between hydrocarbon saturation and resistivity index.

II - The temperature coefficient of electrical conductivity. JPT, Feb.

Wellington SL, Vinegar HJ.1987. X-ray computerized tomography. J. Pet. Techn. V 39 N° 8.

Wilson MD, Pittmann ED.1977. Authigenic clays in sandstones; recognition and influence on reservoir properties and paleoenvironmental analysis. Journal of Sedimentary Petrology, Vol. 47, N° 1, pp. 3-31.

Winsauer WO, Shearing HM Jr, Masson PH, Williams M. 1952. Resistivity of brine saturated sands in relation to pore geometry. *AAPG Bull.,* 36, pp. 253-277.

Worthington PF.1985. The evolution of shaly sand concepts in reservoir evaluation. The Log Analyst, Jan-Feb.

Wright HT, Wooddy LD. 1955. Formation evaluation of the Borregas and Seeligson field, Texas. Symposium on formation evaluation. AIME, October.

Wyllie MRJ. 1957. Fundamentals of Electric Log Interpretation, Academic Press, New York.

Wyllie MRJ, Gregory AR, Gardner GHF. 1958. An experimental investigation of factors afffecting elastic wave velocity in porous media. Geophysics 23, 3, pp. 459-493.

Zinszner B, Meynot C. 1982. Visualisation des propriétés capillaires des roches réservoirs. Oil & Gas Sc. and Tech. Rev. IFP. Vol. 37, pp. 337-361.

Porosity Terms Glossary-Index

———————

Fracture porosity: 313
Porosity located inside the open fracture network.

Interstitial p. see Matrix p.

Intergranular p., Intercrystalline p. see interparticle

Interparticle porosity (intergranular, intercrystalline)**:** *88, 113, 145,* 206, *212, 256, 329, 338*
Porosity located between the rock forming particles (grains, allochems, crystals). This terminology is used (with variations) in most petrological porosity classifications (e. g. [Choquette and Pray 1970]). Although this type of classification accurately locates the evolution of the porosity with respect to the history of the rock, little emphasis is placed on quantifying the dimensions of the porous space but in [Lucia, 1995].

Intragranular p., Intracrystalline p. see. intraparticle

Intraparticle p. (as opposed to interparticle)**:** *88, 113, 207, 256, 329, 338*
Porosity located inside the rock forming particles.

Isolated p. (hydrology) see poorly connected

Kinematic p. (hydrology) see effective p.

Macro/Micro porosity: 18, 58, 69, <u>87</u>, *89,90,91,92*, 102, 112, *113*, 115, 118, 140, 149, <u>153</u>, *154*, <u>156</u>, 211, *212, 332, 333, 338,* 352
The terms micro/macro p. refer to notions which vary widely depending on the method used to study the porosity and on the context of the study.

– *Observation of pores*
   1. Under the naked eye or the naturalist's magnifying glass (pedology, hydrology), a value of 200 µm is sometimes used to mark the boundary between macro and micro porosity.
   2. Under the stereo microscope. Archie's famous classification [1952] used to describe cuttings of carbonate rocks implicitly set the microporosity limit (not visible) at about 20 µm.
   3. Under optical microscope (thin section): Observations of petrographic thin sections: boundary determined by the comparison between the pore size and the section thickness (about 30 µm). In practice: *macroporosity*: pores individually visible under optical microscope (size about 50 µm); *microporosity*: pores not individually visible under optical microscope (size less than 20 mm). Use of the term *mesoporosity* to qualify the intermediate zone may sometimes be justified if pores of this size play an important role in the rock studied.
   4. Under scanning electron microscope (pore cast, etc.): In studies of reservoir rocks, we keep the limits defined for observation in thin section, but it is important to point out that specialists of very fine media such as clays and catalysts use definitions shifted by several orders of magnitude: microporosity corresponds to pores of size less than 2 nanometres (extreme limit of mercury porosimetry systems).

– *Measurements on a porosity spectrum* obtained using the mercury injection porosimetry technique. Separation between two families identifiable on the derived distribution curve. In sedimentary rocks, the boundary is often located between 0.5 and 1 µm of pore threshold equivalent radii. To limit the risks of confusion, it would be best to speak of macro-accessible or micro-accessible porosity.

– *Capillary behaviour of water* in the surface formations (pedological/hydrological applications): The microporosity contains capillary (or pendular [De Marsilly 1986]) water. It is the fraction of water available to plants. It is associated with textural p.: intra-aggregate porosity (in the pedology sense). Macroporosity carries gravity water ("free" water) or funicular water. It is associated with structural p: inter-aggregates (and even intra-aggregates) or clusters (in the pedology sense). It also includes biotic or biogenic porosity: roots and rootlets, worms, insects and other burrowing animals. This porosity may be due to farming (ploughing).

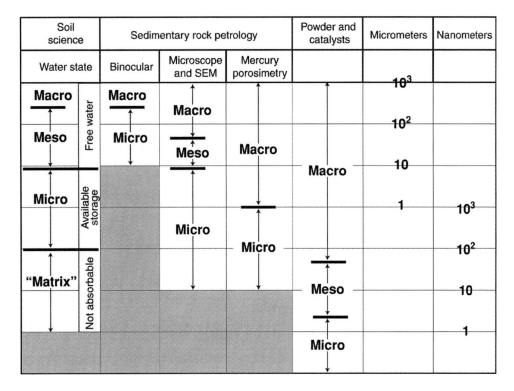

The macro/micro limits for various fields or measurement techniques are roughly schematised on the above table. We can see that these boundaries may vary by several orders of magnitude. We must therefore be vigilant when using these terms and always specify the terminology used.

We must also remember that these classifications may be used jointly. This is the case for instance for the moldic porosity on thin section Fig. 2-2.11 which can be usefully defined as "macro porosity micro accessible".

Matrix porosity:
ambiguous term used in several contexts
- as opposed to fracture, crack, vug p. to express the equivalent of Pore p.:
- as opposed to bulk p, to characterise the porosity of small REV (e.g. measured on routine core analysis plug):
- petrological: porous volume related to the matrix in the sedimentology meaning (matrix as opposed to the detritic phase; matrix as opposed to "granular skeleton"):
- in log analysis: sometimes defines porosity other than that associated with clays.
- in pedology, sometimes corresponds to very fine porosity, accessible only for capillary pressures above the wilting point.

Moldic porosity: *158, 256, 331, 338, 350*
Term used in carbonate petrology related classifications. Porous space created by the partial or total dissolution of a particle. Moldic p. corresponds generally to Vug p.

Neutron porosity: 20
Porosity value calculated from the neutron log. The neutron log responds primarily to the hydrogen concentration in the minerals and in the fluids. (to be used with caution).

NMR porosity: 258
Porosity value calculated from the NMR log. The NMR log responds primarily to the hydrogen concentration in the fluids only.

Non-Fabric Selective porosity (as opposed to Fabric Selective p.)**:**
Pore volume superimposed to the initial rock structure e.g. by fracturing or weathering.

Occluded porosity: 4, 11, 34, 46
Porous volume not connected to the exterior. The criterion of connection to the exterior must be applied absolutely. For a given sample, the occluded porosity may vary significantly depending on the effective stress applied.

Petrophysical porosity: 32
Porosity value measured in the laboratory (generally equivalent to total p.)

Poorly connected p. see connected p.

Pore porosity (as opposed to crack p.)**:** 165, 220, 267
Porous volumes whose dimensions in the three directions have the same order of magnitude.

Primary p. complementary to Secondary p., see this term.

Secondary porosity: 3,
Ambiguous term which should be avoided, possible source of confusion since it is used with highly variable connotations:
– interpretative with respect to a measurement technique: e.g. porosity not taken into account by the sonic log.
– genetic (petrology), porosity created in the rock after early diagenesis. Does not include the notion of non-fabric selective.

Sonic porosity: 241
Porosity value calculated from the sonic log, to be used with caution.

Structural porosity (as opposed to textural p.)**:**
Ambiguous term to be avoided outside the field of soil science, where it is often associated with macroporosity.

Textural porosity:
Related to the initial rock fabric, better to use "fabric selective". In soil science, often equivalent to microporosity.

Total porosity: 5
Fraction of the gross rock volume not occupied by solid material. This porosity is difficult to measure in the presence of clay, organic matter, etc.

Trapped porosity: 4, 59, 62, 63, *340,*
Porosity retaining the non-wetting fluid during imbibition.
Sometimes confused (incorrectly) with "poorly connected" p. or even "disconnected" p. To be reserved for debates on the localisation of fluids in polyphasic experiments. For the quantitative aspect, prefer the use of "residual saturation"

True p. abusively used for total p.

Unusable p. (hydrology) as opposed to effective p.

Usable p. (hydrology) see effective

Well connected p. see connected p.

Vug p., Vuggy porosity: 150, *158*, 167, 193
Porous space formed from subspheric volume (a complement or inversion of the interparticle p.). Basic vuggy p. displays is devoid of inter-pore connection as long as the vugs are not in mutual contact. Sometimes associated with the notion of Poorly connected p. or even of Isolated p., Unusable p. (hydrology). This association may be very misleading since, as soon as there are numerous contacts between vugs, these porous media become highly permeable.

Subject Index

Author Index

Achevé d'imprimer par Corlet Numérique - 14110 Condé-sur-Noireau
N° d'Imprimeur : 45177 - Dépôt légal : novembre 2007 - *Imprimé en France*